Efrén Mezura-Montes (Ed.)

Constraint-Handling in Evolutionary Optimization

Studies in Computational Intelligence, Volume 198

Further volumes of this series can be found on our homepage: springer.com

Vol. 175. Godfrey C. Onwubolu and Donald Davendra (Eds.)
Differential Evolution: A Handbook for Global Permutation-Based Combinatorial Optimization, 2009
ISBN 978-3-540-92150-9

Vol. 176. Beniamino Murgante, Giuseppe Borruso and Alessandra Lapucci (Eds.)
Geocomputation and Urban Planning, 2009
ISBN 978-3-540-89929-7

Vol. 177. Dikai Liu, Lingfeng Wang and Kay Chen Tan (Eds.)
Design and Control of Intelligent Robotic Systems, 2009
ISBN 978-3-540-89932-7

Vol. 178. Swagatam Das, Ajith Abraham and Amit Konar
Metaheuristic Clustering, 2009
ISBN 978-3-540-92172-1

Vol. 179. Mircea Gh. Negoita and Sorin Hintea
Bio-Inspired Technologies for the Hardware of Adaptive Systems, 2009
ISBN 978-3-540-76994-1

Vol. 180. Wojciech Mitkowski and Janusz Kacprzyk (Eds.)
Modelling Dynamics in Processes and Systems, 2009
ISBN 978-3-540-92202-5

Vol. 181. Georgios Miaoulis and Dimitri Plemenos (Eds.)
Intelligent Scene Modelling Information Systems, 2009
ISBN 978-3-540-92901-7

Vol. 182. Andrzej Bargiela and Witold Pedrycz (Eds.)
Human-Centric Information Processing Through Granular Modelling, 2009
ISBN 978-3-540-92915-4

Vol. 183. Marco A.C. Pacheco and Marley M.B.R. Vellasco (Eds.)
Intelligent Systems in Oil Field Development under Uncertainty, 2009
ISBN 978-3-540-92999-4

Vol. 184. Ljupco Kocarev, Zbigniew Galias and Shiguo Lian (Eds.)
Intelligent Computing Based on Chaos, 2009
ISBN 978-3-540-95971-7

Vol. 185. Anthony Brabazon and Michael O'Neill (Eds.)
Natural Computing in Computational Finance, 2009
ISBN 978-3-540-95973-1

Vol. 186. Chi-Keong Goh and Kay Chen Tan
Evolutionary Multi-objective Optimization in Uncertain Environments, 2009
ISBN 978-3-540-95975-5

Vol. 187. Mitsuo Gen, David Green, Osamu Katai, Bob McKay, Akira Namatame, Ruhul A. Sarker and Byoung-Tak Zhang (Eds.)
Intelligent and Evolutionary Systems, 2009
ISBN 978-3-540-95977-9

Vol. 188. Agustín Gutiérrez and Santiago Marco (Eds.)
Biologically Inspired Signal Processing for Chemical Sensing, 2009
ISBN 978-3-642-00175-8

Vol. 189. Sally McClean, Peter Millard, Elia El-Darzi and Chris Nugent (Eds.)
Intelligent Patient Management, 2009
ISBN 978-3-642-00178-9

Vol. 190. K.R. Venugopal, K.G. Srinivasa and L.M. Patnaik
Soft Computing for Data Mining Applications, 2009
ISBN 978-3-642-00192-5

Vol. 191. Zong Woo Geem (Ed.)
Music-Inspired Harmony Search Algorithm, 2009
ISBN 978-3-642-00184-0

Vol. 192. Agus Budiyono, Bambang Riyanto and Endra Joelianto (Eds.)
Intelligent Unmanned Systems: Theory and Applications, 2009
ISBN 978-3-642-00263-2

Vol. 193. Raymond Chiong (Ed.)
Nature-Inspired Algorithms for Optimisation, 2009
ISBN 978-3-642-00266-3

Vol. 194. Ian Dempsey, Michael O'Neill and Anthony Brabazon (Eds.)
Foundations in Grammatical Evolution for Dynamic Environments, 2009
ISBN 978-3-642-00313-4

Vol. 195. Vivek Bannore and Leszek Swierkowski
Iterative-Interpolation Super-Resolution Image Reconstruction:
A Computationally Efficient Technique, 2009
ISBN 978-3-642-00384-4

Vol. 196. Valentina Emilia Balas, János Fodor and Annamária R. Várkonyi-Kóczy (Eds.)
Soft Computing Based Modeling in Intelligent Systems, 2009
ISBN 978-3-642-00447-6

Vol. 197. Mauro Birattari
Tuning Metaheuristics, 2009
ISBN 978-3-642-00482-7

Vol. 198. Efrén Mezura-Montes (Ed.)
Constraint-Handling in Evolutionary Optimization, 2009
ISBN 978-3-642-00618-0

Efrén Mezura-Montes (Ed.)

Constraint-Handling in Evolutionary Optimization

Springer

Efrén Mezura-Montes
Full-time researcher
National Laboratory on Advanced Informatics (LANIA A.C.)
Rébsamen 80, CENTRO
Xalapa, Veracruz, 91000
Mexico
E-mail: emezura@lania.mx

ISBN 978-3-642-00618-0 e-ISBN 978-3-642-00619-7

DOI 10.1007/978-3-642-00619-7

Studies in Computational Intelligence ISSN 1860949X

Library of Congress Control Number: 2009921825

Typeset & Cover Design: Scientific Publishing Services Pvt. Ltd., Chennai, India.

Printed in acid-free paper

9 8 7 6 5 4 3 2 1

springer.com

To Margarita and Diego

Preface

Evolutionary algorithms (EAs), as well as other bio-inspired heuristics, are widely used to solve numerical optimization problems. However, in their original versions, they are limited to unconstrained search spaces i.e they do not include a mechanism to incorporate feasibility information into the fitness function. On the other hand, real-world problems usually have constraints in their models. Therefore, a considerable amount of research has been dedicated to design and implement constraint-handling techniques. The use of (exterior) penalty functions is one of the most popular methods to deal with constrained search spaces when using EAs. However, other alternative methods have been proposed such as: special encodings and operators, decoders, the use of multiobjective concepts, among others.

An efficient and adequate constraint-handling technique is a key element in the design of competitive evolutionary algorithms to solve complex optimization problems. In this way, this subject deserves special research efforts.

After a successful special session on constraint-handling techniques used in evolutionary algorithms within the Congress on Evolutionary Computation (CEC) in 2007, and motivated by the kind invitation made by Dr. Janusz Kacprzyk, I decided to edit a book, with the aim of putting together recent studies on constrained numerical optimization using evolutionary algorithms and other bio-inspired approaches.

The intended audience for this book comprises graduate students, practitioners and researchers interested on alternative techniques to solve numerical optimization problems in presence of constraints.

The book covers six main topics: The first two chapters refer to swarm-intelligence-based approaches. Differential evolution, a very competitive evolutionary algorithm for constrained optimization, is studied in the next three chapters. Two different constraint-handling techniques for evolutionary multiobjective optimization are presented in the two subsequent chapters. Two hybrid approaches, one with a combination of two nature-inspired heuristics and the other with the mix of a genetic algorithm and a local search operator, are detailed in the next two chapters. Finally, a constraint-handling technique designed for a real-world problem and a survey on artificial immune system in constrained optimization are the subjects of the final two chapters.

Angel E. Muñoz-Zavala, Arturo Hernández-Aguirre and Enrique R. Villa-Diharce, present the adaptation of particle swarm optimization (PSO) to solve constrained optimization problems. The search capabilities improve by means of the following modifications: (1) a novel neighborhood structure to slow-down convergence, (2) two perturbation operators applied to the memory of each particle to favor diversity in the swarm and (3) a dynamic tolerance to handle equality constraints.

Guillermo Leguizamón and Carlos A. Coello Coello show an alternative approach to explore the boundary between the feasible and infeasible regions of the search space by means of two perspectives: (1) the use of *ad hoc* operators and (2) a more general operator. The authors couple their approach to two swarm intelligence search engines and a general evolutionary algorithm. These examples show that significant changes to the original version of each algorithm were not required.

Tetsuyuki Takahama and Setsuko Sakai propose an improved version of their ϵDE algorithm to solve constrained optimization problems. A faster reduction of the relaxation for equality constraints in the ϵ constrained method coupled with a more frequent use of gradient-based mutation lead ϵDE to provide even more competitive results in highly constrained problems. Furthermore, the authors present two mechanisms to keep variable values within the valid search space.

Janez Brest presents some modifications to his jDE algorithm to deal with constrained search spaces: (1) The use of the ϵ-constraint method, (2) a population size reduction, (3) the combination of three differential evolution variants, (4) different mechanisms to keep valid variable values and (5) a self-adaptive approach for two DE parameters (F and CR).

Efrén Mezura-Montes and Ana Gabriela Palomeque-Ortiz analyze the behavior of one deterministically-controlled and three self-adapted parameters in differential evolution for constrained optimization. The approach considers two parameters related to the constraint-handling technique. The experimental design analyzes (1) the online-behavior of the algorithm by using two performance measures and (2) the behavior shown by the parameter values.

Gary G. Yen presents a parameterless adaptive penalty function coupled with a distance measure for evolutionary multiobjective constrained optimization. The non-dominated sorting process then uses this modified fitness value. The number of feasible solutions in the population determines the behavior of the process, which may lead the search to either find more feasible solutions or locate the optimal solution.

Tapabrata Ray, Hemant Kumar Singh, Amitay Isaacs and Warren Smith emphasize the importance of maintaining infeasible solutions close to the feasible space in evolutionary multiobjective constrained optimization. The aim is to focus the search precisely on the boundaries of the feasible and infeasible regions.

Heder S. Bernardino, Helio J.C. Barbosa, Afonso C.C. Lemonge and Leonardo G. Fonseca combine the use of an artificial immune system to bias the search to the feasible region and a standard genetic algorithm. A clearing procedure, based on a niching mechanism, helps the search by improving the diversity in the population.

Marcella C. Araujo, Elizabeth F. Wanner, Frederico G. Guimarães and Ricardo H.C. Takahashi, improve a genetic algorithm with a local search operator based on quadratic and linear approximations for the objective function and the constraints of the problem as well. This operator defines a sub-problem with a quadratic objective function and quadratic and/or linear constraints, which is solved with a linear matrix inequality formulation. The aim of the special operator is to improve the satisfaction of constraints.

Akira Oyama presents a constraint-handling technique for aerodynamic and multidisciplinary design optimization. The approach is suitable for problems where the number of evaluations must be kept low due to the cost associated with each one of them. A combination of dominance in the constraints space and a niching mechanism helps the approach to reach the feasible region by requiring a low number of evaluations.

Nareli Cruz-Cortes presents the main proposals for constrained optimization based on an artificial immune system. The suggested taxonomy divides the approaches in "hybrid" (artificial immune systems with genetic algorithms) and "pure" schemes (i.e., those in which only artificial immune system processes and theories are adopted for the search engine).

The themes tackled in this book give evidence of the current research paths regarding constraint-handling in evolutionary optimization, such as the following:

- The generation of special mechanisms to focus the search on the boundaries of the feasible region and the importance of good infeasible solutions in the process.
- Constraint-handling in evolutionary multiobjective optimization.
- Parameter control mechanisms to keep the user from the fine-tuning process.
- Hybrid algorithms, such as global-local search, combination of heuristics-based approaches and the use of mathematical programming methods, in order to improve the search capabilities in constrained search spaces.
- The exploration of novel bio-inspired approaches such as particle swarm optimization, ant colony optimization, artificial immune systems, differential evolution, among others.
- Constraint-handling techniques able to perform well with a low number of objective function evaluations.

I want to thank all the authors for their high-quality contributions and also for their interest in this book. I would also want to thank Professor Janusz Kacprzyk for the opportunity to edit a book in the Studies in Computational Intelligence Series. Many thanks to Dr. Thomas Ditzinger and Ms. Heather

King from Springer for their valuable editorial support. I want to express my gratitude to Dr. Cristina Loyo-Varela, Dr. Cora Beatriz Excelente-Toledo, Dr. Edgard Ivan Benitez-Guerrero, MsC. Pedro Nolasco-Vázquez and all my coworkers at LANIA for their continuous support to my research activities. I appreciate the help of Luis Guillermo Osorio-Hernández and Jorge Isacc Flores-Mendoza in the final edition of this book. The support provided by CONACyT (the mexican council of science and technology) through project No. 79809 is also greatly appreciated. Many thanks to Dr. Carlos A. Coello Coello, my former thesis adviser, for his support and suggestions on the preparation of this book. Finally, I want to thank Margarita and Diego for their encouragement, understanding and patience.

Xalapa, Veracruz, MEXICO
January 2009

Efrén Mezura-Montes

Contents

List of Contributors

Marcella C. Araujo
Departamento de Física,
Universidade Federal de Ouro Preto,
Ouro Preto, MG, Brazil,
marcellaop@gmail.com

Helio J.C. Barbosa
Laboratório Nacional de
Computação Científica,
Av. Getúlio Vargas 333,
25651-075 Petrópolis,
RJ, Brazil,
hcbm@lncc.br

Heder S. Bernardino
Universidade Federal de Juiz
de Fora, Brazil,
hedersb@gmail.com

Janez Brest
Faculty of Electrical
Engineering and Computer Science,
University of Maribor,
Smetanova ul. 17, 2000 Maribor,
Slovenia,
janez.brest@uni-mb.si

Carlos A. Coello Coello
CINVESTAV-IPN, Evolutionary
Computation Group (EVOCINV),
Departamento de Computación,
Av. IPN No. 2508. Col.
San Pedro Zacatenco,
México D.F. 07300, México,
ccoello@cs.cinvestav.mx

Nareli Cruz-Cortés
Center for Computing Research,
National Polytechnic Institute
(CIC-IPN), 07738 Mexico City,
Mexico,
nareli@cic.ipn.mx

Leonardo G. Fonseca
Laboratório Nacional de
Computação Científica,
Av. Getúlio Vargas 333,
25651-075 Petrópolis, RJ, Brazil,
goliatt@lncc.br

Frederico G. Guimarães
Departamento de Engenharia
Elétrica, Universidade Federal de
Minas Gerais, Belo Horizonte,
MG, Brazil,
frederico.guimaraes@yahoo.com.br

Arturo Hernández Aguirre
Center for Research in Mathematics
(CIMAT), Department of
Computer Science A.P. 402,

Guanajuato, Gto. CP. 36240,
México,
artha@cimat.mx

Amitay Isaacs
University of New South Wales,
Australian Defence Force Academy,
Northcott Drive, Canberra,
ACT 2600, Australia.
a.isaacs@adfa.edu.au

Guillermo Leguizamón
LIDIC - Universidad Nacional
de San Luis,
Ejército de Los Andes 950,
San Luis 5700, Argentina,
legui@unsl.edu.ar

Afonso C.C. Lemonge
Universidade Federal
de Juiz de Fora, Brazil,
afonso.lemonge@ufjf.edu.br

Efrén Mezura-Montes
Laboratorio Nacional de Informática
Avanzada (LANIA A.C.),
Rébsamen 80, Centro, Xalapa,
Veracruz, 91000, México,
emezura@lania.mx

Angel E. Muñoz Zavala
Center for Research in
Mathematics (CIMAT),
Department of Computer Science
A.P. 402, Guanajuato,
Gto. CP. 36240,
México, aemz@cimat.mx

Akira Oyama
Institute of Space and
Astronautical Science,
Japan Aerospace Exploration
Agency, 3-1-1 Yoshinodai,
Sagamihara, Kanagawa,
229-8510, Japan,
oyama2@flab.isas.jaxa.jp

Ana Gabriela Palomeque-Ortiz
Laboratorio Nacional de
Informática Avanzada
(LANIA A.C.), Rébsamen 80,
Centro, Xalapa,
Veracruz, 91000, México,
apalomeque@lania.edu.mx

Tapabrata Ray
University of New South Wales,
Australian Defence Force Academy,
Northcott Drive, Canberra,
ACT 2600, Australia.
t.ray@adfa.edu.au

Setsuko Sakai
Faculty of Commercial Sciences,
Hiroshima Shudo University,
Asaminami-ku,
Hiroshima 731-3195 Japan,
setuko@shudo-u.ac.jp

Hemant Kumar Singh
University of New South Wales,
Australian Defence Force Academy,
Northcott Drive, Canberra,
ACT 2600, Australia.
hemant.singh@adfa.edu.au

Warren Smith
University of New South Wales,
Australian Defence Force Academy,
Northcott Drive, Canberra,
ACT 2600, Australia,
w.smith@adfa.edu.au

Tetsuyuki Takahama
Department of Intelligent Systems,
Hiroshima City University,
Asaminami-ku,
Hiroshima 731-3194
Japan,
takahama@its.hiroshima-cu.ac.jp

Ricardo H. C. Takahashi
Departamento de Matemática,
Universidade Federal de Minas Gerais,
Belo Horizonte, MG, Brazil,
taka@mat.ufmg.br

Enrique R. Villa Diharce
Center for Research in Mathematics (CIMAT),
Department of Statistics A.P. 402,
Guanajuato, Gto. CP. 36240,
México, villadi@cimat.mx

Elizabeth F. Wanner
Departamento de Matemática,
Universidade Federal de Ouro Preto,
Ouro Preto, MG, Brazil,
efwanner@iceb.ufop.br

Gary G. Yen
Oklahoma State University,
School of Electrical and Computer Engineering,
Stillwater, OK 74078, USA,
gyen@okstate.edu

Continuous Constrained Optimization with Dynamic Tolerance Using the COPSO Algorithm

Angel E. Muñoz Zavala, Arturo Hernández Aguirre, and Enrique R. Villa Diharce

Abstract. This work introduces a hybrid PSO algorithm which includes perturbation operators to keep population diversity. A new neighborhood structure for Particle Swarm Optimization called Singly-Linked Ring is implemented. The approach proposes a neighborhood similar to the ring structure, but which has an innovative neighbors selection. The objective is to avoid the premature convergence into local optimum. A special technique to handle equality constraints with low side effects on the diversity is the main feature of this contribution. Two perturbation operators are used to improve the exploration, applying the modification only in the particle best population. We show through a number of experiments how, by keeping the selection pressure on a decreasing fraction of the population, COPSO can consistently solve a benchmark of constrained optimization problems.

Keywords: PSO, Constrained Optimization, Neighborhood Structure, COPSO.

1 Introduction

Particle swarm optimization (PSO) algorithm is a population-based optimization technique inspired by the motion of a bird flock. In the PSO model, every particle flies over a real valued n-dimensional space of decision variables $\mathbf{X}$. Each particle keeps track of its position $\mathbf{x}$, velocity $\mathbf{v}$, and remembers the best position ever visited, P_{Best}. The particle with the best P_{Best} value is called the leader, and its position

Angel E. Muñoz Zavala and Arturo Hernández Aguirre
Center for Research in Mathematics (CIMAT), Department of Computer Science
A.P. 402, Guanajuato, Gto. CP. 36240, México
e-mail: aemz@cimat.mx, artha@cimat.mx

Enrique R. Villa Diharce
Center for Research in Mathematics (CIMAT), Department of Statistics
A.P. 402, Guanajuato, Gto. CP. 36240, México
e-mail: villadi@cimat.mx

E. Mezura-Montes (Ed.): Constraint-Handling in Evolution. Optimi., SCI 198, pp. 1–23.
springerlink.com

is called global best, G_{Best}. The next particle's position is computed by adding a velocity term to its current position, as follows:

$$\mathbf{x}_{t+1} = \mathbf{x}_t + \mathbf{v}_{t+1} \tag{1}$$

The velocity term combines the local information of the particle with global information of the flock, in the following way.

$$\mathbf{v}_{t+1} = w * \mathbf{v}_t + \phi_1 * (P_{Best} - \mathbf{x}_t) + \phi_2 * (G_{Best} - \mathbf{x}_t) \tag{2}$$

The equation above reflects the socially exchanged information. It resumes PSO three main features: distributed control, collective behavior, and local interaction with the environment [9, 21].The second term is called the cognitive component, while the last term is called the social component. w is the inertia weight, and ϕ_1 and ϕ_2 are called acceleration coefficients. The inertia weight regulates the mixture of previous velocity with the current one.

When the flock is split into several neighborhoods the particle's velocity is computed with respect to its neighbors. The best P_{Best} value in the neighborhood is called the local best, L_{Best}.

$$\mathbf{v}_{t+1} = w * \mathbf{v}_t + \phi_1 * (P_{Best} - \mathbf{x}_t) + \phi_2 * (L_{Best} - \mathbf{x}_t) \tag{3}$$

Neighborhoods can be interconnected in many different ways, some of the most popular are shown in Figure 1. The star topology is in fact one big neighborhood where every particle is connected to each other, thus enabling the computation of a global best. The ring topology allows neighborhoods, thereby, it is commonly used by the PSO with local best.

PSO is a fast algorithm whose natural behavior is to quickly converge to the best explored local optima. However, attaining flock's convergence to the *global optimum* with high probability implies certain adaptations. The approaches range from modifications to the main PSO equation, to the incorporation of reproduction operators. This chapter introduces the Constrained Optimization via PSO algorithm (COPSO), a hybrid PSO with the following new features:

- Two perturbation operators to keep diversity. Although perturbations are not included in the original PSO model, they are quite common nowadays. The formal analysis of van den Bergh shows that the PSO algorithm will only converge to the best position visited, not the global optimum [40]. Therefore, the quest for high diversity is a sound approach to locate the global optimum since more diversity leads to increasing exploration.
- Singly-linked ring topology. COPSO creates several neighborhoods around a new ring topology that we called the "singly linked ring topology" (SLR). Thus COPSO promotes a flock with several local leaders to improve the exploration capacity [9, 21].
- Constraint handling. PSO lacks an explicit mechanism to bias the search towards the feasible region in constrained search spaces. For selecting a neighborhood

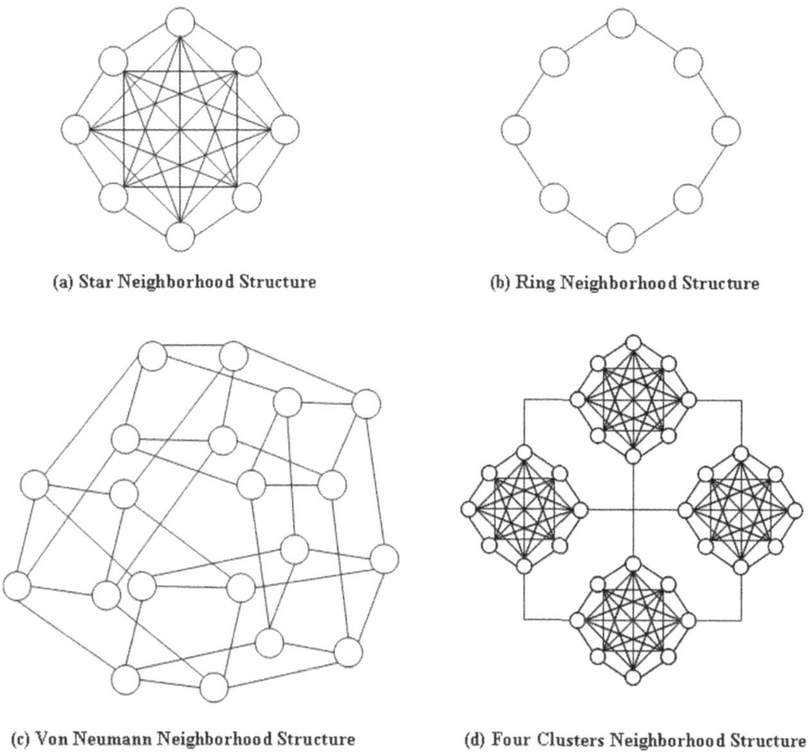

Fig. 1 Neighborhood structures for PSO. A representation of the social networks applied in PSO

leader, COPSO picks the winner through a tournament of feasibility and sum of constraint violations ("superiority of feasible solutions" [7]).

COPSO performs the main PSO algorithm but executes two additional steps which implement the mentioned features: the C-perturbation which is oriented to sustain global exploration by keeping diversity, and the M-perturbation oriented to the local refinement of solutions. The constraint handling technique which selects promising particles is embedded in the function *ParticleBest* (see Figure 2).

The organization of this chapter is the following. A review of diversity control techniques for PSO is presented in Section 2. Next, a review of constraint handling techniques used by PSO algorithms is presented in Section 3. The COPSO algorithm is thoroughly explained in Section 4. The general class of problems of interest is defined in Section 5. In Section 6, COPSO is used to solve a state of the art benchmark of 24 functions. Comparisons against four different approaches are provided (two of them based on PSO). Conclusions are provided in Section 7.

```
X0 = Rand(LL, UL)
V0 = 0
F0 = Fitness ( X0 )
C0 = SCV ( X0 )
PBest= X0      best position of particle
FBest= F0      best fitness value of particle
CBest= C0      SCV value of particle
----------------------------------------------------------------
For i = 1 To maxgenerations
   LBest = LocalBest ( FBest, CBest )
   GBest = TheBest ( FBest, CBest )
   V = Velocity ( V, X, PBest, LBest, GBest)
   X = X + V                                        Stage 1
   F = Fitness ( X )
   C = SCV ( X )
   [PBest, FBest, CBest] = ParticleBest ( PBest, X, FBest, F, CBest, C )
----------------------------------------------------------------
   If (U(0,1) < pC )
      Temp = C-Perturbation (PBest )
      FTemp = Fitness ( Temp )
      CTemp = SCV ( Temp )
      [PBest, FBest, CBest] = ParticleBest ( PBest, Temp, FBest, FTemp, CBest, CTemp )
      TempBest = TheBest ( FBest, CBest )
      If (TempBest < GBest)
         pC = pC*0.99
      Else
         pC = pC*1.01
      End If                                        Stage 2
      GBest = TempBest
      If (i % n/2 = 0 )
         PBest = Temp
         FBest = FTemp
         CBest = CTemp
      End If
   End If
----------------------------------------------------------------
   If (U(0,1) < pM )
      Temp = M-Perturbation (PBest )
      FTemp = Fitness ( Temp )
      CTemp = SCV ( Temp )
      [PBest, FBest, CBest] = ParticleBest ( PBest, Temp, FBest, FTemp, CBest, CTemp )
      TempBest = TheBest ( FBest, CBest )
      If (TempBest < GBest)
         pM = pM*0.99
      Else
         pM = pM*1.01
      End If                                        Stage 3
      GBest = TempBest
      If (i % n/2 = 0 )
         PBest = Temp
         FBest = FTemp
         CBest = CTemp
      End If
   End If
End For
```

Fig. 2 Pseudo-code of *COPSO* algorithm

2 Diversity Control for PSO Algorithm

A natural problem in evolutionary computation is the premature convergence. It means that the evolutionary algorithm could stay trapped in a region containing a local optimum. Premature convergence can be caused by the diversity loss, which occurs when the population reaches a suboptimal state where evolutionary algorithm can no longer produce offspring which outperforms their parents [10]. A way to keep diversity is by sustaining the balance between exploration and exploitation [16].

In PSO the source of diversity, called *variation*, is the difference between the particle's position **x** and P_{Best}, or L_{Best}. Although variation provides diversity, it can only be sustained for a limited number of generations because convergence of the flock is necessary to refine the solution. The need to keep diversity is well known and has been addressed by several authors. Angeline [1], and also Eberhart [8], proposed population breeding. Then, two randomly chosen particles (parents) may reproduce and create offsprings. Lovbjerg [26], and more recently Settles [36] implemented breeding with some success. More investigations on reproduction as source of diversity were recently conducted by S. Das [6]. He adapted the reproduction operator of differential evolution [37,38] to PSO, and reported robust performance in a small set of global optimization problems. W. Zhang [41] proposed to compute the velocity term by taking turns between PSO and differential evolution. At odd generations the individuals are updated by the rules of motion of PSO; at even generations by the differential evolution formalism.

Note how these approaches keep diversity by preventing premature convergence. Other approaches let premature convergence happen but later in the process they try to extinguish it. For instance, in Krink's approach, the particles are clustered and their density used as a measure of crowding [23]. Once such clusters are detected, their density is reduced by bouncing away the particles. Blackwell also investigated a mechanism that repels clustered particles [2].

3 Constraint Handling Techniques for the PSO Algorithm

Real-world optimization problems are subject to a number of equality and inequality constraints, which can be linear or nonlinear. These constraints determine which areas of the search space are feasible and which are infeasible. In addition to these constraints, boundary constraints are usually imposed to the search space [29]. Also, there is the possibility that the feasible space is fragmented and separated by infeasible regions, requiring that both the feasible and infeasible regions be searched.

Several authors have noted how the adoption of a constraint handling technique may cause diversity loss due to the additional selection pressure required to bias the population towards the feasible region [11, 13, 14, 27].

PSO is an unconstrained search technique. Thus, adopting a constraint handling technique into the main PSO algorithm is an open research area. There is a considerable amount of research regarding mechanisms that allow the evolutionary algorithms to deal with equality and inequality constraints. Early approaches did not combine a diversity maintenance strategy with a constraint handling technique. For instance, Parsopoulos used a multi-stage assignment penalty function without diversity control [31]. Hu and Eberhart proposed a feasibility preservation strategy that determines the best particle [17,18]. Both penalty and feasibility preservation strategies were analyzed by G. Coath [5] (whose experiments clearly detect the need of some form of diversity control). He and Prempain used a "fly-back" mechanism that returns an unfeasible particle to its previous feasible position [12]. A more important drawback of this technique is the requirement of an all-feasible initial population. Recent approaches include some form of diversity control. For instance Toscano and Coello [39], use a turbulence operator (a kind of mutation) combined with a superiority of feasible solutions [7]. They succeeded in most problems but faced weak capacity to refine the solutions.

We contrast our method with the best results of the recent proposals reviewed in this section.

4 COPSO: Constrained Optimization via PSO

A brief analysis of the state-of-the-art in PSO to solve constrained optimization problems was presented. Now, we are going to introduce our approach called Constrained Optimization via Particle Swarm Optimization algorithm (COPSO) [15]. COPSO is a local PSO with a `singly-linked` ring neighborhood. COPSO handles constraints adopting a `superiority of feasible solutions` complemented with a `dynamic tolerance` for handling equality constraints. The main components of COPSO are the `C-Perturbation` and `M-Perturbation` operators; these are applied to the variable P_{Best} of the flock [30].

Along the present section, the essential components of COPSO are explained: interaction model, neighborhood structure, diversity mechanism and constraint handling.

4.1 Interaction Model

There are 4 interaction models proposed by Kennedy [19]; these models were defined by omitting components of the velocity formula. The `full` model is composed by the cognition component and the social component. Dropping the social component results in the `cognition-only` model, whereas dropping the cognition component defines the `social-only` model. In a fourth model, `selfless` model, the neighbourhood best is chosen only from the neighbors, without considering the current individual. Experimental results prove that the social-only model consistently found solutions faster than the full model, but the reliability of the social-only model is lower than the full model [4, 19].

Parsopoulos and Vrahatis [32] proposed a new scheme that combine the global and the local PSO variants. The approach calculates two velocity directions, one from a global full model and another from a local full model. Both directions, global and local, are linearly combined through an `influence` parameter. In the same way, Cagnina et all [3] proposed an extended full model using a combination of both the global and the local PSO variants with a constriction factor k (the k parameter affects the whole velocity equation). In COPSO, the velocity equation combines a new global term with the inertia, cognition and social terms.

$$v_{t+1} = w * v_t + \phi_1 * (P_{Best} - x_t) + \phi_2 * (L_{Best} - x_t) + \phi_3 * (G_{Best} - x_t) \tag{4}$$

The inertia weight in COPSO is linearly decremented from 1.0 to 0.5 according to the function evaluations. Thus, at the beginning of the search, the algorithm performs more exploration, and along the process the algorithm is gradually focused to exploitation. The experiments suggest that the best parameters for the model are the following:

$$\phi_1 = 1.0 * U(0,1.0) = U(0,1.0)$$
$$\phi_2 = 1.0 * U(0,1.0) = U(0,1.0)$$
$$\phi_3 = 0.1 * U(0,1.0) = U(0,0.1)$$

where $U(\alpha,\beta)$ is a uniform distribution with random numbers between α and β. Note that the new global term in the velocity equation has a low influence (ϕ_3) but its contribution is important to keep the flock on promising regions. This concept will be understood when we analyse the technique to handle equality constraints, where the incorporation of the global influence is essential to improve the results of some test functions.

4.2 Neighborhood Structure

In the PSO scheme each particle moves following a leader. Particles in the same neighborhood communicate with one another by exchanging information for moving towards a better position. The flow of information through the flock depends on the neighborhood structure. Figure 1 presents a few neighborhood structures developed for PSO.

In a highly connected neighborhood structure, the information about the best particle in the swarm (G_{Best}) is quickly transmitted through the whole flock. This means faster convergence, but also implies a higher risk to converge to a local minimum. Also, Kennedy and Mendes empirically show that the `star` neighborhood attains faster convergence than the other topologies, but it meets the optimal fewer times than any other approach they tested [22]. They suggest trying the `Von Neumann` neighborhood structure, which performed more consistently in their experiments than the topologies commonly found in current practice. The success of the Von Neumann neighborhood in unconstrained optimization is due to the interaction that each particle has with other particles, an average of 5 neighbors. This promotes the exploitation, but unfortunately fails to provide the exploration required by the constrained optimization problems. It is important to note that the conclusions reached by Kennedy and Mendes are valid for unconstrained optimization problems. COPSO works with a new neighborhood structure, the `singly-linked ring`.

The singly-linked ring rises from discovering that the ring and the Von Neumann neighborhoods are double-linked lists; as shown in Figure 3-a. Suppose that every particle is assigned a permanent label which is used to construct the neighborhoods. For each particle k, a neighborhood of size n is composed by the next $n/2$ linked particles, and by $n/2$ previous particles; and the best particle in the neighborhood is the local best of particle k (L_{Best}). For example, in a neighborhood of size $n=2$, particle k has two neighbors, particles $k-1$ and $k+1$. In turn, particles $k-1$ and $k+1$ have particle k as a neighbor. In this way, there is a mutual attraction between consecutive particles, forming overlapped clusters.

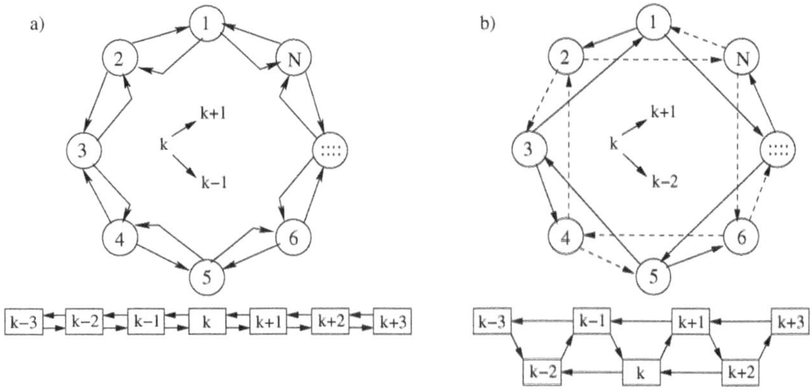

Fig. 3 Ring neighborhood structures for PSO: a)Doubly-Linked Ring (original Ring structure) and b)Singly-Linked Ring

Now, analyse a PSO based on a singly-linked ring, see Figure 3-b, and assume particle k is again the best of flock. This time, particle k has particles $k-2$ and $k+1$ as neighbors (not $k-1$ and $k+1$ as in the double link). Observe that particle $k+1$ has particles $k-1$ and $k+2$ as neighbors, and particle $k-1$ has particles $k-3$ and k as neighbors. Then, k attracts $k-1$ but $k-1$ only attracts k through particle $k+1$. Therefore, the particle in between cancels the mutual attraction. Besides, the information through the whole swarm is transmitted faster than in the original ring topology. Therefore, the singly-linked ring keeps the exploration of the search space, and increases the exploitation of the best solutions [15].

For each particle i, the members of a neighborhood of size k are selected by the next algorithm.

COPSO uses a singly-linked ring with a neighborhood of size $n=4$. For each particle k, its neighbors are the particles: $k+1$, $k-2$, $k+3$ and $k-4$.

Fig. 4 Procedure to find the neighbors for $i-th$ particle in a singly-linked ring structure

```
% Find k particles for neighborhood N_i
N_i = ∅
Step = 1
Switch = 1
Repeat
   N_i = N_i ∪ P_(i+Switch*Step)
   Step = Step + 1
   Switch = -Switch
Until Size(N_i) = k
```

4.3 Diversity Mechanism

In his PhD thesis, van den Bergh gives a theorem that specifies under which conditions an algorithm can be considered a global optimization method [40]. The theorem implies that a general algorithm, without *a priori* knowledge, must be able to generate an infinite number of samples distributed throughout the whole search space in order to guarantee that it will find the global optimum with asymptotic probability 1.

This can be achieved by periodically adding randomized particles to the swarm. Nevertheless, resetting the position of the particles is not a trivial task; a bad decision affects directly in the exploitation of the best solutions [40]. We propose, based on the observation that the P_{Best} particles drive the swarm, perturbing the P_{Best} population instead.

Before explaining the perturbation operators, it is necessary to introduce the main algorithm of COPSO for better understanding. The complete pseudocode of COPSO algorithm is shown in Figure 2.

Flying the particles is the main task of PSO, see Stage 1. Variables LL and UL are the lower and upper limits of the search space. Function *LocalBest* returns the best neighbor for each particle. Function *ParticleBest* updates the P_{Best} population. Function $TheBest$ updates the G_{Best} particle. The function SCV computes the *Sum of Constraint Violations*, that is, total value of unfeasible constraints. The number of particles is n, d is the dimension of the space, and i is the generation index. The perturbations are applied to P_{Best} in the next two stages.

The goal of the second stage is to add a perturbation generated from the linear combination of three random vectors. This perturbation is preferred over other operators because it preserves the distribution of the population (also used for reproduction by the differential evolution algorithm [33]). In COPSO this perturbation is called C-Perturbation. It is applied to the members of P_{Best} to yield a set of temporal particles $Temp$. Then each member of $Temp$ is compared with its corresponding father and P_{Best} is updated with the child if it wins the tournament. Figure 5 shows the pseudo-code of the **C-Perturbation** operator.

In the third stage every vector is perturbed again so a particle could be deviated from its current direction as responding to external, maybe more promising, stimuli. This perturbation is implemented by adding small random numbers (from a uniform distribution) to every design variable. The perturbation, called M-Perturbation, is

Fig. 5 Pseudo-code of **C-Perturbation**

```
For k = 0 To n
  For j = 0 To d
    r = U(0, 1)
    p1 = Random(n)
    p2 = Random(n)
    Temp[k, j] = PBest[k, j] + r (PBest[p1, j] - PBest[p2, j])
  End For
End For
```

applied to every member of P_{Best} to yield a set of temporal particles *Temp*. Then each member of *Temp* is compared with its corresponding father and P_{Best} is updated with the child if it wins the tournament. Figure 6 shows the pseudo-code of the **M-Perturbation** operator. The perturbation is added to every dimension of the decision vector with probability $1/d$ (d is the dimension of the decision variable vector).

Fig. 6 Pseudo-code of **M-Perturbation**

```
For k = 0 To n
  For j = 0 To d
    r = U(0, 1)
    If r ≤ 1/d Then
      Temp[k, j] = Rand(LL, UL)
    Else
      Temp[k, j] = PBest[k, j]
  End For
End For
```

In Figure 7 the position P_{Best} is relocated to a new "best" after the perturbation operations. Notice this change is made to the particle's memory of best visited location. When PSO takes turn to perform its computations, it finds everything as left in the previous generation, except that the memory P_{Best} may store a better position.

The cooperation between PSO and the perturbation operators have been carefully analyzed by the authors through out the many experiments conducted. The PSO

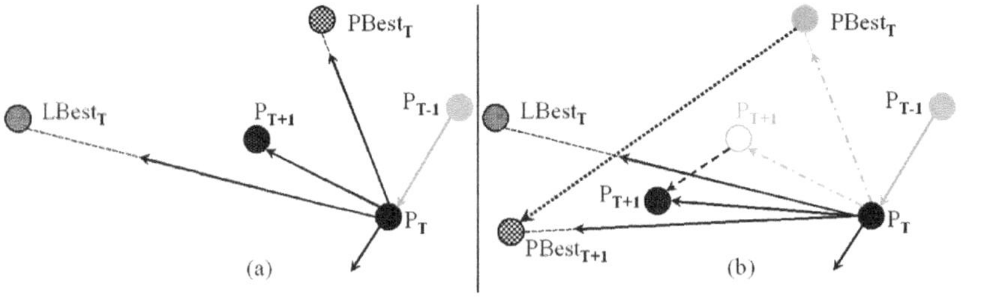

Fig. 7 C-perturbation and M-perturbation on PBest. (a) PBest in stage 1. (b) PBest after stages 2 and 3

stage performs very efficiently at refining solutions in a local space, but exploration is performed by the perturbation operators. The perturbation operators adapt their activity along generations as follows: the operator is invoked with probability $p_C = p_M = 1$ when the flock is flown for the first time. This probability is decremented by a factor of 1% whenever the perturbation stage fails to improve the G_{Best} value, otherwise, it is incremented by a factor of 1%.

In his work, van den Bergh proposed two algorithms which periodically add particles randomly to the flock for accomplishing his theorem that guarantees convergence to the global optimum [40]. Following that line of thought in COPSO, a *Temp* population is created after applying the perturbation operators, and then used (as if they were random particles) to update the whole P_{Best} population, saving the G_{Best} particle. This is performed every $n/2$ generations (n=flock size), subject to the current probabilities p_C and p_M of the perturbation operators. Figure 2 shows clearly this feature.

The perturbation of P_{Best} is what makes COPSO different. Other approaches perturb the position of the particle and later P_{Best} is updated accordingly. COPSO, however, applies the perturbations to the memory P_{Best} and let the motion of the particles to PSO.

4.4 Constraint Handling Approach

K. Deb introduced the `superiority of feasible solutions` selection based on the idea that any individual in a constrained search space must first comply with the constraints and then with the function value [7]. COPSO adopted such popular tournament selection whose rules have been included in the functions *LocalBest*, *ParticleBest* and *TheBest*:

1. Given two feasible particles, pick the one with better function value;
2. From a pair of feasible and infeasible particles, pick the feasible one;
3. If both particles are infeasible, pick the particle with the lowest sum of constraint violation.

The sum of constraint violations is, of course, the total value by which unsatisfied constraints are violated (computed by function **SCV** in Figure 2). The superiority of feasible solutions does not require tuning parameters or applying special operators. Just a simple comparison is used to choose the best individual. Our approach applies this method, allowing feasible and infeasible solutions in the swarm. It enriches the information about the search space, especially at boundaries. Nevertheless, for handling equality constraints, it is not enough just converting them into inequality constraints of the form $|h_j| \leq \varepsilon$, where ε is called the tolerance. COPSO uses a *Dynamic Tolerance* for handling equality constraints.

Dynamic Tolerance: An initial tolerance value of $\varepsilon = 1.0$ is decremented by 10% to a specified *target value* τ. The tolerance value is decreased whenever a percentage of feasible particles (PFP) is attained. The initial value of PFP is 100%, and it is updated at every generation with the following equation:

$$PFP = \left(1 - \frac{generation}{maxgeneration}\right)\% \qquad (5)$$

In addition, when COPSO has performed 90% of the function evaluations and the tolerance value has not reached the specified target value ($\varepsilon > \tau$), the current tolerance value is replaced with that target value ($\varepsilon = \tau$). For the last 10% of the function evaluations, the tolerance value is kept fixed ($\varepsilon = \tau$); thus, the particles have additional time to achieve convergence.

In brief, at the beginning of the process the particles are allowed to be feasible in a large number. Whenever the tolerance is decremented some particles might become infeasible, resulting in the lost of promising regions. The incorporation of the G_{Best} component to the velocity equation helps to keep up the flock into the promising regions. The efficiency of the dynamic tolerance is shown in Section 6, where a set of test problems with equality constraints is solved by COPSO.

5 Problem Statement

We are interested in the general nonlinear programming problem in which we want to:

$$\text{Find } \mathbf{x} \text{ which optimizes } f(\mathbf{x})$$

subject to:

$$g_i(\mathbf{x}) \leq 0, \quad i = 1, \ldots, I$$

$$h_j(\mathbf{x}) = 0, \quad j = 1, \ldots, E$$

where $\mathbf{x}$ is the vector of solutions $\mathbf{x} = [x_1, x_2, \ldots, x_d]^T$, I is the number of inequality constraints and E is the number of equality constraints (in both cases, constraints could be linear or non-linear). For an inequality constraint that satisfies $g_i(\mathbf{x}) = 0$, then we will say that it is active at $\mathbf{x}$. All equality constraints h_j (regardless of the value of $\mathbf{x}$ used) are considered active at all points of $\mathcal{F}$ ($\mathcal{F}$ = feasible region).

6 Experiments on Benchmark Functions

For all experiments, COPSO used the parameters mentioned in Section 4. These parameters agree with van den Bergh as to guarantee convergent trajectories [40].

$$\begin{aligned}\frac{1}{2}(c_1+c_2)-1 &< w \\ \frac{1}{2}(1+1)-1 &< 0.5 \\ 0 &< 0.5\end{aligned} \qquad (6)$$

COPSO uses a flock size of 75 members. Total number of function evaluations is 350,000 (suggested by Runnarson and Yao [34]). When required, a specific number of function evaluations is performed to develop a fair comparison against other algorithms. In our experiements, the tolerance value (ε) decreased from 1.0 to a target value of 1E-06 for all equality constraints. The other algorithms apply a fixed tolerance of ε =1E-04 for transforming equality constraints to inequality constraints. A PC computer with Windows XP and C++ Builder Compiler, Pentium-4 processor at 3.00GHz, 1.00 GB of RAM, was used to perform the experiments. A well-know benchmark of 24 functions was used to compare COPSO against the state-of-the-art algorithms.

The formal definition of all problems and their optimal values are available in the following document:
`http://www.cimat.mx/reportes/enlinea/I-07-04.pdf`

6.1 The Benchmark Problems

Mezura extended to 24 functions the original 13 benchmark functions of Runnarson and Yao [28]. The maximization problems were transformed to minimization problems to generalize the comparison (*max* f(x) = *min* -f(x)). The basic statistics for 30 runs are shown in Table 1.

COPSO solved 22 out of 24 benchmark problems. Also, the median of the 30 runs reaches the optimal solution in 20 out of 24 problems, and it is very near in test problem g23. COPSO was unable to find feasible solutions for test problems g20 and g22. These problems were used in the Special Session on Constrained Real-Parameter Optimization at the CEC 2006. The technical report [24] coments that the best known solution of test problem g20 is a little infeasible and no feasible solution has been found so far. About test problem g22, we believe that COPSO was unable to find a feasible solution because this problem is subject to 19 equality constraints; it is the test problem with more equality constraints in the benchmark. In test problem g17, COPSO finds a better optimum than the optimum reported, due that an error value of ε =1E-06 allows to attain better solutions. Finally, COPSO always found the optimal solution in 17 out of 24 test problems, along the 30 runs.

Several benchmark problems required less than 350,000 function evaluations to find the optimal value. The Table 2 shows the COPSO's number of objective function evaluations required to approximate the best-known optimum within a margin

Table 1 The results of COPSO on the benchmark

TF	Optimal	Best	Median	Mean	Worst	S. D.
g01	-15.000000	-15.000000	-15.000000	-15.000000	-15.000000	0
g02	-0.803619	-0.803619	-0.803619	-0.799859	-0.787296	5.0161E-03
g03	-1.000000	-1.000005	-1.000005	-1.000005	-1.000005	0
g04	-30665.539	-30665.538672	-30665.538672	-30665.538672	-30665.538672	0
g05	5126.4981	5126.498096	5126.498096	5126.498096	5126.498096	0
g06	-6961.8138	-6961.813876	-6961.813876	-6961.813876	-6961.813876	0
g07	24.306209	24.306209	24.306209	24.306209	24.306218	1.6111E-06
g08	-0.095825	-0.095825	-0.095825	-0.095825	-0.095825	0
g09	680.630057	680.630057	680.630057	680.630057	680.630057	0
g10	7049.248	7049.248020	7049.248020	7049.248074	7049.249209	2.2285E-04
g11	0.750000	0.749999	0.749999	0.749999	0.749999	0
g12	-1.000000	-1.000000	-1.000000	-1.000000	-1.000000	0
g13	0.053950	0.053950	0.053950	0.053950	0.053950	0
g14	-47.761	-47.761128	-47.761128	-47.761128	-47.761128	0
g15	961.715	961.715171	961.715171	961.715171	961.715171	0
g16	-1.905	-1.905155	-1.905155	-1.905155	-1.905155	0
g17	8853.539	8853.539831	8927.597675	8908.678674	8952.490404	34.119
g18	-0.8660	-0.866025	-0.866025	-0.866025	-0.866025	0
g19	32.386	32.348678	32.348678	32.348680	32.348705	5.3077E-06
g20	0.204979	*0.204389	*0.224929	*0.224840	*0.238456	1.0289E-02
g21	193.785	193.785033	193.785033	193.785033	193.785033	0
g22	236.430975	*186.911502	*7970.676246	*8608.014343	*19226.181834	6.7078E+03
g23	-400.000551	-400.000551	-399.968046	-396.336901	-364.913882	9.1364
g24	-5.508	-5.508013	-5.508013	-5.508013	-5.508013	0

$*$ Infeasible solution

of 1E-4. As mentioned, the top value is 350,000 function evaluations. In test problems g03, g05, g11, g13, g14, g15, g17, g20, g21, g22, and g23, the number of function evaluations required is reported when a tolerance value of ε=1E-6 was reached, since these bechmark problems are subject to equality constraints.

In the Table 2, we also show the number of feasible runs. A run that finds at least one feasible solution in less than 350,000 fitness evaluations is called feasible. The Table 2 shows the number of successful runs (when the best value found is within 1E-4 of the optimal the run is successful). The experiments show that just 3 of the 22 problems which found the optimal solution, require more than 200,000 evaluations to reach the optimal region. The convergence reported by COPSO suggest that the algorithm would perform efficienly with just 150,000 function evaluations. In Table 3 we show the results of COPSO with just 150,000 function evaluations.

COPSO showed a similar behavior with both 350,000 and 150,000 function evaluations. Just in test function g23, COPSO did not reach the optimal with 150,000

Table 2 Fitness function evaluations to reach the optimum within 1E-04

TF	Best	Median	Mean	Worst	S.D.	Feasible Runs	Successful Runs
g01	80776	90343	89506.30	96669	3567.79	30	30
g02	87419	93359	93958.82	99654	3932.39	30	17
g03	97892	106180	107512.90	122540	6503.81	30	30
g04	93147	103308	102903.10	110915	3868.52	30	30
g05	149493	165915	165510.33	188040	9201.72	30	30
g06	95944	109795	111170.83	130293	8250.71	30	30
g07	114709	138767	140822.77	208751	20003.28	30	30
g08	2270	4282	4131.53	5433	744.80	30	30
g09	94593	103857	104695.03	119718	6881.20	30	30
g10	109243	135735	139707.50	193426	17682.16	30	30
g11	89734	112467	111577.33	127650	8560.88	30	30
g12	482	6158	5884.50	9928	2281.72	30	30
g13	149727	160964	159213.40	168800	5582.01	30	30
g14	138471	149104	148658.67	165292	4830.97	30	30
g15	127670	135323	135892.67	147268	4407.20	30	30
g16	65872	75451	75250.20	83087	4696.44	30	30
g17	221036	232612	231250.50	236434	5346.64	30	8
g18	97157	107690	107851.40	124217	6664.79	30	30
g19	109150	122279	125989.53	167921	12745.50	30	30
g20	NR	NR	NR	NR	NR	0	0
g21	206559	221373	220885.73	233325	7495.97	30	30
g22	NR	NR	NR	NR	NR	0	0
g23	260154	274395	274390.67	291456	10256.16	30	6
g24	11081	18278	25991.47	63338	15855.57	30	30

NR: Optimal not reached

function evaluations. In some test problems the other basic statistics (median, mean and worst values) show an insignificant reduction of performance.

6.2 COPSO and Dynamic Tolerance for Handling Equality Constraints

For showing the efficiency of the dynamic tolerance to handle equality constraints, COPSO solves the test problems g03, g05, g11, g13, g14, g15, g17, g20, g21, g22, and g23, applying a constant tolerance value $\varepsilon = 1\text{E-06}$ along the whole process. Table 4 presents the results of COPSO without dynamic tolerance and 350,000 fitness function evaluations.

The results presented in Table 4, shows that COPSO without dynamic tolerance was able to find the optimal solution in only 4 out of 11 test problems (compare versus Table 1). The median of the 30 runs is near to the optimal region only

Table 3 The results of COPSO on the benchmark with 150,000 function evaluations

TF	Optimal	Best	Median	Mean	Worst	S. D.
g01	-15.000000	-15.000000	-15.000000	-15.000000	-15.000000	0
g02	-0.803619	-0.803619	-0.794897	-0.796831	-0.778322	7.0699E-03
g03	-1.000000	-1.000005	-1.000005	-1.000005	-1.000005	0
g04	-30665.539	-30665.538672	-30665.538672	-30665.538672	-30665.538672	0
g05	5126.4981	5126.498096	5126.498096	5126.498096	5126.498096	0
g06	-6961.8138	-6961.813876	-6961.813876	-6961.813876	-6961.813876	0
g07	24.306209	24.306209	24.306216	24.306371	24.307563	3.7682E-04
g08	-0.095825	-0.095825	-0.095825	-0.095825	-0.095825	0
g09	680.630057	680.630057	680.630057	680.630057	680.630057	0
g10	7049.248	7049.248020	7049.256651	7049.329736	7049.956260	1.5754E-01
g11	0.750000	0.749999	0.749999	0.749999	0.749999	0
g12	-1.000000	-1.000000	-1.000000	-1.000000	-1.000000	0
g13	0.053950	0.053950	0.053950	0.053950	0.053950	0
g14	-47.761	-47.761128	-47.761128	-47.761128	-47.761124	8.1367E-07
g15	961.715	961.715171	961.715171	961.715171	961.715171	0
g16	-1.905	-1.905155	-1.905155	-1.905155	-1.905155	0
g17	8853.539	8853.539831	8927.597682	8922.847388	8953.504502	24.38
g18	-0.8660	-0.866025	-0.866025	-0.866025	-0.866025	7.2927E-08
g19	32.386	32.348679	32.349428	32.351628	32.364675	4.3035E-03
g20	0.204979	*0.230695	*0.237481	*0.237140	*0.239278	1.6854E-03
g21	193.785	193.785033	193.785054	193.785111	193.785657	1.3518E-04
g22	236.430975	*202.731972	*3752.915089	*6624.040387	*19316.081466	6.3870E+03
g23	-400.000551	-399.135562	-389.925796	-384.098457	-336.566384	17.315
g24	-5.508	-5.508013	-5.508013	-5.508013	-5.508013	0

∗ Infeasible solution

Table 4 The results of COPSO without Dynamic Tolerance

TF	Optimal	Best	Median	Mean	Worst	S. D.
g03	-1.000000	-0.978109	-0.223987	-0.258190	-0.000886	2.0407E-01
g05	5126.4981	5126.498096	5149.681308	5180.149212	5557.763291	9.3208E+01
g11	0.750000	0.749999	0.779451	0.783479	0.859523	3.3474E-02
g13	0.053950	0.749138	0.961657	0.946136	0.998083	5.7773E-02
g14	-47.761	-47.758761	-47.219714	-46.970898	-45.085320	6.8130E-01
g15	961.715	961.715171	961.782214	962.058256	964.752897	6.3404E-01
g17	8853.539	8933.066289	8945.915638	8960.000781	9156.003831	5.2685E+01
g20	0.204979	*0.205469	*0.205484	*0.205483	*0.205485	2.6263E-06
g21	193.785	193.785034	193.785034	227.231279	327.462518	4.9553E+01
g22	236.430975	*382.880649	*5967.141521	*7665.320210	*18753.721473	6.2330E+03
g23	-400.000551	-393.740205	-256.575290	-246.415279	-35.961120	9.4751E+01

∗ Infeasible solution

in 2 out of 11 test problems. The mean and worst values are far away of the optimal region in all the test problems subject to equality constraints.

6.3 Comparison of COPSO vs CPSO

Firstly, we compare COPSO and the Constrained Particle Swarm Optimization algorithm (CPSO) developed by Cagnina et all [3]. Since both approaches are based on PSO with local-global model, this comparison is quite fair.

CPSO handles constraints through a superiority of feasible solutions, and keeps diversity by adding mutations to the velocity vector. Also, CPSO uses a Gaussian update equation proposed by Kennedy [20]. The comparison is shown in Table 5. CPSO performed 340,000 fitness function evaluations, 10,000 less than COPSO, but it is not significative for the comparison. COPSO's performance is better than CPSO on test problems g02, g05, g06, g07, g10 and g13. In 12 of 13 benchmark problems, the worst solution of COPSO is better than the mean solution of CPSO, along 30 runs. The comparison was just performed for the first 13 benchmark problems because is the available information of CPSO. Hu and Eberhart [17], and Zhang [42] reported the first solutions to these 13 benchmark problems with very limited success.

Table 5 Comparison of COPSO and CPSO in the benchmark problems

		Best Result		Mean Result		Worst Result	
TF	**Optimal**	**COPSO**	**CPSO**	**COPSO**	**CPSO**	**COPSO**	**CPSO**
g01	-15.000000	-15.000000	-15.000000	-15.000000	-15.000000	-15.000000	*-134.2191
g02	-0.803619	-0.803619	-0.801388	-0.799859	-0.7653	-0.787296	-0.0917
g03	-1.000000	-1.000005	-1.000	-1.000005	-1.0000	-1.000005	-1.0000
g04	-30665.539	-30665.538672	-30665.659	-30665.538672	-30665.6564	-30665.538672	-25555.6267
g05	5126.4981	5126.498096	5126.497	5126.498096	5327.9569	5126.498096	*2300.5443
g06	-6961.8138	-6961.813876	*-6961.825	-6961.813876	-6859.0759	-6961.813876	64827.5545
g07	24.306209	24.306209	24.400	24.306209	31.4854	24.306218	4063.5252
g08	-0.095825	-0.095825	-0.095825	-0.095825	-0.0958	-0.095825	0.0006
g09	680.630057	680.630057	680.636	680.630057	682.3973	680.630057	18484.7591
g10	7049.248	7049.248020	7052.8523	7049.248020	8533.6999	7049.249209	13123.4656
g11	0.750000	0.749999	0.749	0.749999	0.7505	0.749999	*0.4466
g12	-1.000000	-1.000000	-1.000	-1.000000	-1.000	-1.000000	*-9386
g13	0.053950	0.053950	0.054237	0.053950	1.4139	0.053950	0.9675

$*$ Infeasible solution

6.4 Comparison of COPSO vs ISRES

Runarsson and Yao first proposed the Stochastic Ranking algorithm for constrained optimization [34], and later they developed an improved version called "Improved Stochastic Ranking Evolution Strategy", (ISRES) [35] (still one major representant of the state of the art). Experiments for test problems g14 through g24 were developed using the ISRES's code available at Runarsson's page. The parameters used were the same suggested by the authors [35]. The comparison is shown in Table 6. Both algorithms performed the same number of fitness function evaluations,

Table 6 Comparison of COPSO and ISRES on the benchmark problems

		Best Result		Mean Result		Worst Result	
TF	Optimal	COPSO	ISRES	COPSO	ISRES	COPSO	ISRES
g01	-15.000000	-15.000000	-15.000000	-15.000000	-15.000000	-15.000000	-15.000000
g02	-0.803619	-0.803619	-0.803619	-0.799859	-0.782715	-0.787296	-0.723591
g03	-1.000000	-1.000005	-1.001	-1.000005	-1.001	-1.000005	-1.001
g04	-30665.539	-30665.53867	-30665.539	-30665.53867	-30665.539	-30665.53867	-30665.539
g05	5126.4981	5126.498096	5126.497	5126.498096	5126.497	5126.498096	5126.497
g06	-6961.8138	-6961.813876	-6961.814	-6961.813876	-6961.814	-6961.813876	-6961.814
g07	24.306209	24.306209	24.306	24.306209	24.306	24.306218	24.306
g08	-0.095825	-0.095825	-0.095825	-0.095825	-0.095825	-0.095825	-0.095825
g09	680.630057	680.630057	680.630	680.630057	680.630	680.630057	680.630
g10	7049.248	7049.248020	7049.248	7049.248020	7049.25	7049.249209	7049.27
g11	0.750000	0.749999	0.750	0.749999	0.750	0.749999	0.750
g12	-1.000000	-1.000000	-1.000000	-1.000000	-1.000000	-1.000000	-1.000000
g13	0.053950	0.053950	0.053942	0.053950	0.066770	0.053950	0.438803
g14	-47.761	-47.761128	-47.761129	-47.761128	-47.759250	-47.761128	-47.735569
g15	961.715	961.715171	961.715171	961.715171	961.715171	961.715171	961.715171
g16	-1.905	-1.905155	-1.905155	-1.905155	-1.905155	-1.905155	-1.905155
g17	8853.539	8853.539831	8889.9003	8908.678674	8889.9442	8952.490404	8890.9516
g18	-0.8660	-0.866025	-0.866025	-0.866025	-0.866025	-0.866025	-0.866025
g19	32.386	32.348678	32.348689	32.348680	32.374095	32.348705	32.644735
g20	0.204979	*0.204389	NA	*0.224840	NA	*0.238456	NA
g21	193.785033	193.785033	193.785034	193.785033	220.056989	193.785033	325.144812
g22	236.430975	*186.911502	NA	*8608.014343	NA	*19226.181834	NA
g23	-400.000551	-400.000551	-400.000551	-396.336901	-321.342939	-364.913882	-47.842844
g24	-5.508	-5.508013	-5.508013	-5.508013	-5.508013	-5.508013	-5.508013

* Infeasible solution, NA Not available

350,000. Note that ISRES and COPSO find the best values in the same problems, except in test problem g17 where ISRES was unable to find an optimal solution. Also, COPSO average is closer to the optimum value and is better than ISRES in problems g02, g13, g14, g19, g21 and g23. But ISRES is better in problem g17, where it finds an average solution lower than COPSO. Both COPSO and ISRES were unable to find feasible solutions for test problems g20 and g22.

6.5 *Comparison of COPSO vs SMES*

In his Ph.D. thesis, Mezura proposed the extended benchmark of 24 test problems, and an approach to solve it, the "Simple Multimember Evolutionary Strategy" (SMES), which worked reasonable well on the first 13 problems but had a weak performance on the new problems (g14 through g23), mainly due to reduced exploration [27]. In Table 7 we show the comparison of COPSO and SMES. In this case, COPSO and SMES performed 240,000 fitness function evaluations.

It can be seen that COPSO is clearly better than SMES in problems g02, g05, g07, g10, g13, g14, g15, g17, g19, g21 and g23. Although the best values reported for the rest of the problems are comparable, COPSO outperforms SMES in the average results for problems g02, g05, g06, g07, g09, g10, g13, g14, g15, g17, g18, g19, g21 and g23. COPSO and SMES were unable to find feasible solutions for test problems

Table 7 Results of COPSO and SMES for benchmark problems

		Best Result		Mean Result		Worst Result	
TF	Optimal	COPSO	SMES	COPSO	SMES	COPSO	SMES
g01	-15.000000	-15.000000	-15.000000	-15.000000	-15.000000	-15.000000	-15.000000
g02	-0.803619	-0.803619	-0.803601	-0.801137	-0.785238	-0.792607	-0.751322
g03	-1.000000	-1.000005	-1.000000	-1.000005	-1.000000	-1.000005	-1.000000
g04	-30665.539	-30665.53867	-30665.539	-30665.53867	-30665.539	-30665.53867	-30665.539
g05	5126.4981	5126.498096	5126.599	5126.498096	5174.492	5126.498096	5304.167
g06	-6961.8138	-6961.813876	-6961.814	-6961.813876	-6961.284	-6961.813876	-6952.482
g07	24.306209	24.306209	24.327	24.306209	24.475	24.306212	24.843
g08	-0.095825	-0.095825	-0.095825	-0.095825	-0.095825	-0.095825	-0.095825
g09	680.630057	680.630057	680.632	680.630057	680.643	680.630057	680.719
g10	7049.248	7049.248020	7051.903	7049.252037	7253.047	7049.320549	7638.366
g11	0.750000	0.749999	0.750000	0.749999	0.750000	0.749999	0.750000
g12	-1.000000	-1.000000	-1.000000	-1.000000	-1.000000	-1.000000	-1.000000
g13	0.053950	0.053950	0.053986	0.053950	0.166385	0.053950	0.468294
g14	-47.761	-47.761128	-47.535	-47.761128	-47.368	-47.761118	-47.053
g15	961.715	961.715171	*961.698	961.715171	963.922	961.715171	967.787
g16	-1.905	-1.905155	-1.905	-1.905155	-1.905	-1.905155	-1.905
g17	8853.539	8853.539831	*8890.1826	8900.046859	*8954.1364	8928.146735	*9163.6767
g18	-0.8660	-0.866025	-0.866	-0.866025	-0.716	-0.866025	-0.648
g19	32.386	32.348679	34.223	32.348776	37.208	32.349921	41.251
g20	0.204979	*0.233685	*0.211364	*0.237738	*0.251130	*0.240781	*0.304414
g21	193.785033	193.785033	*347.9809	193.785033	*678.3924	193.785107	*985.7821
g22	236.430975	*202.238522	*2340.6166	*8801.959136	*9438.2549	*19831.711035	*17671.5351
g23	-400.000551	-399.98688	*-1470.1525	-394.098457	*-363.5082	-333.566384	*177.2526
g24	-5.508	-5.508013	-5.508	-5.508013	-5.508	-5.508013	-5.507

$*$ Infeasible solution

g20 and g22. But, COPSO finds feasible solutions for test problems g17, g21 and g23, where SMES could not find feasible solutions in any single run.

6.6 Comparison of COPSO vs DOPSO Using 50,000 Fitness Function Evaluations

Finally, we compare COPSO and the Dynamic-Objective Particle Swarm Optimization algorithm (DOPSO) developed by Lu and Chen [25]. DOPSO handles constraints through a bi-objective unconstrained optimization problem, where one objective is to find the feasible region, and the other one is to optimize the original objective function. Also, DOPSO uses a restricted velocity PSO (RVPSO) [25], which modifies the velocity equation replacing the inertia term $(w * v_t)$ by $w * (G_{Best} - P_{Best})$. DOPSO performed $50,000$ fitness function evaluations. Thereby, a new experiment with just $50,000$ function evaluations is performed by COPSO for a fair comparison. Table 8 presents the results of this comparison.

DOPSO finds better solutions than the optimal reported for test problems g03, g05, g11, and g13 because it uses a tolerance value of ε =1E-03 (remember that COPSO reaches a tolerance value of ε =1E-06). In general, COPSO's performance is better than DOPSO on test problems g01, g02, g05 and g13. But, DOPSO is better than COPSO in test problem g10.

Table 8 Comparison of COPSO and DOPSO in the benchmark problems

		Best Result		Mean Result		Worst Result	
TF	Optimal	COPSO	DOPSO	COPSO	DOPSO	COPSO	DOPSO
g01	-15.000000	-14.999999	-15	-14.999999	-14.4187	-14.999999	-12.4531
g02	-0.803619	-0.803614	-0.664028	-0.792865	-0.413257	-0.778276	-0.259980
g03	-1.000000	-1.000005	-1.0050	-1.000004	-1.0025	-1.000003	-0.9334
g04	-30665.539	-30665.538672	-30665.539	-30665.538672	-30665.539	-30665.538672	-30665.539
g05	5126.4981	5126.498030	5126.4842	5126.520522	5241.0549	5126.634522	5708.2250
g06	-6961.8138	-6961.813876	-6961.81388	-6961.813876	-6961.81388	-6961.813876	-6961.8138
g07	24.306209	24.306802	24.306	24.316905	24.317	24.345260	24.385
g08	-0.095825	-0.095825	-0.095825	-0.095825	-0.095825	-0.095825	-0.095825
g09	680.630057	680.630057	680.630	680.630059	680.630	680.630080	680.630
g10	7049.248	7049.278315	7049.2480	7050.506423	7049.2701	7058.453341	7049.5969
g11	0.750000	0.749999	0.749	0.749999	0.749	0.749999	0.749
g12	-1.000000	-1.000000	-1	-1.000000	-1	-1.000000	-1
g13	0.053950	0.053949	0.0538666	0.053963	0.6811235	0.054012	2.0428924

7 Final Remarks and Future Research

This chapter reviews and compares a new algorithm called COPSO. It has shown high performance in constrained optimization problems of linear or nonlinear nature. The experimental results are highly competitive with respect to the state-of-the-art algorithms. Three important contributions of COPSO are worth to mention: A new neighborhood structure for PSO, the incorporation of perturbation operators without modifying the essence of the PSO, and a dynamic tolerance to handle equality constraints.

The first contribution is the singly-linked neighborhood structure. It slows down the convergence of the flock, breaking the double-link that exists between the particles in the ring neighborhood structure. COPSO implements a singly-linked ring with a neighborhood of size $n = 4$, but a general algorithm to build neighborhoods of size n is given.

Another relevant idea developed by COPSO is the perturbation of the P_{Best} memory as a source of variation and therefore diversity. Two perturbation operators, C-perturbation and M-perturbation are applied to the $PBest$. It is equivalent to perturb the particle's memory, but not its behavior (as it is performed by other approaches, that tend to destroy the flock's organization capacity).

The last property of COPSO is the adoption of a dynamic tolerance to handle equality constraints. The effectiveness of the approach was demonstrated through a specific set of experiments reported in Table 4.

The performance of COPSO is highly robust, reporting the best results on the benchmark problems when limited to only 50,000 fitness function evaluations (see Table 8).

Future research should address the advantage meant by the different neighborhood topologies in constrained optimization (most researchers have approached global optimization problems). Which topology is best to maintain the flock's diversity? Or provides fast convergence to the constrained optimum ?

Acknowledgements. The first author acknowledges support from National Council of Science and Technology, CONACYT, to pursue graduate studies at the Center for Research in Mathematics. The second author acknowledges support from CONACYT through project No. 51667-Y.

References

1. Angeline, P.J.: Evolutionary Optimization versus Particle Swarm Optimization: philosophy and performance differences. In: Porto, V.W., Waagen, D. (eds.) EP 1998. LNCS, vol. 1447, pp. 601–610. Springer, Heidelberg (1998)
2. Blackwell, T.M.: Particle swarms and population diversity. Soft Computing 9(11), 793–802 (2005)
3. Cagnina, L.C., Esquivel, S.C., Coello, C.A.: A particle swarm optimizer for constrained numerical optimization. In: Proceedings of the 9th International Conference - Parallel problem Solving from Nature, PPSN IX, pp. 910–919 (2006)
4. Carlisle, A., Dozier, G.: Adapting Particle Swarm Optimization to Dynamic Environments. In: Proceedings of the International Conference on Artificial Intelligence, ICAI 2000, pp. 429–434 (2000)
5. Coath, G., Halgamuge, S.K.: A comparison of Constraint-Handling Methods for the Application of Particle Swarm Optimization to Constrained Nonlinear Optimization Problems. In: Proceedings of the 2003 Congress on Evolutionary Computation, pp. 2419–2425. IEEE Press, Los Alamitos (2003)
6. Das, S., Konar, A., Chakraborty, U.K.: Improving particle swarm optimization with differentially perturbed velocity. In: Proceedings of the 2005 Conference on Genetic and Evolutionary Computation, pp. 177–184. ACM Press, New York (2005)
7. Deb, K.: An efficient constraint handling method for genetic algorithms. Computer Methods in Appplied Mechanics and Engineering 186(2-4), 311–338 (2000)
8. Eberhart, R., Shi, Y.: Comparison between genetic algorithms and particle swarm optimization. In: Porto, V.W., Waagen, D. (eds.) EP 1998. LNCS, vol. 1447, pp. 611–616. Springer, Heidelberg (1998)
9. Eberhart, R., Dobbins, R., Simpson, P.: Computational Intelligence PC Tools. Academic Press Professional, London (1996)
10. Fogel, D.: An Introduction to Simulated Evolutionary Optimization. IEEE Transaction on Neural Networks 5(1), 3–14 (1994)
11. Hamida, S.B., Petrowski, A.: The need for improving the exploration operators for constrained optimization problems. In: Proceedings of the Congress on Evolutionary Computation, pp. 1176–1183. IEEE Press, Los Alamitos (2000)
12. He, S., Prempain, E., Wu, Q.H.: An Improved Particle Swarm Optimizer for Mechanical Design Optimization Problems. Engineering Optimization 36(5), 585–605 (2004)
13. Hernandez-Aguirre, A., Botello, S., Coello, C.: PASSSS: An implementation of a novel diversity strategy to handle constraints. In: Proceedings of the 2004 Congress on Evolutionary Computation CEC 2004, pp. 403–410. IEEE Press, Los Alamitos (2004)
14. Hernandez-Aguirre, A., Botello, S., Coello, C., Lizarraga, G., Mezura, E.: Handling constraints using multiobjective optimization concepts. International Journal for Numerical Methods in Engineering 59(13), 1989–2017 (2004)
15. Hernández, A., Muñoz, A., Villa, E., Botello, S.: COPSO: Constrained Optimization via PSO Algorithm. Technical Report of the Computer Sciences Department, Centro de Investigación en Matemáticas, Guanajuato, México (2007), `http://www.cimat.mx/reportes/enlinea/I-07-04.pdf`

16. Holland, J.: Adaptation in Natural and Artificial Systems. University of Michigan Press, Ann Arbor (1975)
17. Hu, X., Eberhart, R.: Solving Constrained Nonlinear Optimization Problems with Particle Swarm Optimization. In: Proceedings of the 6th World Multiconference on Systemics, Cybernetics and Informatics, SCI 2002, p. IIIS (2002)
18. Hu, X., Eberhart, R., Shi, Y.: Engineering optimization with particle swarm. In: Proceedings of the IEEE Swarm Intelligence Symposium, pp. 53–57. IEEE Press, Los Alamitos (2003)
19. Kennedy, J.: The Particle Swarm: Social Adaptation of Knowledge. In: Proceedings of the IEEE International Conference on Evolutionary Computation, pp. 303–308. IEEE Press, Los Alamitos (1997)
20. Kennedy, J.: Bare Bones Particle Swarms. In: IEEE Swarm Intelligence Symposium, pp. 80–87. IEEE Press, Los Alamitos (2003)
21. Kennedy, J., Eberhart, R.: The Particle Swarm: Social Adaptation in Information-Processing Systems. McGraw-Hill, London (1999)
22. Kennedy, J., Mendes, R.: Population Structure and Particle Swarm Performance. In: Proceedings of the 2002 Congress on Evolutionary Computation, pp. 1671–1676. IEEE Press, Los Alamitos (2002)
23. Krink, T., Vesterstrom, J.S., Riget, J.: Particle Swarm Optimization with Spatial Particle Extension. In: Proceedings of the 2002 Congress on Evolutionary Computation, pp. 1474–1479. IEEE Press, Los Alamitos (2002)
24. Liang, J., Runarsson, T., Mezura-Montes, E., Clerc, M., Suganthan, P., Coello, C., Deb, K.: Problem Definitions and Evaluation Criteria for the CEC 2006. Special Session on Constrained Real-Parameter Optimization, Technical Report (2006)
25. Lu, H., Chen, W.: Dynamic-objective particle swarm optimization for constrained optimization problems. Journal of Combinatorial Optimization 12(4), 408–418 (2006)
26. Lvbjerg, M., Rasmussen, T., Krink, T.: Hybrid particle swarm optimiser with breeding and subpopulations. In: Proceedings of the 2001 Genetic and Evolutionary Computation Conference (2001)
27. Mezura, E.: Alternatives to Handle Constraints in Evolutionary Optimization. CINVESTAV-IPN, D.F., Mexico (2004)
28. Mezura, E., Coello, C.: Identifying on-line behavior and some sources of difficulty in two competitive approaches for constrained optimization. In: Proceedings of the Conference on Evolutionary Computation, CEC 2005, pp. 56–63. IEEE Press, Los Alamitos (2005)
29. Michalewicz, Z.: Genetic Algorithms + Data Structures = Evolution Programs. Springer, Germany (1994)
30. Muñoz, A., Hernández, A., Villa, E.: Constrained optimization via particle evolutionary swarm optimization algorithm (PESO). In: Proceedings of the Genetic and Evolutionary Computation Conference, GECCO 2005, pp. 209–216. Association for Computing Machinery (2005)
31. Parsopoulos, K., Vrahatis, M.: Particle swarm optimization method for constrained optimization problems. Intelligent Technologies - Theory and Application: New Trends in Intelligent Technologies 76, 214–220 (2002)
32. Parsopoulos, K., Vrahatis, M.: Unified Particle Swarm Optimization for Solving Constrained Engineering Optimization Problems. In: Wang, L., Chen, K., S. Ong, Y. (eds.) ICNC 2005. LNCS, vol. 3612, pp. 582–591. Springer, Heidelberg (2005)
33. Price, K.V., Storn, R.M., Lampinen, J.A.: Differential Evolution: A practical approach to global optimization. Springer, Berlin (2005)
34. Runarsson, T.P., Yao, X.: Stochastic Ranking for Constrained Evolutionary Optimization. IEEE Transactions on Evolutionary Computation 4(3), 284–294 (2000)

35. Runarsson, T.P., Yao, X.: Search Biases in Constrained Evolutionary Optimization. IEEE Transactions on Systems, Man and Cybernetics - Part C: Applications and Reviews 35(2), 233–243 (2005)
36. Settles, M., Soule, T.: Breeding Swarms: A GA/PSO Hybrid. In: Proceedings of the 2005 Conference on Genetic and Evolutionary Computation, pp. 161–168. ACM Press, New York (2005)
37. Storn, R.: System Design by Constraint Adaptation and Differential Evolution. IEEE Transactions on Evolutionary Computation 3(1), 22–34 (1999)
38. Storn, R., Price, K.: Differential Evolution - A Simple and Efficient Adaptive Scheme for Global Optimization over Continuous Spaces. Technical Report TR-95-012, International Computer Science Institute (1995)
39. Toscano, G., Coello, C.: A Constraint-Handling Mechanism for Particle Swarm Optimization. In: Proceedings of the 2004 Congress on Evolutionary Computation, pp. 1396–1403. IEEE Press, Los Alamitos (2004)
40. van den Bergh, F.: An Analysis of Particle Swarm Optimizers. University of Pretoria, South Africa (2002)
41. Zhang, J., Xie, F.: DEPSO: Hybrid Particle Swarm with Differential Evolution Operator. In: Proceedings of IEEE International Conference on Systems, Man and Cybernetics, pp. 3816–3821. IEEE Press, Los Alamitos (2003)
42. Zhang, W., Xie, X., Bi, D.: Handling boundary constraints for numerical optimization by Particle Swarm flying in periodic search space. In: Proceedings of the 2004 Congress on Evolutionary Computation, pp. 2307–2311. IEEE Press, Los Alamitos (2004)

[illegible]

Boundary Search for Constrained Numerical Optimization Problems

Guillermo Leguizamón and Carlos Coello Coello

Abstract. The necessity of approaching the boundary between the feasible and infeasible search space for many constrained optimization problems is a paramount challenge for every constraint-handling technique. It is true that many of the state-of-the-art constraint-handling techniques performs well when facing constrained problems. However, it is a common situation that reaching the boundary between the feasible and infeasible search space could bc a difficult task for some particular problems. Firstly, this chapter shows a general overview of the constraint-handling techniques based on a boundary approach and emphasizing a recent proposal applying a more general boundary operator. In addition, the chapter includes some particular considerations related to the implementation aspects of the boundary approach when facing problems with one o more constraints. Another important issue also considered here is about the implementation of this approach when taking into account different search engines. On this direction, some basic examples are depicted as guidelines for possible implementations under well-known metaheuristics as Evolutionary Algorithms (EAs), Particle Swarm Optimization (PSO), and Ant Colony Optimization (ACO). To validate the boundary approach implemented under the above metaheuristics, an experimental study is presented in which well-known problems were considered. Finally, a brief summary of the chapter and some ideas for future works are given which could help the researchers interested in developing advanced constraint-handling techniques.

Keywords: Constraint-Handling Techniques, Boundary Search, Particle Swarm Optimization, Ant Colony Optimization.

Guillermo Leguizamón
LIDIC - Universidad Nacional de San Luis Ejército de Los Andes 950 San Luis 5700, Argentina
e-mail: `legui@unsl.edu.ar`

Carlos A. Coello Coello
CINVESTAV-IPN Evolutionary Computation Group (EVOCINV) Departamento de Computación Av. IPN No. 2508. Col. San Pedro Zacatenco México D.F. 07300, México
e-mail: `ccoello@cs.cinvestav.mx`

E. Mezura-Montes (Ed.): Constraint-Handling in Evolution. Optimi., SCI 198, pp. 25–49.
springerlink.com

1 Introduction

The boundary search can be considered as alternative approach when facing problems with active constraints (see Sect. 2) at the optimal or high quality solutions. This is mainly observed for that problems that include at least one equality constraint. However, there exist many optimization problems without any equality constraint for which many of their constraints are active for the best feasible solutions. Clearly the more appropriate situation for the boundary approach is when the problem has only one equality constraint. In addition, the boundary search could be used as a complementary mechanism of another constraint-handling technique to rapidly reach or force the exploration towards the regions around the boundary between the feasible and infeasible search space.

Specific operators (or *ad hoc* boundary operators) could be the right candidates to search only on the boundary region between the feasible and infeasible search space. However, it is not always possible to design specific boundary operators for each problem constraint. Furthermore, there exist only a few examples of this kind of boundary operators in the literature. Michalewicz et al. [17] wrote one of the first papers on boundary search through the use of evolutionary algorithms for constrained numerical optimization problems. The efficiency of this approach was shown by using two constrained optimization problems: Keane's function (also known as $G02$) [10] and another function with one equality constraint (also known as $G03$). For solving these problems the authors proposed two genetic operators which generate offspring lying on the boundary between the feasible and infeasible search space. Similarly, Schoenauer and Michalewicz [19] proposed several evolutionary operators capable of exploring a general surface of dimension $n-1$ (n is the number of variables). The design of these operators, tested on three problems, depends on the surface representation: curves-based, plane-based, and parametric representation. Although not using an *ad hoc* boundary operator, Wu et al. [23] proposed a GA for the optimization of a water distribution system, which is a highly constrained optimization problem. The proposed approach co-evolves and self-adapts two penalty factors in order to guide and preserve the search towards the boundary of the feasible search space.

On the other hand, Gottlieb [9] introduces and remarks the use of the the boundary approach for a combinatorial optimization problem, more precisely, the multiple knapsack problem. By the same year Leguizamón and Michalewicz proposed for the same problem, an Ant System which biases the search boundary region and gives encouraging results [15]. The maximum independent set problem is also considered under the same approach and many instances of this problem were solved optimally [7].

The reduction of the search space is one of the most relevant characteristics of the boundary search approach since the exploration considers only the boundary of the feasible search space. However, many of the test cases considered in the former works only include problems with one constraint for which it is possible to define *ad hoc* genetic operators that fit perfectly the boundary of the feasible region. However, this sort of approach is impractical in an arbitrary problem with many constraints, and it is therefore necessary to define a more general approach for boundary search which can be as robust as possible to deal with different types of constraints. More recently, Leguizamón and Coello [13, 14] proposed a boundary approach that focuses the search on the boundary region by considering a sort of more general boundary operator applicable to any type of constraints. The experimental reports show the applicability of the boundary approach using ant colony based algorithms as a search engine.

The next section of this chapter shows a general overview of the constraint-handling techniques. Sect. 3 describes the boundary search approach and two alternatives for exploring the boundary between the feasible and infeasible search space: *ad hoc* operators and a general operator. In the last case, emphasizing a more recent proposal according to Leguizamón and Coello [13, 14]. In order to visualize some considerations about specific implementation aspects of the boundary approach, Sect. 4 displays the pseudocode by taking into account different search engines. On this direction, some basic examples are depicted as guidelines for possible implementations under well-known metaheuristics (MHs) as Evolutionary Algorithms [1, 6, 8], Particle Swarm Optimization [6, 11, 12], and Ant Colony Optimization [4–6] which have been used as alternative search engines to successfully implement constraint-handling techniques. Section 5 shows the results of the application of EAs, ACO, and PSO to a well-known testbed of numerical optimization problems. Finally, a brief summary and some ideas for future work are given which could help the researchers interested in developing advanced constraint-handling techniques.

2 A General Overview of Constraint-Handling Techniques

The general nonlinear programming problem whose aim is to find $\mathbf{x}$ so as to optimize:

$$f(\mathbf{x}) \quad \mathbf{x} = (x_1, x_2, \ldots, x_n) \in \mathbb{R}^n$$

where $\mathbf{x} \in \mathscr{F} \subset \mathscr{S}$. The set $\mathscr{S} \subset \mathbb{R}^n$ defines the search space and sets $\mathscr{F} \subset \mathscr{S}$ and $\mathscr{U} = \mathscr{S} - \mathscr{F}$ define the *feasible* and *infeasible* search spaces, respectively. The search space $\mathscr{S}$ is defined as an n-dimensional rectangle in $\mathbb{R}^n$ (domains of variables defined by their lower and upper bounds):

$$l(i) \leq x_i \leq u(i) \text{ for } 1 \leq i \leq n$$

whereas the feasible set $\mathscr{F}$ is defined by the intersection of $\mathscr{S}$ and a set of additional $m \geq 0$ constraints:

$$g_j(\mathbf{x}) \leq 0, \text{ for } j = 1, \ldots, q \text{ and } h_j(\mathbf{x}) = 0 \text{ for } j = q+1, \ldots, m.$$

At any point $\mathbf{x} \in \mathscr{F}$, the constraints g_k that satisfy $g_k(\mathbf{x}) = 0$ are called the active constraints at $\mathbf{x}$. Equality constraints h_j are active at all points of $\mathscr{F}$.

The most common way of extending MHs (e.g., EAs, PSO) to optimize constrained problems has been through the use of penalty functions, which are the oldest and more widely used approaches. However, due to the well-known difficulties associated with them, researchers in MHs (mainly in EAs) have proposed different ways to automate the definition of good penalty factors, which remains as the main drawback of using penalty functions. Additionally, several researchers have developed a considerable amount of alternative approaches to handle constraints, mainly to deal with specific features of some complex optimization problems in which it is difficult to estimate good penalty factors or to even generate a single feasible solution.

A comprehensive survey of constraint-handling techniques that have been adopted over the years to handle all sorts of constraints (linear, non-linear, equality, and inequality) in EAs can be found in Coello Coello [3]. This survey covers extensively: a) penalty functions in several of their variations that have been used with EAs (i.e., static, dynamic, annealing, adaptive, co-evolutionary, and death penalties); b) the use of special representations and genetic operators (e.g., operators that preserve feasibility at all times and decoders that transform the shape of the search space); c) repair algorithms, which are normally used in combinatorial optimization problems in which the traditional genetic operators tend to generate infeasible solutions all (or at least most of) the time. Thus, "repair" refers, in this context, to make valid (or feasible) these individuals through the application of a certain (normally heuristic) procedure; d) techniques that handle objectives and constraints separately; and e) discusses approaches that use hybrids with other techniques such as Lagrangian multipliers or fuzzy logic as well as other more novel approaches.

Although the Coello Coello's survey is mainly concerned with constraint-handling techniques from the perspective of EAs, the concepts depicted conform a general framework to be applied with other search engines. Examples of the more recent applications of using novels MHs (e.g., DE, PSO, ACO, etc.) for constraint handling techniques can be found at the web site from EVOCINV [2] which includes upto date references to the more representatives constraint-handling techniques implemented under different search engines.

It is worth remarking that many problems formulated as at the beginning of this section, include active constraints at the best known or optimal solutions. For example, for problems with at least one equality constraint h_j, the respective optimal solution will lay on the region defined by $h_j(\mathbf{x}) = 0$. Furthermore, for many problems, the best solutions may lay on the boundary between the feasible and infeasible search space of some inequality constrains, i.e., the region defined by $g_j(\mathbf{x}) = 0$. When those conditions are met for a particular problem, the design of *ad hoc* operators or approaches that explore the search space focusing on the boundary region (according either to the equality and/or inequality constraints) can be a suitable alternative for including in a specific search engine or metaheuristic.

3 The Boundary Search Approach

In the following we first explain how the boundary region can be approached given a specific search space; more precisely, a subset of the n-dimensional space $\mathbb{R}^n$. Then, we also describe the manner in which this search space could be explored assuming a hypothetical search engine and exploration operators, as well as the properties that they should satisfy. Afterwards, we present in detail the proposed technique that takes advantage of the boundary approach to explore some specific regions of the boundary of the feasible search space by considering *ad hoc* and a general boundary operators.

Definition 1. Given a constrained numerical optimization problem, the respective constraints determine a feasible search space $\mathscr{F}$. In addition, these problem constraints determine a set $\mathscr{F}_{\mathscr{B}} \subseteq \mathscr{F}$ which represent the points in $\mathscr{F}$ which make active at least one of the problem constraints ($\mathscr{F}_{\mathscr{B}} = \mathscr{F}$ when the problem includes only one equality constraint).

To appropriately define boundary operators, we must take into account that set $\mathscr{F}_{\mathscr{B}}$ must be closed under the application of a boundary operator. Let us suppose that Ω_r is a r-ary boundary operator, i.e., it takes r points in set $\mathscr{F}_{\mathscr{B}}$; then the resulting point must be in $\mathscr{F}_{\mathscr{B}}$. In other words, an r-ary boundary operator can generally be defined as $\Omega_r : (\mathscr{F}_{\mathscr{B}})^r \rightarrow \mathscr{F}_{\mathscr{B}}$. For example, when considering a "boundary" *crossover operator* that takes two parents to generate one child, it can be defined in the boundary context as $\Omega_2 : \mathscr{F}_{\mathscr{B}} \times \mathscr{F}_{\mathscr{B}} \rightarrow \mathscr{F}_{\mathscr{B}}$. Fig. 1 display a set of points laying on the boundary region with respect to a problem constraint and the application of boundary operators that take respectively 2, 3, and 4 points as argument. Clearly, Ω_r operator could be any operator (e.g., genetic operators as in EAs) or equation (e.g., velocity and position updating of the particles in PSO) used in different MHs used for sampling new points in the search space.

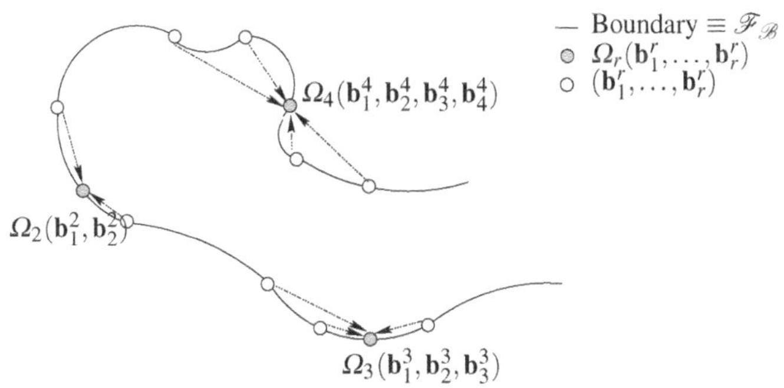

Fig. 1 This figure shows a set of points laying on $\mathscr{F}_{\mathscr{B}}$ (white circles) and the respective points sampled on $\mathscr{F}_{\mathscr{B}}$ (filled circles) after the application of the boundary operator Ω_r, for $r = 2, 3, 4$. Notice that $\mathscr{F}_{\mathscr{B}}$ is closed under the application of Ω_r

It is worth noting that the definition of an Ω_r operator which makes set $\mathscr{F}_{\mathscr{B}}$ closed under its application could be a difficult or at least impossible task for most of the usual problem constraints. The most paradigmatic case is the proposal of Michalewicz et al. [17] where different boundary operators where defined which perfectly fit in our definition of the Ω_r operator: *Geometrical Crossover, Spherical*

Crossover, and *ad- hoc* mutation operators. However, their application is limited to couple of problems with specific characteristics. In order to mitigate this drawback, the design of more general operators to explore $\mathscr{F}_{\mathscr{B}}$ could be an interesting approach when considering a wider set of problems to be tackled under the boundary approach. It must be noted that the above classification, i.e., *ad hoc* and general operators, is not intended to be a general rule, instead, it only represents the authors' point of view in the way the boundary approach could be conceived.

In the following we present two possibles alternative to define an Ω_r operator, either by *ad hoc* operators or a general operator. The main difference among these two types of operators is that *ad hoc* operators are defined to operate on the *phenotype space* where as the general operator does it on the *genotype space* (Sect. 3.2 describes these concepts in more detail).

3.1 *Ad hoc* Boundary Operators

As mentioned in a previous section, the work of Michalewicz et al. [17] represents one of the first intents to define specific *ad hoc* operators to explore the boundary between the feasible and infeasible search space. Although this approach to explore the boundary region can be useful and efficient, it in not always possible to define a specific one for any problem constraint. Indeed, for most of the problem constraints, to find an adequate operator could be as difficult as solving the original problem. As a manner of showing the way in which the boundary region can be explored, we will describe in the following a classical example, the Keane's problem [10]. The reason for showing this alternative through an example is because an *ad hoc* operator completely depends on the shape of the involved constraints.

3.1.1 Exploration of the Boundary Region under *ad-hoc* Operators

The exploration of the boundary search space under *ad hoc* operators can be visualized more clearly for a particular problem since its definition depends on the particular constraint considered. For our example, we have chosen a very well-known constraint optimization problem widely used as a benchmark to test different constraint-handling techniques: the Keane's problem. This problem, also known as $G02$ (see [16]) includes a non-linear objective function and two inequality constraints. More precisely, an optimal solution for $G02$ aims at maximizing:

$$f(\mathbf{x}) = \left|\frac{\sum_{i=1}^{n} cos^4(x_i) - 2\prod_{i=1}^{n} cos^2(x_i)}{\sqrt{\sum_{i=1}^{n} i x_i^2}}\right|$$

subject to:

$$g_1(\mathbf{x}) = 0.75 - \prod_{i=1}^{n} x_i \leq 0$$
$$g_2(\mathbf{x}) = \sum_{i=1}^{n} x_i - 7.5n \leq 0$$

where $n = 20$ and $\mathscr{S} = \{\mathbf{x} \in \mathbb{R}^{20} | 0 \leq x_i \leq 10, \text{ for } i = 1 \ldots 20\}$. The best known solution is at $\mathbf{x}^* = (3.16237443645701, 3.12819975856112, 3.09481384891456, 3.06140284777302, 3.02793443337239, 2.99385691314995, 2.95870651588255, 2.92182183591092, 0.49455118612682, 0.48849305858571, 0.48250798063845, 0.47695629293225, 0.47108462715587, 0.46594074852233, 0.46157984137635, 0.45721400967989, 0.45237696886802, 0.44805875597713, 0.44435772435707, 0.44019839654132)$ where $f(\mathbf{x}^*) = 0.80619$ and constraint g_1 is close to being active. Fig. 2 shows a plotting of the $G02$'s objective function for $n = 2$.

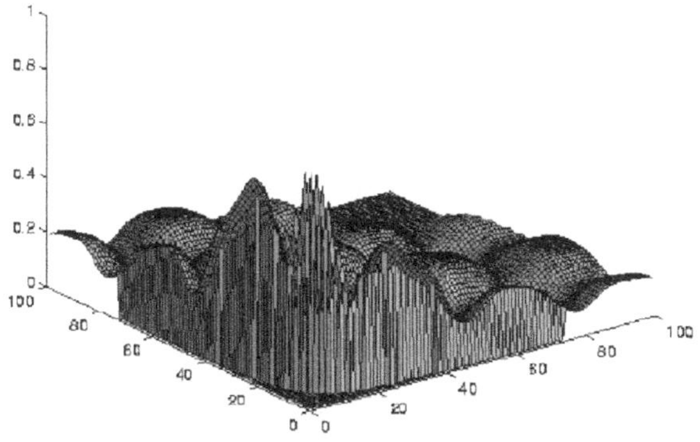

Fig. 2 Keane's function with $n = 2$. Infeasible solutions were assigned value zero

Despite of problem $G02$ has two constraints, one of them is just dismissed when solving this problem since: a) the second constraint is satisfied for all solutions laying on the boundary of the first constraint and b) the first constraint is close to being active at best known solution (see [17]). For this problem, the space $\mathscr{F}_{\mathscr{B}} = \{\mathbf{x} \in \mathscr{S} | 0.75 - \prod_{i=1}^{20} x_i = 0\}$

The search engine used in Michalewicz et al. [17] is an EA where the main components are described as follows:

i. **Initialization:** Randomly choose a positive variable for x_i and use it inverse as a variable for x_{i+1}. The last variable is either 0.75 (when n is odd) or multiplied by 0.75 (if n is even).
ii. **Mutation:** is a unary operator ($r = 1$) defined by

$$\Omega_1(\mathbf{x}) = (x_1, \ldots, q \times x_i, \ldots, \frac{1}{q} \times x_j, \ldots, x_n),$$

where q is a random factor restricted to respect the bounds on the variables and $1 \leq i, j \leq 20$ randomly chosen, with $i \neq j$.
iii. **Crossover:** is a binary operator ($r = 2$) called *geometrical crossover* and defined according to our notation by

$$\Omega_2(\mathbf{x}, \mathbf{y}) = (x_1^{\alpha} y_1^{1-\alpha}, \ldots, x_n^{\alpha} y_n^{1-\alpha}), \text{ with } 1 \leq \alpha \leq 1,$$

where α is a random number.

By using the above initialization procedure, all points in the initial population will lay on $\mathcal{F}_{\mathcal{B}}$. Similarly, $\mathcal{F}_{\mathcal{B}}$ is closed under the application of *ad hoc* operators Ω_1 and Ω_2, therefore, all points generated will also lay on the boundary. It is worth remarking that the application of boundary operators for problem $G02$ produced at least two very important results in the area of constraint-handling techniques. First of all, good quality solutions were formerly obtained by using a boundary operator and second, showed the usefulness and potential of the boundary approach for certain types of numerical optimization problems.

3.2 A General Boundary Operator

We describe here an alternative[1] general boundary approach (proposed in [13, 14]) which is based on the notion that each point $\mathbf{b}$ of the boundary region can be represented by means of two different points $\mathbf{x}$ and $\mathbf{y}$, where $\mathbf{x}$ is some feasible point and $\mathbf{y}$ is some infeasible one, i.e., $(\mathbf{x},\mathbf{y})$ can represent one point lying on the boundary by applying a "binary search" on the straight line connecting the points $\mathbf{x}$ and $\mathbf{y}$ (when considering an equality constraint, $\mathbf{z} \in \mathcal{F}$ iff $h(\mathbf{z}) \leq 0$; otherwise, $\mathbf{z} \in \mathcal{U}$). Fig. 3 shows a hypothetical search space including the feasible (shadowed area) and infeasible regions. We can identify four points lying on the boundary $\mathbf{b}_1$, $\mathbf{b}_2$, $\mathbf{b}_3$, and $\mathbf{b}_4$ which are respectively obtained from $(\mathbf{x}_1,\mathbf{y}_1)$, $(\mathbf{x}_2,\mathbf{y}_2)$, $(\mathbf{x}_3,\mathbf{y}_3)$, and $(\mathbf{x}_4,\mathbf{y}_4)$.

The binary search applied to each pair of points $(\mathbf{x},\mathbf{y})$ is achieved following the steps described in function BS (see Algorithm 1). For example, a possible application of this process can be seen in Fig. 3 where we adopt the pair of points $(\mathbf{x}_3,\mathbf{y}_3)$ from which we obtain the point $\mathbf{b}_3$, which lies on the boundary. The first step (labeled (1)) indicates that the first mid point found is feasible. Consequently, the left side of the straight line ($\mathbf{x}_3$) is moved to point $\mathbf{p}_1$. In the next step (labeled (2)) we consider the points $\mathbf{p}_1$ and $\mathbf{y}_3$ as extreme points for which the mid point is the infeasible point $\mathbf{p}_2$. Thus, the new feasible point or right extreme of the line is now the point $\mathbf{p}_2$. Finally, the last point generated is $\mathbf{b}_3$ which can be either lying on or close to the boundary. Condition ((dist_to_boundary($\mathbf{m}$) $\leq \delta$) AND Feasible($\mathbf{m}$)) defines a threshold to stop the process of approaching the boundary. However, the second part of this condition (i.e., "Feasible($\mathbf{m}$)") it is only applied when considering an inequality constraint. In this way, function *BS* guarantees that $\mathbf{m}$ is in the feasible side regarding the corresponding inequality constraint under consideration. It is worth noticing that parameters $\mathbf{x}$ and $\mathbf{y}$ are local to BS, i.e., function BS behaves as a decoder of the pair of feasible and infeasible points passed as parameters. Therefore, the number of "mid_points_between" $\mathbf{x}$ and $\mathbf{y}$ before approaching the boundary within a distance less that δ is given by $log_2(d)$ where $d = (dist(\mathbf{x},\mathbf{y})))/\delta$. Thus, the closer to the boundary, the larger $log_2(d)$.

[1] It is possible that other general operators can be visualized to implement under the boundary search approach.

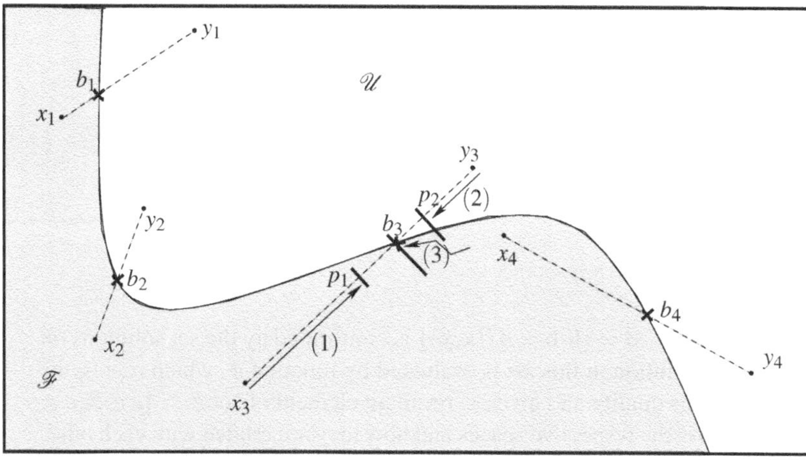

Fig. 3 Given one feasible and one infeasible point, the respective point lying on the boundary can be easily reached by using a simple binary search. In this way, the each point on the boundary can be reached from at least a pair of points $(\mathbf{x}, \mathbf{y})$ with $\mathbf{x} \in \mathscr{F}$ and $\mathbf{y} \in \mathscr{U}$

3.2.1 Exploring the Boundary Region under a General Operator

So far, we have shown how a point lying on the boundary $\mathbf{b}$ can be represented through a pair of points $(\mathbf{x}, \mathbf{y})$ with $\mathbf{x} \in \mathscr{F}$ and $\mathbf{y} \in \mathscr{U}$. Now we need to consider the exploration of the search space which, according to our proposal, can be defined as $\mathscr{G} = \{(\mathbf{x}, \mathbf{y}) | \mathbf{x} \in \mathscr{F} \subset \mathbb{R}^n \wedge \mathbf{y} \in \mathscr{U} \subset \mathbb{R}^n\}$, that is, the set of pair of points $(\mathbf{x}, \mathbf{y})$ as described above. This space can be considered a *genotype space* as known in the area of evolutionary computation. Since each point from $\mathscr{G}$ represents a point on the boundary, it is necessary the application of the decoder represented by function *BS* (see Algorithm 1) to obtain the respective *phenotype*, i.e., the "gene expression" of

Algorithm 1. BS(**x**,**y**: real vector): real vector

```
m: real vector;
repeat
    m = mid_point_between(x, y);
    if Is_on_Boundary(m) then
        return m; { m is a point lying on the boundary }
    end if
    if Feasible(m) then
        x = m;
    else
        y = m;
    end if
until (dist_to_boundary(m) ≤ δ) AND (Feasible(m));
return m; {The closest point to the boundary according to δ }
```

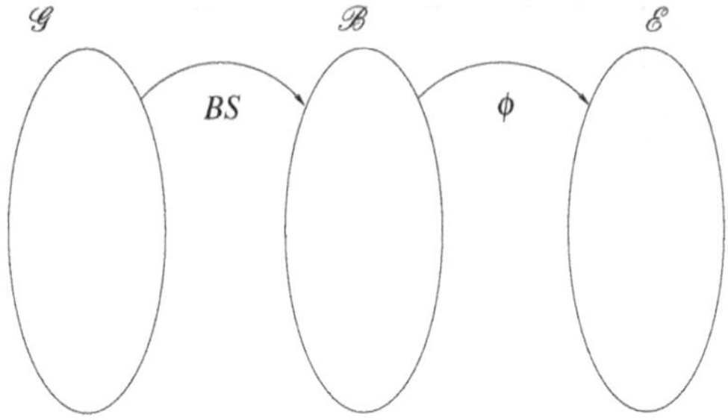

Fig. 4 The search or genotype space ($\mathscr{G}$), phenotype space ($\mathscr{B}$), and space $\mathscr{E}$, and the respective connection through the decoder *BS* and function evaluation ϕ

$(\mathbf{x},\mathbf{y}) \in \mathscr{G}$. Thus, the set $\mathscr{B} = \{\mathbf{b}|\mathbf{b} = BS(\mathbf{x},\mathbf{y})\}$ is conformed by the set solutions on the boundary. Each solution in this set is evaluated by function ϕ, which represents a measure of solutions quality and gives as result an element of set $\mathscr{E} = \{e \in \mathbb{R}|e = \phi(\mathbf{b})\}$. Fig. 4 displays the respective spaces and how they are related with each other by the application of functions *BS* and ϕ, respectively.

From the above described, is clear that the search engine must deal with the exploration of space $\mathscr{G}$. Fig. 5 shows a set of three hypothetical points $\{(\mathbf{x}_1,\mathbf{y}_1), (\mathbf{x}_2,\mathbf{y}_2), (\mathbf{x}_3,\mathbf{y}_3)\}$ in $\mathscr{G}$, a problem constraint, and the respective points $\{\mathbf{b}_1,\mathbf{b}_2,\mathbf{b}_3\}$ on the boundary. The application of the general Ω_3 operator on $(\mathbf{b}_1,\mathbf{b}_2,\mathbf{b}_3)$ gives as result a point $\mathbf{b}$ on $\mathscr{F}_{\mathscr{B}}$. To obtain this point on the boundary, an operator χ is applied respectively on the points on $\mathscr{F}$ and $\mathscr{U}$ to obtain a new point on $\mathscr{G}$, i.e., $(\mathbf{x},\mathbf{y})$, from which a new point on the boundary is obtained as displayed in the following:

$$\begin{aligned}\Omega_3(\mathbf{b}_1,\mathbf{b}_2,\mathbf{b}_3) &= \Omega_3(BS(\mathbf{x}_1,\mathbf{y}_1),BS(\mathbf{x}_2,\mathbf{y}_2),BS(\mathbf{x}_3,\mathbf{y}_3))\\ &= BS(\chi_3(\mathbf{x}_1,\mathbf{x}_2,\mathbf{x}_3),\chi_3(\mathbf{y}_1,\mathbf{y}_2,\mathbf{y}_3))\\ &= BS(\mathbf{x},\mathbf{y})\\ &= \mathbf{b}\end{aligned}$$

Indeed, operator χ could be any exploration operator which will depend on the search engine used to explore space $\mathscr{G}$. For example, from the perspective of evolutionary algorithms, it can be created an initial population of individuals where each one of them represents an element of set $\mathscr{G}$. Therefore, suitable operators to be chosen could be any qualified crossover and/or mutation operators for floating-point representations. A similar approach can be adopted if using another search engine suitable for exploring continuous spaces, e.g., particle swarm optimization, ant colony optimization, differential evolution, immune systems, etc. Sect. 4 describes through three MHs for which the boundary approach can be easily implemented with just a few changes when considering different search engines.

3.3 *Focusing on the Problem Constraints*

It is important to remember that we are assuming active constraints at the global optimum to proceed with this method which focuses the exploration on the boundary region. However, either using an *ad hoc* or general operator (as the proposed here),

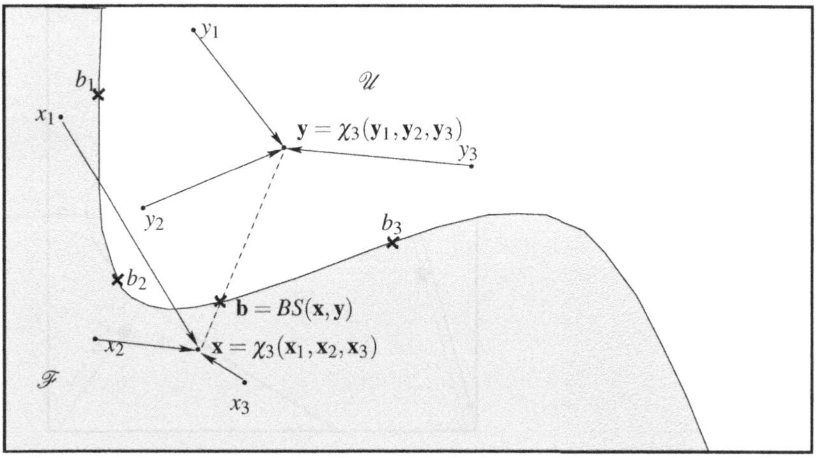

Fig. 5 A set of hypothetical points $\{(\mathbf{x}_1, \mathbf{y}_1), (\mathbf{x}_2, \mathbf{y}_2), (\mathbf{x}_3, \mathbf{y}_3)\}$ in $\mathscr{G}$, a problem constraint, and the set respective points $\{\mathbf{b}_1, \mathbf{b}_2, \mathbf{b}_3\}$ on the boundary. The application of the general 3-ary operator on $(\mathbf{b}_1, \mathbf{b}_2, \mathbf{b}_3)$ gives as result a point $\mathbf{b}$ on $\mathscr{F}_{\mathscr{B}}$. In fact, the operator Ω_r is a combination of operators (χ) that respectively works on space $\mathscr{F}$ and $\mathscr{U}$, in addition to the decoding function *BS*

the main difficulty of a boundary operator is concerned with problems with more that one constraint.

Certainly, the simplest case to apply the boundary approach is when the problem has only one constraint which could be either an equality or an inequality constraint. Let us suppose that the problem includes only one constraint, let us say h. Then, the search engine should proceed by sampling: a) when applying an *ad hoc* operator, a set of solutions laying on the boundary and after that, applying the specific *ad hoc* operators to explore $\mathscr{F}_{\mathscr{B}}$ directly or b) when applying the general operator, a set of pair of points on the *genotype space* $\mathscr{G}$ which each one of them is mapped via function *BS* in to the boundary region, after that, $\mathscr{F}_{\mathscr{B}}$ is indirectly explored through the exploration of space $\mathscr{G}$. In both cases, all solutions generated will be feasible. Two examples of case (a) are certainly given in the Michalewicz et al.'s proposal [17]. For the second case (b), Fig. 5 display a hypothetical problem with one constraint, some points on space $\mathscr{G}$, and the modification of this points which gives, via function *BS*, a new point on the boundary.

On the other hand, when facing the typical situation in which we have more than one constraint, it is necessary to define an appropriate policy to explore the boundary as efficiently as possible since space $\mathscr{F}_{\mathscr{B}}$ will be now determined by a set of constraints rather than one. In this case, it will be not possible to define any type of boundary operators that make closed $\mathscr{F}_{\mathscr{B}}$ under their application.

In the following we focus in some alternatives to manage this situation considering only the use of the general operator. In fact, the same approach can be applied when considering *ad hoc* operators, however, we believe that this is unlikely due to the difficulty to define them for any type of constraint. Therefore, one

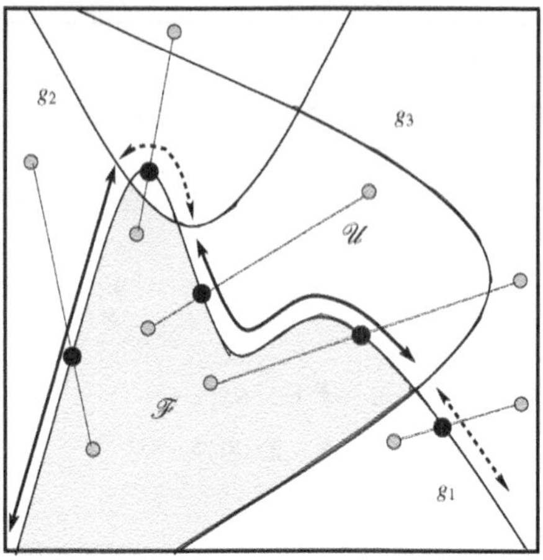

Fig. 6 Feasible search space defined by 3 inequality constraints. The search proceeds on the boundary of constraint g_1, however, some points on the boundary of g_1 are infeasible when considering the whole set of the problem constrains

possibility is to explore in turn the boundary of each problem constraint. The selection of the constraints to search for can be determined using different methods. If the problem includes at least one equality constraint, such equality constraints are the most appropriate candidates to be selected first. However, a possible search engine could keep focused on a particular constraints over the whole run or may be change from one problem constraint to another depending on a particular condition. In our previous work [13] we defined a simple condition based on a parameter called t_c which counts the number of iterations the algorithm focuses in a particular constraint. However, more complex condition could be considered, for example, taking into account the population diversity or the degree in which some problem constraints are being violated. For the last case, the scheme proposed by Schoenauer and Xanthakis [20] could be adapted and applied when focusing on the boundary region. The proposed scheme consists on a multi-steps evolutionary process based on behavioral memory that considers each problem constraint in turn. The process starts from the first constraint. When the current constraint j is processed, the solutions that violate constraints $j-1,\ldots,1$ are given a zero fitness. Simultaneously, when constraint j is satisfied (according to a particular threshold), constraint $j+1$ is then processed. The process continues until all constraints have been considered.

As an illustrative example when facing a problem with more than one constraint, Fig. 6 shows a hypothetical search space determined by three inequality constraints. Let us suppose that the search proceeds starting on constraint g_1. If the visited points are on the boundary of $\mathscr{F}$, these points will also satisfy the remaining problem constraints (filled line in Fig. 6). However, the exploration of the boundary with respect to constraint g_1 will eventually produce points violating constraints g_2 and g_3 (dotted line in 6). One of the simplest methods to deal with this situation could be for example, the application of a penalty function for the infeasible solutions. In addition, if g_1 is active at the global optimum, the method will focus the search on the boundary in order to restrict the explored regions of the whole search space. Note however, that other (more sophisticated) constraint-handling techniques can also be adopted. For example, it could be considered the inclusion of the Stochastic Ranking approach [18] to make the comparisons among the solutions generated [14] and thus avoiding the inclusion and tuning of any penalty factor for solutions evaluation. As a manner of showing some concrete examples of the possible application of the boundary approach, in the next section, we focus on its implementation from the perspective of three different search engines: Evolutionary Algorithms, Particle Swarm Optimization, and Ant Colony Optimization.

4 Implementation Issues

This section is aimed to explain in some detail the implementation the boundary approach under the general boundary operator by using three search engines: EAs, PSO, and ACO. Since their implementation under *ad hoc* operators is straightforward, i.e., they do not produce any important change on the respective search engine, we have decided not to include the respective implementation.

The selection of the three mentioned search engines does not follow any kind or priority of one over the remaining ones. In first place, EAs can be considered the more popular MHs used in optimization and particularly in numerical constrained optimization problems. Second, PSO is a more recent MHs which lately have been successfully applied to solve many of the state-of-the-art benchmarks for numerical optimization. Finally, we consider ACO as a possible alternative which was chosen for two main reasons: 1) it was the first search engine used to test the boundary approach using a general operator with encouraging results and 2) more advanced version of the ACO metaheuristic have been recently developed which can successfully be applied to problems defined over continuous domains with or without constraints.

Before giving any detail about the respective algorithms, is worth noticing that all the algorithms were designed including the Stochastic Ranking as complementary handling technique, i.e., the solutions are ranked based on the sorting procedure given in Alg. 2 which receives as argument an structure T containing a set of solutions on $\mathscr{F}_{\mathscr{B}}$ and the respective objective value. Thus, $T.\mathbf{x}_i$ and $T.e_i$ represent respectively the solution i and its objective value. It should be noticed that I_j and I_{j+1} are indexes that point to the structure T. In addition, it is also important to remark that the algorithms described in the following are designed for the general case, i.e., when the problem includes more than one constraint. However, for the simplest case (problems with only one constraint) the designed algorithms are still applicable by modifying a few lines of code as will be explained for each particular search engine considered. On the other hand, each algorithm includes references to different structures called $T_{\mathscr{F}}$, $T_{\mathscr{U}}$, and $T_{\mathscr{B}}$ which respectively represent a population of solutions in spaces $\mathscr{F}$, $\mathscr{U}$, and $\mathscr{F}_{\mathscr{B}}$ (the same applies to the auxiliary structures called A and A'). Similarly, an structure $T_{\mathscr{E}}$ is used to save the objective value for the respective solutions in $T_{\mathscr{B}}$. Finally, variable 'ctr' indicates the current constraint under consideration, i.e., indicates that the algorithm is currently focusing the

Algorithm 2. A general outline of the stochastic ranking algorithm using a bubble-sort like algorithm as defined in [18]. P_f represents the probability of using only the objective function for comparisons when ranking solutions in the infeasible regions of the search space (a value of $0.4 < P_f < 0.5$ was reported as the most appropriate). Parameters N and λ represent respectively the maximum number of sweeps and number of solutions that are ranked by comparing adjacent solutions in at least λ sweeps, and $rnd \in U(0,1)$.

```
procedure Sort(var T)
I_j = j, ∀j{1,...,λ}
for i in 1 : N do
    for j in 1 : λ − 1 do
        if (T.x_{I_j} == T.x_{I_{j+1}}) || (rnd < P_f)) then
            if (T.x_{I_j} > T.x_{I_{j+1}}) then
                swap(I_j, I_{j+1})
            end if
        else
            if (T.x_{I_j} > T.x_{I_{j+1}}) then
                swap(I_j, I_{j+1})
            end if
        end if
    end for
    if no swap done then
        break
    end if
end for
```

exploration on the boundary of constraint 'ctr'. Additional specific structures used by each search engine will be explained when necessary.

Similarly, there exist a set of common functions used through the three algorithms which are described in the following:

- init(): is in charge of obtaining the initial population of points in space $\mathscr{G}$.
- evaluate(): assigns the respective objective value.
- Boundary(): applies function BS() to all the pair of points in $(T_{\mathscr{F}}, T_{\mathscr{U}})$ and returns the respective decoding of those points (the returning structure is usually saved in $T_{\mathscr{B}}$).
- change_constraint(): returns a boolean value indicating the decision of focusing the search on a different problem constraint.
- Re_init(): when a change of constraint occurs, this function reinitialize the points in structure $T_{\mathscr{F}}$ and $T_{\mathscr{U}}$ when necessary (e.g., it could be useful a simple perturbation operator here).
- get_next_constraint(): implements the policy for the selection of the next constraint to be considered for exploration. As indicated in the algorithms described further in this section, a possible policy could be the *Round-Robin* policy, however, others (more informed) policies are also possible.
- First$_k$(): returns the first k solutions found in the structure given as parameter.

- Sort(): applies the Alg. 2 to further make the selection of the set first k solutions.
- Update(): selects the best current solutions in $(T_{\mathscr{F}} \oplus A_{\mathscr{F}}, T_{\mathscr{U}} \oplus A_{\mathscr{F}})^2$.

4.1 Boundary by Means of EAs

The application of EAs to solve any optimization problems mainly includes a procedure to obtain the initial population, a selection operator, and a set of genetic operators. Alg. 3 describes the more important components of an EA used to solve constrained numerical optimization problems according to the boundary approach. Lines 3 to 5 are aimed to obtain the initial population on space $\mathscr{G}$, the respective points on the boundary and their objective values. It can be observed that the selection process (function Select()) is applied considering $T_{\mathscr{F}}$ and $T_{\mathscr{U}}$, and the respective objective values in $T_{\mathscr{E}}$. This process produces two intermediate structures $A'_{\mathscr{F}}$ and $A'_{\mathscr{U}}$, i.e., selected points on $\mathscr{G}$ which undergo genetic operators (in two independent steps) as can be observed in lines 14 and 15. The resulting structures are then used as decoders to obtain the respective new points on the boundary (line 17) saved in structure $A_{\mathscr{B}}$. The next step consists in operate on the ranking (see the SR based function Sort()) of the union of $A_{\mathscr{B}}$ and $T_{\mathscr{B}}$. From this ranking, only the respective best k solutions from space $\mathscr{G}$ will survive for the next generation ('Update()' is used to keep the respective best points on space $\mathscr{G}$).

When the problem has only one constraint (or only one is considered as in problem $G02$) there is no need to change from one to another constraint, therefore a few code lines can be dismissed, not necessarily dropped, from the general algorithm. In line 2, the 'initial_constraint' is the only problem constraint. The set of lines 7 through 16 are dropped and replaced by lines 13 through 15.

4.2 Boundary by Means of PSO

Differently to EA$_{\mathscr{B}}$, a PSO algorithm includes some other additional structures in addition to that used to keep the population or swarm. This particular structures are those representing the respective particles' velocities (called here $V_{\mathscr{F}}$ for space $\mathscr{F}$ and $V_{\mathscr{U}}$ for space $\mathscr{U}$) which let the algorithm explore the respective feasible and infeasible regions, i.e., the decoder space $\mathscr{G}$. Alg. 4 gives a general outline of a PSO implementing the boundary approach for solving constrained optimization problems. This algorithm follows the principle of PSO design known as "Local Best PSO". For that reason, two proper functions of this PSO version are added, 'Set_the_best_personal_position()' and 'Set_the_best_local_position()'. It can be noticed that they are first applied on $T_{\mathscr{F}}$ and then, on $T_{\mathscr{U}}$. In both cases, the selection on the local and best positions take into account the objective values $T_{\mathscr{E}}$ corresponding to the points they represent on $T_{\mathscr{B}}$.

[2] Operator $\oplus$ is defined as follows: $A \oplus B = (a_1, \ldots, a_N) \oplus (b_1, \ldots, b_M) = (a_1, \ldots, a_N, b_{N+1}, \ldots, b_{N+M})$ taking into account $T_{\mathscr{B}}$ and the respective objective values.

Algorithm 3. A general outline of EA$_{\mathscr{B}}$

1: $t = 0$
2: ctr= initial_constraint // 'ctr' represents the problem constraint under consideration
3: init($T_{\mathscr{F}}$,$T_{\mathscr{U}}$, ctr);
4: $T_{\mathscr{B}}$ =Boundary($T_{\mathscr{F}}$, $T_{\mathscr{U}}$)
5: $T_{\mathscr{E}}$ =evaluate($T_{\mathscr{B}}$)
6: **while** (stop condition not met) **do**
7:     **if** (change_constraint()) **then**
8:         ctr=get_next_ctr(ctr) // The search continues considering another problem constraint,
9:             // e.g., following a *Round-Robin* policy.
10:         $A_{\mathscr{F}}$ =Reinit($T_{\mathscr{F}}$, ctr);
11:         $A_{\mathscr{U}}$ =Reinit($T_{\mathscr{U}}$, ctr);
12:     **else**
13:         ($A'_{\mathscr{F}}$,$A'_{\mathscr{U}}$) =Select($T_{\mathscr{F}}$, $T_{\mathscr{U}}$, $T_{\mathscr{E}}$);
14:         $A_{\mathscr{F}}$ =Genetic_ops($A_{\mathscr{F}}$, ctr);
15:         $A_{\mathscr{U}}$ =Genetic_ops($A_{\mathscr{U}}$, ctr);
16:     **end if**
17:     $A_{\mathscr{B}}$ =Boundary($A_{\mathscr{F}}$,$A_{\mathscr{U}}$)
18:     $T_{\mathscr{B}}$ =First$_k$(Sort($T_{\mathscr{B}} \oplus A_{\mathscr{B}}$))
19:     $T_{\mathscr{E}}$ =evaluate($T_{\mathscr{B}}$)
20:     Update($T_{\mathscr{F}}$,$T_{\mathscr{U}}$,$T_{\mathscr{E}}$); { According to the new $T_{\mathscr{B}}$}
21:     $t = t + 1$
22: **end while**

The exploration stage for PSO$_{\mathscr{B}}$ is accomplished from lines 21 to 28. It can be noticed that the application of two specific PSO steps: 'Update_velocity()' and 'Update_position()'. These two functions are applied on structures $V_{\mathscr{F}}$ and $T_{\mathscr{F}}$ for the feasible part of space $\mathscr{G}$, and structures $V_{\mathscr{U}}$ and $T_{\mathscr{U}}$ for the infeasible one. In the case of 'Update_position()', it returns the modified point position that can be assigned as in lines 23 and 27. After the obtaining of the new solutions in $A_{\mathscr{F}}$ and $A_{\mathscr{U}}$, the process follows the same steps as for EA$_{\mathscr{B}}$. Finally, when the problem has only one constraint (or only one is considered as in problem $G02$) there is no need to change from one to another constraint, therefore a few code lines can be dismissed, not necessarily dropped, from the general algorithm. In line 2, the 'initial_constraint' represents the only problem constraint. The set of lines 15 through 29 are replaced by lines 21 through 28.

4.3 Boundary by Means of ACO

The last search engine is one based on the ACO metaheuristic. Particularly, we have chosen a recent and advanced version of an ACO algorithm for continuous problems proposed by Socha and Dorigo [22] which is called ACO$_{\mathbb{R}}$ where the solutions are built by using a probability density distribution (PDF). At step i each ant generates a random number according to a mixture of normal kernels of PDFs $P^i(x_i)$ defined

Algorithm 4. A general outline of lbest PSO$_{\mathscr{B}}$

```
t = 0
ctr= initial_constraint // 'ctr' represents the problem constraint under consideration
init(T_F, T_U, ctr); // Initializes feasible and infeasible swarms
T_B = Boundary(T_F, T_U)
T_E = evaluate(T_B)
while (stop condition not met) do
    for each particle i in T_F do
        Set_the_best_personal_position(T_F(i), T_E(i))
        Set_the_best_local_position(T_F(i), T_E(i))
    end for
    for each particle i in T_U do
        Set_the_best_personal_position(T_U(i), T_E(i))
        Set_the_best_local_position(T_U(i), T_E(i))
    end for
    if (change_constraint()) then
        ctr= get_next_ctr(ctr) // The search continues considering another problem constraint,
                               // e.g., following a Round-Robin policy.
    T_F = Reinit(T_F, ctr);
    T_U = Reinit(T_U, ctr);
else
    for each particle i in T_F do
        Update_velocity(V_F(i), ctr);
        A_F(i) = Update_position(T_F(i), V_F(i), ctr);
    end for
    for each particle i in T_U do
        Update_velocity(V_U(i), ctr);
        A_U(i) = Update_position(T_U(i), V_U(i), ctr);
    end for
end if
A_B = Boundary(A_F, A_U)
T_B = First_k(Sort(T_B ⊕ A_B))
T_E = evaluate(T_B)
Update(T_F, T_U, T_E); { According to the new T_B}
t = t + 1
end while
```

on the interval $a_i \leq x_i \leq b_i$, i.e., a multimodal PDF aimed at considering several subregions of that interval at the same time. These ideas are extensively presented and details concerning implementation issues are given in Socha and Dorigo [22] through algorithm ACO$_{\mathbb{R}}$ which represents the former ideas proposed by Socha [21] regarding continuous domains. The authors presented an experimental study that considers the application of ACO$_{\mathbb{R}}$ to a test suite of several unconstrained continuous optimization problems. Alg. 5 displays the main components of the ACO$_{\mathscr{B}}$, an ACO$_{\mathbb{R}}$ algorithm that includes the boundary approach for constrained optimization problems. The initialization process includes, in addition, a structure ω which represents a set of weights used as part of the mixture of normal kernels (or PDFs

Algorithm 5. A general outline of $\text{ACO}_{\mathcal{B}}$ algorithm

1: $t = 0$
2: ctr= initial_constraint // 'ctr' represents the problem constraint under consideration
3: init($T_{\mathcal{F}}$,$T_{\mathcal{U}}$, ω, ctr);
4: $T_{\mathcal{B}}$ =Boundary($T_{\mathcal{F}}$,$T_{\mathcal{U}}$)
5: $T_{\mathcal{E}}$ =evaluate($T_{\mathcal{B}}$)
6: **while** (stop condition not met) **do**
7:     **if** (change_constraint()) **then**
8:         ctr= get_next_ctr(ctr) // The search continues considering another problem constraint,
9:         // e.g., following a *Round-Robin* policy.
10:         $A_{\mathcal{F}}$ =Reinit($T_{\mathcal{F}}$, ω, ctr);
11:         $A_{\mathcal{U}}$ =Reinit($T_{\mathcal{U}}$, ω, ctr);
12:     **else**
13:         $A_{\mathcal{F}}$ =BuildSols($T_{\mathcal{F}}$, ω, ctr);
14:         $A_{\mathcal{U}}$ =BuildSols($T_{\mathcal{U}}$, ω, ctr);
15:     **end if**
16:     $A_{\mathcal{B}}$ =Boundary($A_{\mathcal{F}}$,$A_{\mathcal{U}}$)
17:     $T_{\mathcal{B}}$ =First$_k$(Sort($T_{\mathcal{B}} \oplus A_{\mathcal{B}}$))
18:     $T_{\mathcal{E}}$ =evaluate($T_{\mathcal{B}}$)
19:     Update($T_{\mathcal{F}}$,$T_{\mathcal{U}}$,$T_{\mathcal{E}}$); { According to the new $T_{\mathcal{B}}$}
20:     $t = t + 1$
21: **end while**

$P^i(x_i)$). Function 'Build_sol()' samples a new set of solutions according the to the respective points in $\mathcal{G}$ and their ranking. Again, structure ω is involved on the sampling process which uses the previous solutions to build an updated model of the PDFs (more details can be found in the description of $\text{ACO}_{\mathbb{R}}$ in Socha and Dorigo [22] as well as in Leguizamón and Coello [13] where the adaptation of $\text{ACO}_{\mathbb{R}}$ for constrained optimization problems is presented). After the sampling of the new solutions in $A_{\mathcal{F}}$ and $A_{\mathcal{U}}$, the process follows the same steps as for $\text{EA}_{\mathcal{B}}$ and $\text{PSO}_{\mathcal{B}}$.

Similarly to the above the search engines, when the problem has only one constraint (or only one is considered as in problem $G02$) there is no need to change from one to another constraint, therefore a few code lines can be dismissed, not necessarily dropped, from the general algorithm. In line 2, the 'initial_constraint' represents the only problem constraint. The set of lines 6 through 15 are replaced by lines 13 through 14.

5 Experimental Study

This section presents a (brief) experimental study involving the applications of the boundary approach under the three metaheuristics described in the previous section.

Table 1 Set of well-known nonlinear problems

Problem	Opt/BK
G01	-15.00
G02	1.0
G03	0.80619
G04	-30665.539
G05	5126.4981
G06	-6961.8138
G07	24.306209
G09	680.630
G10	7049.2083
G11	0.75
G13	0.053950
G14	-47.764
G15	961.715
G17	8853.539
G21	193.7783
G23	-400.0025
G24	-5.5079
G25	16.73889

The problems considered are conformed by a subset of a well-known testbed nonlinear problems used in literarure [16]. Table 1 displays the selected problems and the respectives optimal or best known values (column Opt/Bk).

The study is divided in two parts. The first part (Section 5.2) shows the results of the application of one of the three metaheuritics, namely the ACO approach, by considering two different algorithms. One of them, the original $ACO_{\mathbb{R}}$ enhanced with the stochastic ranking technique to handle constraints ($ACO_{\mathscr{N}\mathscr{B}}$). The other one, is the $ACO_{\mathbb{R}}$, but including the boundary approach and stochastich ranking ($ACO_{\mathscr{B}}$). The objective of this study is to show the benefit of the boundary search when included as an alternative constraint handling technique. To do that, we considered just a simple version of the original $ACO_{\mathbb{R}}$ and standard parameter setting for the stochastic ranking in order to better visualize the performance of the algorithm when the search focuses on the boundary between the feasible and infeasible search space. The second part (6) is aimed to compare the performance the boundary approach when implemented under ACO, PSO, and EAs metaheuristics. The respective algorithms are: $ACO_{\mathscr{B}}$, $PSO_{\mathscr{B}}$, and $EA_{\mathscr{B}}$.

5.1 Parameter Setting

$t_{max} = 10000$, $t_c = 200$, only on active constraints, for all the algorithms

$PSO_{\mathscr{B}}$

$$v_i^j = w \cdot v_i^j + c1 \cdot r1 \cdot (x_i^{lb} - x_i^j) + c2 \cdot r2 \cdot (x_i^{gb} - x_i^j)$$
$$x_i^j = x_i^j + v_i^j$$

$c1 = 0.5$, $c2 = 0.5$, $w = 0.85$, $NP = 50$ (NP is k in Algorithm 3, line 18)

$\text{EA}_{\mathscr{B}}$

$p_c = 0.65$, $p_m = 0.01$, $\mu = 50$ (μ is k in Algorithm 4, line 31), $\lambda = 50$, $(\mu + \lambda)$, real vector representation for the individuals, arithmetic crossover, and simple mutation which produces a little change on the value of a partciular dimension vector.

$\text{ACO}_{\mathscr{B}}$

$\xi = 0.85$, $q = 0.1$, $NK = 50$ (NP is k in Algorithm 5, line 17), number of ants set to 50.

5.2 Performance Comparison of $\text{ACO}_{\mathscr{B}}$ and $\text{ACO}_{\mathscr{N}\mathscr{B}}$

In this section, we compare the performance of $\text{ACO}_{\mathscr{B}}$ and $\text{ACO}_{\mathscr{N}\mathscr{B}}$. Table 2 shows the results obtained for all the problems studied. Each column shows respectively for each algorithm, the best found value (BF), average and standard deviation ($\text{Avg}_{\pm}std$), and number feasible solutions found out of 30 runs. Numbers in boldface (column BF) indicate that the optimal solution was obtained for the respective algorithm. Preliminary results from $\text{ACO}_{\mathscr{N}\mathscr{B}}$ were obtained by setting $q \in \{0.0001, 0.01, 0.1\}$ and fixing $\xi = 85$. $\text{ACO}_{\mathscr{N}\mathscr{B}}$ showed a very similar behavior (not results) to $\text{ACO}_{\mathscr{B}}$ with respect to parameter q. Even more, the use of a varying q showed to be the more representative for $\text{ACO}_{\mathscr{N}\mathscr{B}}$ considering the overall performance (quality and number of feasible solutions found).

It can be seen that $\text{ACO}_{\mathscr{N}\mathscr{B}}$ gives an important number of feasible solutions for some of the problems, however was not able obtain feasible solutions for all the problems (e.g., for $G21$ and $G23$ no feasible solutions were found at all). Taking into account the solved problems by $\text{ACO}_{\mathscr{N}\mathscr{B}}$ ($G01$, $G06$, $G07$, $G09$, $G11$, $G24$, and $G25$) its performance is inferior to the $\text{ACO}_{\mathscr{B}}$ for some of the above problems when considering the number of feasible solutions and/or average values (see problems $G01$, $G06$, $G11$, $G24$, and $G25$). For some of the remaining problems, $\text{ACO}_{\mathscr{N}\mathscr{B}}$ performs fairly well giving results very close to the optimal ones, however, the main difference with $\text{ACO}_{\mathscr{B}}$ is on the average values which shows a less robust algorithm. In addition, for problems $G05$, $G13$, and $G17$; $\text{ACO}_{\mathscr{N}\mathscr{B}}$ showed the worst performance regarding the solution quality. On the other hand, it can be seen that for problems $G10$, $G17$, and $G21$, $\text{ACO}_{\mathscr{B}}$ obtained values very close to the optimal ones, whereas, for problem $G23$, $\text{ACO}_{\mathscr{B}}$ showed a poor performance with respect to the solution quality.

It is worth remarking that $\text{ACO}_{\mathscr{N}\mathscr{B}}$ is a very simple adaptation of the original $\text{ACO}_{\mathbb{R}}$ to handling constraints. Despite of that, the results of $\text{ACO}_{\mathscr{N}\mathscr{B}}$ shows the potential of the ACO approach for continuous problems. Particularly this potential is exploited here by incorporating a boundary approach.

Table 2 BF in bold means that the optimal value was found for the respective problem. It should be noticed that the results correspond to the setting $q = 0.1$ and $\xi = 0.85$. NA stands for "Not Available"

	ACOBoundary $\equiv \mathscr{F}_{\mathscr{B}}$			$\text{ACO}_{\mathscr{N}\mathscr{B}}$		
Prob.	BF	$\text{Avg}_{\pm\sigma}$	#Fea	BF	$\text{Avg}_{\pm\sigma}$	#Fea
G01	**-15.00**	$-15.00_{\pm 0}$	30	**-15.00**	$-14.10_{\pm 1.198}$	24
G02	**0.80619**	$0.776871_{\pm 0.025}$	30	0.728079	$0.431599_{\pm 0.1145}$	30
G03	**1.00**	$1.00_{\pm 0}$	30	1.00	$1.00_{\pm 0.0}$	29
G04	**-30665.539**	$-30665.539_{\pm 0}$	30	-30665.58	$-30665.57_{\pm 0.005}$	30
G05	**5126.49**	$5135.99_{\pm 14.54}$	27	5231.96	$5231.96_{\pm 0}$	1
G06	**-6961.814**	$-6961.814_{\pm 0}$	30	**-6961.814**	$-6961.814_{\pm 0}$	2
G07	**24.306**	$24.5370_{\pm 0.240}$	30	24.320	$25.041_{\pm 0.62}$	29
G09	**680.630**	$680.630_{\pm 0}$	30	**680.630**	$680.635_{\pm 0.003}$	30
G10	7049.32	$7155.99_{\pm 94.92}$	30	7049.33	$7659.54_{\pm 596.20}$	29
G11	**0.75**	$0.75_{\pm 0}$	30	**0.75**	$0.75_{\pm 0}$	8
G13	**0.053950**	$0.053960_{\pm 0}$	26	0.097069	$0.557918_{\pm 0.2844}$	11
G14	**-47.760**	$-47.686_{\pm 0.08}$	30	-47.497	$-46.315_{\pm 0.8131}$	10
G15	**961.7151**	$961.7157_{\pm 0}$	30	961.7324	$961.8092_{\pm 0.125}$	3
G17	8863.67	$8958_{\pm 37.64}$	30	9011.91	$9018.82_{\pm 5.582}$	26
G21	193.7860	$193.8794_{\pm 0.09}$	20	NA	NA	NA
G23	-303.54	$22.54_{\pm 145.84}$	17	NA	NA	NA
G24	**5.5080**	$5.5080_{\pm 0}$	30	**5.5080**	$5.5080_{\pm 0}$	28
G25	**16.73889**	$16.73889_{\pm 0}$	30	**16.73889**	$16.73889_{\pm 0}$	7

6 Comparison of ACO, PSO, and EAs under the Boundary Approach

This section shows the performance of $\text{ACO}_{\mathscr{B}}$, $\text{PSO}_{\mathscr{B}}$, and $\text{EA}_{\mathscr{B}}$. Table 3 displays the Best Found (BF) value, and the respective average and deviation values ($\text{Avg}_{\pm}$) for each one of the algorithms implemented. Numbers in boldface means that the optimal (or best known) value was found. The number of feasible solutions found by the respective algorithms are not showed, however the three algorithms found very similar values to those values displayed in Table 2 for $\text{ACO}_{\mathscr{B}}$.

Clearly, there are a subset of problems for which the three algorithms performs identically (see $G01$, $G03$, $G04$, $G06$, $G09$, $G11$, $G24$, and $G25$). For the problem $G05$, the performance is identical with respect to the best value found, however the average values are different but not statistically significant. On the other hand, the results for problems $G02$, $G07$, and $G14$ show that $\text{ACO}_{\mathscr{B}}$ outperformed $\text{PSO}_{\mathscr{B}}$ and $\text{EA}_{\mathscr{B}}$ when considering BF value, however, $\text{EA}_{\mathscr{B}}$ is more robust for this problem (see columns BF and $\text{Avg}_{\pm\sigma}$). In the case of $G07$ it should be noticed that the three algorithms perform similarly (the average value were not statistically significant).

There exist other particular cases that are analyzed in the following. The results for problem $G10$ show that $\text{ACO}_{\mathscr{B}}$ and $\text{EA}_{\mathscr{B}}$ outperformed $\text{PSO}_{\mathscr{B}}$ considering both, best found and average values. However, $\text{ACO}_{\mathscr{B}}$ outperformed $\text{EA}_{\mathscr{B}}$ in terms of the

Table 3 Performance of $\mathrm{ACO}_{\mathscr{B}}$, $\mathrm{PSO}_{\mathscr{B}}$, and $\mathrm{EA}_{\mathscr{B}}$ on the benchmark problems. BF in boldface means that the optimal value was found for the respective problem. NA stands for "Not Available". The respective parameter setting is described in Section 5.1

	$\mathrm{ACO}_{\mathscr{B}}$		$\mathrm{PSO}_{\mathscr{B}}$		$\mathrm{EA}_{\mathscr{B}}$	
Prob.	BF	Avg$_{\pm\sigma}$	BF	Avg$_{\pm\sigma}$	BF	Avg$_{\pm\sigma}$
G01	**-15.00**	-15.00$_{\pm 0}$	**-15.00**	-14.10$_{\pm 1.198}$	**-15.00**	-15.00$_{\pm 0}$
G02	**0.80619**	0.776871$_{\pm 0.025}$	0.80512	0.5939$_{\pm 0.06661}$	0.803550	0.803352$_{\pm 0.0001}$
G03	**1.00**	1.00$_{\pm 0}$	**1.00**	1.00$_{\pm 0.0}$	**1.00**	1.00$_{\pm 0.0}$
G04	**-30665.539**	-30665.539$_{\pm 0}$	**-30665.539**	-30665.539$_{\pm 0}$	**-30665.539**	-30665.529$_{\pm 0.01}$
G05	**5126.49**	5135.99$_{\pm 14.54}$	**5126.49**	5130.74$_{\pm 5.44}$	5126.64	5130.12$_{\pm 3.84}$
G06	**-6961.814**	-6961.814$_{\pm 0}$	**-6961.814**	-6961.814$_{\pm 0}$	**-6961.814**	-6961.814$_{\pm 0}$
G07	**24.306**	24.5370$_{\pm 0.240}$	24.375	25.053$_{\pm 0.728}$	24.372	24.546$_{\pm 0.128}$
G09	**680.630**	680.630$_{\pm 0}$	**680.630**	680.635$_{\pm 0.003}$	**680.630**	680.633$_{\pm 0.0}$
G10	7049.3261	7155.9948$_{\pm 94.924}$	7093.0151	8040.5537$_{\pm 972.585}$	7049.333	7659.540$_{\pm 596.20}$
G11	**0.75**	0.75$_{\pm 0}$	**0.75**	0.75$_{\pm 0}$	**0.75**	0.75$_{\pm 0}$
G13	**0.053950**	0.053960$_{\pm 0}$	0.053961	0.063768$_{\pm 0.0103}$	0.053984	0.069971$_{\pm 0.048}$
G14	**-47.760**	-47.686$_{\pm 0.08}$	-47.129	-43.778$_{\pm 1.68}$	47.6724	47.6708$_{\pm 0.0022}$
G15	**961.7151**	961.7157$_{\pm 0}$	**961.7151**	962.1973$_{\pm 1.1407}$	**961.7151**	961.7149$_{\pm 0.0002}$
G17	8863.67	8958$_{\pm 37.64}$	8859.9541	9019.2519$_{\pm 125.776}$	8866.86	9004.288$_{\pm 91.11}$
G21	193.786	193.879$_{\pm 0.09}$	196.392	201.034$_{\pm 3.49}$	193.804	193.880$_{\pm 0.085}$
G23	-303.54	22.54$_{\pm 145.84}$	NA	NA	-177.16	2.67$_{\pm 160.70}$
G24	**5.5080**	5.5080$_{\pm 0}$	**5.5080**	5.5080$_{\pm 0}$	**5.5080**	5.5080$_{\pm 0}$
G25	**16.73889**	16.73889$_{\pm 0}$	**16.73889**	16.73889$_{\pm 0}$	**16.73889**	16.73889$_{\pm 0}$

average values (statistically significant). For problem $G13$ the above described situation it can also be observed, except that $\mathrm{ACO}_{\mathscr{B}}$ found the optima value for this particular problem. A different situation is for problem $G17$ for which $\mathrm{PSO}_{\mathscr{B}}$ found the best solution, however, $\mathrm{ACO}_{\mathscr{B}}$ and $\mathrm{EA}_{\mathscr{B}}$ showed not statistically significant average values, but better and statistically significant with respect to $\mathrm{PSO}_{\mathscr{B}}$. Considering the average value, the three algorithms showed similar performance for problem $G21$, except that $\mathrm{ACO}_{\mathscr{B}}$ achieved the best value. Finally, problem $G23$, no one of the three algorithms perform well. $\mathrm{PSO}_{\mathscr{B}}$ was not capable of finding any feasible solution, whereas, $\mathrm{ACO}_{\mathscr{B}}$ and $\mathrm{PSO}_{\mathscr{B}}$ found a few feasible solutions of bad quality.

7 Summary and Future Work

In this chapter we have presented the boundary approach as an alternative approach to explore the boundary between the feasible and infeasible search space for constrained optimization problems. The boundary approach was presented under two perspective, either by using *ad hoc* operators as presented in [17], and a more general operator as proposed in previous works [13, 14] as possible alternatives to be applied when facing constrained problems. In addition, three different search machines (EAs, PSO, and ACO) were considered to show how the boundary approach could be implemented without producing major modifications on the original algorithms. Furthermore, it is worth noticing that the boundary approach is an interesting mechanism that could be applied to many constrained optimization problems, particularly this observation is true for the 'general boundary operator' described in detail here and proposed in earlier works.

As presented in this chapter, the boundary approach can be used alone (as a constraint-handling techniques itself) or in combination with a complementary constraint-handling technique like penalty functions or any other like Stochastic Ranking which is used here to display three algorithms due to his simplicity and for being an efficient and very well known technique. The three algorithms (i.e., $EA_{\mathscr{B}}$, $PSO_{\mathscr{B}}$, and $ACO_{\mathscr{B}}$) presented here as a guidelines show that the boundary approach is flexible enough to be implemented under different search machines.

Finally, is important to remark that the boundary approach (seen as having the possibility of using boundary operators) can be considered when implementing a more general constraint-handling technique under some search engine. Particularly when facing problems with an several constraints. For example, the general operator (as defined here) needs two points, one from the feasible region and the other one form the infeasible one. Let us suppose that a metaheuristic implements a constraint-handling technique to explore the whole space $\mathscr{F}$. However, the boundary of a particular constraint could be approached by considering points from both, the feasible and infeasible one with respect to that constraint. By this way, the generation of solutions laying on the boundary could help to quickly reach (in a controlled manner) the boundary region. Nevertheless, there are still place to consider alternative ways of implementing a *general boundary operator* different form the proposed in this work.

Acknowledgements. The first author acknowledges support from Universidad Nacional de San Luis and the ANPCYT (National Agency for Promotion of Science and Technology). The second author acknowledges support from CONACyT project no. 45683-Y.

References

1. Bäck, T., Fogel, D.B., Michalewicz, Z. (eds.): Handbook of Evolutionary Computation. Oxford University Press/ Institute of Physics Publishing, New York/Bristol (1997)
2. Coello Coello, C.: List of references on constraint-handling techniques used with evolutionary algorithms, `http://www.cs.cinvestav.mx/-constraint/`
3. Coello Coello, C.: Theoretical and numerical constraint-handling techniques used with evolutionary algorithms: A survey of the state of the art. Computer Methods in Applied Mechanics and Engineering 191(11-12), 1245–1287 (2002)
4. Corne, D., Dorigo, M., Glover, F. (eds.): New Ideas in Optimization. McGraw-Hill International, New York (1999)
5. Dorigo, M., Stützle, T.: Ant Colony Optimization. MIT Press, Cambridge (2004)
6. Engelbrecht, A.P.: Fundamentals of Computational Swarm Intelligence. John Wiley & Sons, Ltd, Chichester (2005)
7. Schütz, M., Leguizamón, G., Michalewicz, Z.: An ant system for the maximum independent set problem. In: Proceedings of the 2001 Argentinian Congress on Computer Science, El Calafate, Argentina, pp. 1027–1040 (2001)

8. Glover, F., Kochenberg, G.A. (eds.): Handbook of Metaheuristics. Kluwer Academic Publishers, London (2003)
9. Gottlieb, J.: Evolutionary algorithms for multidimensional knapsack problems: the relevance of the boundary of the feasible region. In: Banzhaf, W., Daida, J., Eiben, A.E., Garzon, M.H., Honavar, V., Jakiela, M., Smith, R.E. (eds.) Proc. of the Genetic and Evolutionary Computation Conf. GECCO 1999, p. 787. Morgan Kaufmann, San Francisco (1999)
10. Keane, A.J.: Experiences with optimizers in structural design. In: Parmee, I.C. (ed.) Proceedings of the Conference on Adaptive Computing in Engineering Design and Control, Plymouth, UK, pp. 14–27. University of Plymouth (1994)
11. Kennedy, J., Eberhart, R.: Particle swarm optimization. In: IEEE International Conference on Neural Networks, Perth, Australia, vol. IV, pp. 1942–1948. IEEE Service Center (1995)
12. Kennedy, J., Eberhart, R.C., Shi, Y.: Swarm Intelligence. Morgan Kaufmann, San Francisco (2001)
13. Leguizamón, G., Coello Coello, C.: Boundary search for constrained numerical optimization problems in aco algorithms. In: Dorigo, M., Gambardella, L.M., Birattari, M., Martinoli, A., Poli, R., Stützle, T. (eds.) ANTS 2006. LNCS, vol. 4150, pp. 108–119. Springer, Heidelberg (2006)
14. Leguizamón, G., Coello Coello, C.: A Boundary Search based ACO Algorithm Coupled with Stochastic Ranking. In: 2007 IEEE Congress on Evolutionary Computation (CEC 2007), Singapore, September 2007, pp. 165–172. IEEE Press, Los Alamitos (2007)
15. Leguizamón, G., Michalewicz, Z.: A New Version of Ant System for Subset Problems. In: Proceedings of the 1999 Congress on Evolutionary Computation, pp. 1459–1464. IEEE Press, Piscataway (1999)
16. Liang, J.J., Runarsson, T.P., Mezura-Montes, E., Clerc, M., Suganthan, P.N., Coello Coello, C., Deb, K.: Problem Definitions and Evaluation Criteria for the CEC. Technical report, Special Session on Constrained Real-Parameter Optimization, School of Electrical and Electronic Engineering Nanyang Technological University, Singapore (2006), http://www.ntu.edu.sg/home5/lian0012/cec2006/technical_report.pdf
17. Michalewicz, Z., Nazhiyath, G., Michalewicz, M.: A note on usefulness of geometrical crossover for numerical optimization problems. In: Fogel, L.J., Angeline, P.J., Bäck, T. (eds.) Evolutionary Programming V: Proc. of the Fifth Annual Conf. on Evolutionary Programming, pp. 305–311. MIT Press, Cambridge (1996)
18. Runarsson, T.P., Yao, X.: Stochastic ranking for constrained evolutionary optimization. IEEE Transactions on Evolutionary Computation 4(3), 284–294 (2000)
19. Schoenauer, M., Michalewicz, Z.: Evolutionary computation at the edge of feasibility. In: Ebeling, W., Rechenberg, I., Voigt, H.-M., Schwefel, H.-P. (eds.) PPSN 1996. LNCS, vol. 1141, pp. 245–254. Springer, Heidelberg (1996)
20. Schoenauer, M., Xanthakis, S.: Constrained GA optimization. In: Forrest, S. (ed.) Proc. of the Fifth Int. Conf. on Genetic Algorithms, San Mateo, CA, pp. 573–580. Morgan Kaufmann, San Francisco (1993)
21. Socha, K.: ACO for continuos and mixed-variable optimization. In: Dorigo, M., Birattari, M., Blum, C., Gambardella, L.M., Mondada, F., Stützle, T. (eds.) ANTS 2004. LNCS, vol. 3172, pp. 25–36. Springer, Heidelberg (2004)

22. Socha, K., Dorigo, M.: Ant colony optimization for continuous domains. European Journal of Operational Research 185(3), 1115–1173 (2008), http://dx.doi.org/10.1016/j.ejor.2006.06.046
23. Wu, Z.Y., Simpson, A.R.: A self-adaptive boundary search genetic algorithm and its application to water distribution systems. Journal of Hidraulic Research 40(2), 191–203 (2002)

Solving Difficult Constrained Optimization Problems by the ε Constrained Differential Evolution with Gradient-Based Mutation

Tetsuyuki Takahama and Setsuko Sakai

Abstract. While research on constrained optimization using evolutionary algorithms has been actively pursued, it has had to face the problem that the ability to solve multi-modal problems is insufficient, that the ability to solve problems with equality constraints is inadequate, and that the stability and efficiency of searches is low. We have proposed the εDE, defined by applying the ε constrained method to differential evolution (DE). It is shown that the εDE is a fast and stable algorithm that is robust to multi-modal problems and it can solve problems with many equality constraints by introducing a gradient-based mutation which finds a feasible point using the gradient of constraints. In this chapter, an improved εDE is proposed, in which faster reduction of the relaxation of equality constraints in the ε constrained method and higher gradient-based mutation rate are adopted in order to solve problems with many equality constraints and to find feasible solutions faster and very stably. Also, cutting off and reflecting back solutions outside of search space are adopted to improve the efficiency in finding optimal solutions. The improved εDE realizes stable and efficient searches, and can solve difficult constrained optimization problems with equality constraints. The advantage of the improved εDE is shown by applying it to twenty four constrained problems of various types.

Keywords: differential evolution, constrained optimization, ε constrained method, ε-level comparison, gradient-based mutation, pseudoinverse.

1 Introduction

An evolutionary algorithm (EA) is the term commonly used for algorithms based on principles of evolution, and includes genetic algorithms (GAs), evolution

Tetsuyuki Takahama
Department of Intelligent Systems, Hiroshima City University, Asaminami-ku, Hiroshima 731-3194 Japan
e-mail: takahama@its.hiroshima-cu.ac.jp

Setsuko Sakai
Faculty of Commercial Sciences, Hiroshima Shudo University, Asaminami-ku, Hiroshima 731-3195 Japan
e-mail: setuko@shudo-u.ac.jp

E. Mezura-Montes (Ed.): Constraint-Handling in Evolution. Optimi., SCI 198, pp. 51–72.
springerlink.com

strategies (ESs), evolutionary and genetic programming, and so on. EAs are direct search methods that use only the value of an objective function and essentially solve unconstrained optimization problems. However, optimization problems in the real world are often constrained optimization problems where objective functions are optimized under given constraints.

There are many studies on solving constrained optimization problems using evolutionary algorithms (EAs) [5,7,11,12]. However there are some difficulties in these studies:

1. The ability to solve multi-modal problems is insufficient. EAs for constrained optimization can locate a feasible region and find feasible solutions in unimodal problems. When multi-modal problems that have many local solutions in a feasible region are solved, even if the EAs can locate the feasible region, they are sometimes trapped to a local solution and cannot search for an optimal solution. Thus a method that is robust to multi-modal problems is required.
2. Obtained solutions in problems with equality constraints are inadequate. Many EAs for constrained optimization cannot directly solve problems with equality constraints. To overcome such problems, the definition of the problems needs to be changed by converting equality constraints into relaxed inequality constraints. As a result, the feasibility of the obtained solutions is inadequate. Also, it is difficult for them to solve problems with many equality constraints.
3. The stability and efficiency of searches is low. Even when solving the same problem, EAs sometimes find good solutions but sometimes find only very bad solutions. The initial search points are usually generated randomly and the search process includes many stochastic operations in EAs. Sometimes EAs cannot overcome the effect of randomness in the search process of some problems. Thus, the stability of the search becomes low. Also, many EAs need rank-based selection or replacement, stochastic selection and mutations based on Gaussian or Cauchy distributions that incur high computational costs. Thus, the efficiency of search also becomes low.

To overcome these problems, we have proposed the εDE [36], defined by applying the ε constrained method [31] to differential evolution (DE) [19, 20]. DE is an evolutionary algorithm for global optimization that incorporates both crossover and mutation into a simple operation, which is realized by selecting a parent and adding scaled difference between two other parents to the parent. By incorporating DE, problem 1. and 3. can be solved; DE is a simple, fast and stable search algorithm that is robust to multi-modal problems. The εDE is stable because it uses a simple and stable selection and replacement mechanism excluding stochastic selection and replacement [17]. The εDE is also efficient because it uses a simple arithmetic operation and does not use any rank-based operations or mutations based on Gaussian and Cauchy distributions. Problem 2. can be solved by using a simple way of controlling the relaxation of equality constraints for the ε constrained method to directly solve problems with equality constraints. Also, problems with many equality constraints can be solved by introducing a gradient-based mutation [32] that finds a feasible point from an infeasible point using the gradient of constraints at the infeasible point.

In this chapter, we propose the improved εDE, in which the following two ideas are introduced:

- Treatment of points outside of search space
  In DE, new points are sometimes generated outside of the search space. In the εDE, boundary conditions, which specify the search space, were converted to constraints. The points outside of space were treated in the same manner as points inside of the space, and they are evaluated wastefully. To avoid this situation, we introduce two ways of handling points outside of the space: cutting off the points on the boundary of the space and reflecting the points back to the inside of the space.
- Improvement of efficiency to find feasible solutions
  It is difficult for the εDE to find feasible points in earlier generation, because it searches for infeasible points having better objective values until the relaxation of equality constraints is reduced completely or becomes zero. In [32], the elitism where more feasible points are preserved as feasible elites was proposed to find feasible solutions faster. However, the elitism sometimes leads premature convergence of search points. In this study, we show that the combination of faster reduction of the relaxation of equality constraints and higher gradient-based mutation rate enables to solve problems with many equality constraints and to find feasible solutions faster and more stable.

The advantage of the improved εDE is shown by applying it to twenty four constrained problems of various types proposed in CEC2006 [10] and comparing the results to those obtained by the εDE.

The rest of this chapter is organized as follows: Section 2 describes previous works. Section 3 briefly describes the ε constrained method. Section 4 describes DE and the improved εDE. Section 5 presents experimental results of various benchmark problems. Finally, Section 6 concludes with a brief summary of this study and some remarks.

2 Previous Works

There exist many studies on solving constrained optimization problems using evolutionary algorithms [11] and particle swarm optimization [4]. These studies can be classified into several categories according to the way the constraints are treated as follows:

1. Constraints are only used to judge whether a search point is feasible or not [8]. Death penalty method is in this category. The searching process begins with one or more feasible points and continues to search for new points within the feasible region. When a new search point is generated and the point is not feasible, the point is repaired or discarded. However, generating initial feasible points is difficult and computationally demanding when the feasible region is very small.

2. The constraint violation, which is the sum of the violation of all constraint functions, is combined with the objective function. The penalty function method is in this category [13]. An extended objective function is defined by adding the constraint violation to the objective function as a penalty. The optimization of the objective function and the constraint violation is realized by the optimization of the extended objective function. The main difficulty of the penalty function method is the difficulty of selecting an appropriate value of the penalty coefficient that adjusts the strength of the penalty.
3. The constraint violation and the objective function are used separately. In this category, both the constraint violation and the objective function are optimized by a lexicographic order in which the constraint violation precedes the objective function. Takahama and Sakai proposed the α constrained method [22, 24] and the ε constrained method [31], which adopt a lexicographic ordering with relaxation of the constraints. Deb [6] proposed a method in which the extended objective function that realizes the lexicographic ordering is used. Runarsson and Yao [16] proposed the stochastic ranking method in which the stochastic lexicographic order, which ignores the constraint violation with some probability, is used. These methods were successfully applied to various problems.
4. The constraints and the objective function are optimized by multiobjective optimization methods. In this category, the constrained optimization problems are solved as the multiobjective optimization problems in which the objective function and the constraint functions are objectives to be optimized [1, 9, 15, 21]. However in many cases, solving multiobjective optimization problems is a more difficult and expensive task than solving single objective optimization problems.

The ε constrained methods can convert algorithms for unconstrained problems to algorithms for constrained problems using the ε-level comparison, which compares the search points based on the constraint violation of them. The ε constrained method is in the promising category 3. and is proposed based on the α constrained method. The α constrained method was applied to Powell's direct search method in [22, 24], the nonlinear simplex method proposed by Nelder and Mead in [23, 26, 29], a genetic algorithm (GA) using linear ranking selection in [25, 27] and particle swarm optimization (PSO) in [28, 30]. The ε constrained method was applied to PSO in [31, 34, 35], GA in [33] and differential evolution (DE) in [32, 36].

3 The ε Constrained Method

In this section, constrained optimization problems are defined and the ε constrained method is explained.

3.1 Constrained Optimization Problems

Constrained optimization problems, especially nonlinear optimization problems, where objective functions are minimized under given constraints, are very important and frequently appear in the real world. In this study, the following optimization problem (P) with inequality constraints, equality constraints, upper bound constraints and lower bound constraints will be discussed.

$$\begin{aligned}
\text{(P) minimize } & f(\mathbf{x}) \\
\text{subject to } & g_j(\mathbf{x}) \leq 0,\ j = 1,\ldots,q \\
& h_j(\mathbf{x}) = 0,\ j = q+1,\ldots,m \\
& l_i \leq x_i \leq u_i,\ i = 1,\ldots,n,
\end{aligned} \tag{1}$$

where $\mathbf{x} = (x_1, x_2, \cdots, x_n)^T$ is an n dimensional vector, $f(\mathbf{x})$ is an objective function, $g_j(\mathbf{x}) \leq 0$ and $h_j(\mathbf{x}) = 0$ are q inequality constraints and $m-q$ equality constraints, respectively. Functions f, g_j and h_j are linear or nonlinear real-valued functions. Values u_i and l_i are the upper bound and the lower bound of x_i, respectively. Also, let the feasible space in which every point satisfies all constraints be denoted by $\mathscr{F}$ and the search space in which every point satisfies the upper and lower bound constraints be denoted by $\mathscr{S}$ ($\supset \mathscr{F}$).

3.2 Constraint Violation and ε-Level Comparison

In the ε constrained method, constraint violation $\phi(\mathbf{x})$ is defined. The constraint violation can be given by the maximum of all constraints or the sum of all constraints.

$$\phi(\mathbf{x}) = \max\{\max_j\{0, g_j(\mathbf{x})\}, \max_j |h_j(\mathbf{x})|\} \tag{2}$$

$$\phi(\mathbf{x}) = \sum_j ||max\{0, g_j(\mathbf{x})\}||^p + \sum_j ||h_j(\mathbf{x})||^p \tag{3}$$

where p is a positive number. Usually, Eq. (3) with $p = 1$ is used.

The *ε-level comparison* is defined as an order relation on the set of $(f(\mathbf{x}), \phi(\mathbf{x}))$. If the constraint violation of a point is greater than 0, the point is not feasible and its worth is low. The ε-level comparisons are defined by a lexicographic order in which $\phi(\mathbf{x})$ precedes $f(\mathbf{x})$, because the feasibility of $\mathbf{x}$ is more important than the minimization of $f(\mathbf{x})$.

Let f_i and ϕ_i be the function value and the constraint violation at a point $\mathbf{x}_i$ ($i = 1, 2$), respectively. Then, for any ε satisfying $\varepsilon \geq 0$, the ε-level comparisons $<_\varepsilon$ and $\leq_\varepsilon$ between (f_1, ϕ_1) and (f_2, ϕ_2) are defined as follows:

$$(f_1, \phi_1) <_\varepsilon (f_2, \phi_2) \Leftrightarrow \begin{cases} f_1 < f_2, & \text{if } \phi_1, \phi_2 \leq \varepsilon \\ f_1 < f_2, & \text{if } \phi_1 = \phi_2 \\ \phi_1 < \phi_2, & \text{otherwise} \end{cases} \tag{4}$$

$$(f_1, \phi_1) \leq_\varepsilon (f_2, \phi_2) \Leftrightarrow \begin{cases} f_1 \leq f_2, & \text{if } \phi_1, \phi_2 \leq \varepsilon \\ f_1 \leq f_2, & \text{if } \phi_1 = \phi_2 \\ \phi_1 < \phi_2, & \text{otherwise} \end{cases} \tag{5}$$

In case of ε=∞, the ε-level comparisons $<_\infty$ and $\leq_\infty$ are equivalent to the ordinal comparisons $<$ and $\leq$ between function values. Also, in case of $\varepsilon = 0$, $<_0$ and $\leq_0$ are equivalent to the lexicographic order in which the constraint violation $\phi(\mathbf{x})$ precedes the function value $f(\mathbf{x})$.

3.3 *The Properties of the ε Constrained Method*

The ε constrained method converts a constrained optimization problem into an unconstrained problem by replacing the order relation in direct search methods with the ε-level comparison. An optimization problem solved by the ε constrained method, that is, a problem in which the ordinary comparison is replaced with the ε-level comparison, ($\mathrm{P}_{\leq_\varepsilon}$), is defined as follows:

$$(\mathrm{P}_{\leq_\varepsilon}) \ \text{minimize}_{\leq_\varepsilon} \ f(\mathbf{x}), \tag{6}$$

where minimize$_{\leq_\varepsilon}$ means the minimization based on the ε-level comparison $\leq_\varepsilon$. Also, a problem (P^ε) is defined that the constraints of (P), that is, $\phi(\mathbf{x}) = 0$, is relaxed and replaced with $\phi(\mathbf{x}) \leq \varepsilon$:

$$\begin{aligned} (\mathrm{P}^\varepsilon) \ &\text{minimize } f(\mathbf{x}) \\ &\text{subject to } \phi(\mathbf{x}) \leq \varepsilon \end{aligned} \tag{7}$$

It is obvious that (P^0) is equivalent to (P).

For the three types of problems, (P^ε), ($\mathrm{P}_{\leq_\varepsilon}$) and (P), the following theorems are given based on the α constrained method [22, 24, 31].

Theorem 1. If an optimal solution of (P^0) exists, any optimal solution of ($\mathrm{P}_{\leq_\varepsilon}$) is an optimal solution of (P^ε).

Theorem 2. If an optimal solution of (P) exists, any optimal solution of ($\mathrm{P}_{\leq_0}$) is an optimal solution of (P).

Theorem 3. Let $\{\varepsilon_n\}$ be a strictly decreasing non-negative sequence which converges to 0. Let $f(\mathbf{x})$ and $\phi(\mathbf{x})$ be continuous functions of $\mathbf{x}$. Assume that an optimal solution $\mathbf{x}^*$ of (P^0) exists and an optimal solution $\hat{\mathbf{x}}_n$ of ($\mathrm{P}_{\leq_{\varepsilon_n}}$) exists for any ε_n. Then, any accumulation point to the sequence $\{\hat{\mathbf{x}}_n\}$ is an optimal solution of (P^0).

Theorem 1 and 2 show that a constrained optimization problem can be transformed into an equivalent unconstrained optimization problem by using the ε-level comparison. So, if the ε-level comparison is incorporated into an existing unconstrained optimization method, constrained optimization problems can be solved. Thus, it is thought that the ε constrained method is *an algorithm transformation method* which can convert an algorithm for unconstrained optimization into an algorithm for constrained optimization. Theorem 3 shows that, in the ε constrained method, an optimal solution of (P) can be given by converging ε to 0 as well as by increasing the penalty coefficient to infinity in the penalty method.

4 The εDE

In this section, we first describe differential evolution. Then, we describe the εDE, which is the integration of the ε constrained method and DE, with gradient-based mutation and the improved εDE.

4.1 Differential Evolution

Differential evolution is a variant of ES proposed by Storn and Price [19,20]. DE is a stochastic direct search method using population or multiple search points. DE has been successfully applied to the optimization problems including non-linear, non-differentiable, non-convex and multi-modal functions [2]. It has been shown that DE is fast and robust to these functions.

The main feature of DE is that DE uses simple arithmetic operations to avoid the control of Gaussian mutation in ES. In general, the mutation process must be adaptive to the step size of the Gaussian mutation, because the ideal step size depends on the gene or element that is mutated and the state of the evolution process. DE adopts the sum of a base vector and the scaled difference vectors as the mutation operation instead of Gaussian mutation. The base vector is selected from the population. The difference vectors are formed by the differences between a pair of vectors randomly selected from the population. As search area formed by the population contracts and expands over generations, the step size in each dimension, which is given by the difference vectors, adapts automatically.

There are some variants of DE that have been proposed, such as DE/best /1/bin and DE/rand/1/exp [14]. The variants are classified using the notation DE/*base*/*num*/*cross*. "*base*" indicates the method of selecting a base vector. For example, DE/rand/*num*/*cross* selects the base vector at random from the population. DE/best/*num*/*cross* selects the best vector in the population. "*num*" indicates the number of difference vectors used to perturb the base vector. "*cross*" indicates the crossover mechanism used to create a trial vector. For example, DE/*base*/*num*/bin shows that crossover is controlled by binomial crossover using constant crossover rate. DE/*base*/*num*/exp shows that crossover is controlled by exponential crossover which is a one-point crossover using exponentially decreasing the crossover rate.

In DE, initial vectors (individuals) are randomly generated within the search space and form an initial population. Each individual contains n genes as decision variables or a decision vector. At each generation or iteration, all individuals are selected as target vectors (parents). Each target vector is processed as follows: The mutation process begins by choosing $1+2\,num$ vectors from the population except for the target vector in the processing. The first vector is a base vector. All subsequent vectors are paired to create *num* difference vectors. The difference vectors are scaled by a scaling factor F and added to the base vector. The resulting vector is then recombined with the target vector. The probability of recombination at an element is controlled by a crossover rate CR. This crossover process produces a trial vector (child). Finally, for survivor selection, the trial vector is accepted for the next generation if the trial vector is better than the parent.

4.2 The εDE with Gradient-Based Mutation

The εDE is an algorithm where the ε constrained method is applied into DE, or ordinary comparisons in DE are replaced with the ε-level comparisons. The gradient-based mutation is explained here, and the algorithm of the εDE will be described later.

The gradient-based mutation is an operation similar to the gradient-based repair method proposed by Chootinan and Chen [3]. The vector of constraint functions $C(\mathbf{x})$, the vector of constraint violations $\Delta C(\mathbf{x})$ and the increment of a point $\mathbf{x}$ to satisfy constraints $\Delta\mathbf{x}$ are defined as follows:

$$
\begin{aligned}
C(\mathbf{x}) &= (g_1(\mathbf{x}) \cdots g_q(\mathbf{x})\, h_{q+1}(\mathbf{x}) \cdots h_m(\mathbf{x}))^T && (8)\\
\Delta C(\mathbf{x}) &= (\Delta g_1(\mathbf{x}) \cdots \Delta g_q(\mathbf{x})\, h_{q+1}(\mathbf{x}) \cdots h_m(\mathbf{x}))^T && (9)\\
\nabla C(\mathbf{x})\Delta\mathbf{x} &= -\Delta C(\mathbf{x}) && (10)\\
\Delta\mathbf{x} &= -\nabla C(\mathbf{x})^{-1}\Delta C(\mathbf{x}) && (11)
\end{aligned}
$$

where $\Delta g_j(\mathbf{x}) = \max\{0, g_j(\mathbf{x})\}$. And $\nabla C(\mathbf{x})$ is the gradient matrix of $C(\mathbf{x})$.

$$
\nabla C(\mathbf{x}) = \begin{pmatrix}
\frac{\partial g_1(\mathbf{x})}{\partial x_1} & \frac{\partial g_1(\mathbf{x})}{\partial x_2} & \cdots & \frac{\partial g_1(\mathbf{x})}{\partial x_n} \\
\vdots & \vdots & \cdots & \vdots \\
\frac{\partial g_q(\mathbf{x})}{\partial x_1} & \frac{\partial g_q(\mathbf{x})}{\partial x_2} & \cdots & \frac{\partial g_q(\mathbf{x})}{\partial x_n} \\
\frac{\partial h_{q+1}(\mathbf{x})}{\partial x_1} & \frac{\partial h_{q+1}(\mathbf{x})}{\partial x_2} & \cdots & \frac{\partial h_{q+1}(\mathbf{x})}{\partial x_n} \\
\vdots & \vdots & \cdots & \vdots \\
\frac{\partial h_m(\mathbf{x})}{\partial x_1} & \frac{\partial h_m(\mathbf{x})}{\partial x_2} & \cdots & \frac{\partial h_m(\mathbf{x})}{\partial x_n}
\end{pmatrix} \tag{12}
$$

The gradient matrix $\nabla C(\mathbf{x})$ can be obtained numerically by calculating $C(\mathbf{x})$ repeatedly with changing the value of each decision variable x_i to $x_i+\eta$:

$$
\nabla C(\mathbf{x}) = \frac{1}{\eta}\left(C(\mathbf{x}+\eta\, e_1) - C(\mathbf{x}), \cdots, C(\mathbf{x}+\eta\, e_n) - C(\mathbf{x})\right) \tag{13}
$$

where $\mathbf{e}_i$ is a unit vector of which the i-th element is 1 and the other elements are 0 and η is a very small amount.

Although the $\nabla C(\mathbf{x})$ is not invertible in general, the Moore-Penrose inverse or pseudoinverse $\nabla C(\mathbf{x})^+$ [18], which gives an approximate or best (least squares) solution to a system of linear equations, can be used instead in Eq. (11). A computationally simple and accurate way to get the pseudoinverse is by using singular value decomposition. The singular value decomposition of a (m,n) matrix A can be described as follows:

$$A = U\Sigma V^T \tag{14}$$

where U is a unitary (m,m) matrix, V^T is the conjugate transpose of a (n,n) unitary matrix V, and Σ is a (m,n) matrix with nonnegative numbers on the diagonal and zeros off the diagonal. The pseudoinverse A^+ of A can be obtained as follows:

$$A^+ = V\Sigma^+ U^T \tag{15}$$

where Σ^+ is the pseudoinverse of a diagonal matrix Σ and can be obtained by inverting each non-zero element on the diagonal.

After $\Delta\mathbf{x}$ is obtained, a mutated vector can be obtained as follows:

$$\mathbf{x}^{\text{new}} = \mathbf{x} + \Delta\mathbf{x} \tag{16}$$

This mutation or repair operation is executed with gradient-based mutation probability P_g. In [3], only non-zero elements of $\Delta C(\mathbf{x})$ are repaired and the repair operation is repeated with some probability while amount of repair is not small. In the εDE, however, non-zero inequality constraints and all equality constraints are considered to keep the feasibility of equality constraints. The mutation operation is repeated fixed times R_g while the point is not ε-feasible; When $\phi(\mathbf{x}) \leq \varepsilon(t)$, the point $\mathbf{x}$ is ε-feasible.

4.3 Controlling the ε-Level

Usually, the ε-level does not need to be controlled. Many constrained problems can be solved based on the lexicographic order where the ε-level is constantly 0. However for problems with equality constraints, the ε-level should be controlled properly to obtain high quality solutions.

In this study, a simple way of controlling the ε-level is used according to Eq. (17). The initial ε-level $\varepsilon(0)$ is the constraint violation of the top θ-th individual in the initial search points. The ε-level is updated until the number of iterations t becomes the control generation T_c. After the number of iterations exceeds T_c, the ε-level is set to 0 to obtain solutions with minimum constraint violation.

$$\begin{aligned} \varepsilon(0) &= \phi(\mathbf{x}_\theta) \\ \varepsilon(t) &= \begin{cases} \varepsilon(0)(1-\frac{t}{T_c})^{cp}, & 0<t<T_c, \\ 0, & t \geq T_c \end{cases} \end{aligned} \tag{17}$$

where $\mathbf{x}_\theta$ is the top θ-th individual and $\theta = 0.2N$ in usual, and cp is a parameter to control the speed of reducing relaxation of constraints.

4.4 The Improved εDE

In order to improve the efficiency of the εDE to find optimal solutions, operations to move a point outside of search space into the inside of the space are required. There are some ways to realize the movement: generating solutions again, cutting off the solutions on the boundary, and reflecting points back to the inside of the boundary [9]. In this study, reflecting back and cutting off operations are used:

- Reflecting back

$$x_{ij} = \begin{cases} l_i + (l_i - x_{ij}) - \left\lfloor \frac{l_i - x_{ij}}{u_i - l_i} \right\rfloor (u_i - l_i) & (x_{ij} < l_i) \\ u_i - (x_{ij} - u_i) + \left\lfloor \frac{x_{ij} - u_i}{u_i - l_i} \right\rfloor (u_i - l_i) & (x_{ij} > u_i) \\ x_{ij} & (\text{otherwise}) \end{cases} \tag{18}$$

where $\lfloor z \rfloor$ is the maximum integer smaller than or equal to z. This operation is applied when a new point is generated by DE operations.

- Cutting off

$$x_{ij} = \begin{cases} l_i & (x_{ij} < l_i) \\ u_i & (x_{ij} > u_i) \\ x_{ij} & (\text{otherwise}) \end{cases} \tag{19}$$

This operation is applied when a new point is generated by gradient-based mutation.

In order to find feasible solutions faster, faster reduction of the relaxation of equality constraints in the ε constrained method is adopted by using a large value for control parameter cp in Eq.(17). In order to find feasible solutions more stable, higher gradient-based mutation rate is adopted by using a large value for gradient-based mutation rate P_g. However, it should be noted that too fast reduction, or too large cp and too high mutation rate, or too large P_g tend to lead too fast convergence of search process and increase the probability that the search process is trapped in a local minimum.

The algorithm of the improved εDE based on DE/rand/1/exp variant, which is used in this study, is as follows:

The pseudo code of the improved εDE is shown in Fig. 1.

```
εDE/rand/1/exp()
{
  P=Generate N individuals {x^i} in the search space S randomly;
  ε=ε(0);
  for(t=1; t≤T_max; t++) {
    for(i=1; i≤N; i++) {
      (p1,p2,p3)=select randomly from [1,N] s.t. p1≠p2≠p3≠i;
      x^new=x^i ∈ P;
      j=select randomly from [1,n];
      k=1;
      do {
        x_j^new=x_j^p1+F(x_j^p2 − x_j^p3);
        j=(j+1)%n;
        k++;
      } while(k≤n && u(0,1)<CR);
      move x^new inside of S by reflecting back;
      if(φ(x^i)>ε(t) && u(0,1)<P_g)
        for(s=0; s<R_g; s++) {
          x^new=x^new − ∇C(x^new)^+ΔC(x^new);
          move x^new inside of S by cutting off;
          if(φ(x^new)≤ε(t)) break;
        }
      if((f(x^new),φ(x^new)) <_ε (f(x^i),φ(x^i)))
        x^i=x^new;
    }
    ε=ε(t);
  }
}
```

where $\varepsilon(t)$ is the ε-level control function, F is a scaling factor, CR is a crossover rate, and $u(0,1)$ is a uniform random number generator in $[0,1]$. Underlined parts are modifications from original DE.

Fig. 1 Pseudo code of the improved εDE

5 Experimental Results

In order to show the performance of the improved εDE, twenty four benchmark problems defined in "Problem Definitions and Evaluation Criteria for the CEC 2006 Special Session on Constrained Real-Parameter Optimization" [10], which can be obtained online, are solved. The improved εDE can solve problems with equality constraints directly. However, according to the definition of [10], problems with equality constraints are relaxed, that is, all equality constraints $h_j(\mathbf{x}) = 0,\ j = q+1,\cdots,m$ are replaced by inequalities:

$$|h_j(\mathbf{x})| \leq \delta,\ \delta > 0 \tag{20}$$

where $\delta = 0.001$.

The main features of the test problems are shown in Table 1. The table shows the name of test problem in the column labeled "Prob.", the number of decision variables in "n" and the form of the objective function in "Form of f". The columns labeled LI, NI, LE and NE show the number of linear inequality constraints, nonlinear inequality constraints, linear equality constraints and nonlinear equality constraints, respectively. The number of active constraints at the optimal solution is shown in "active", where "-" means the number of active constraints is unknown. The estimated ratio between the feasible region and the search space, or $\rho = |\mathscr{F}|/|\mathscr{S}|$ is shown in "ρ". The value of ρ is calculated by generating 10,000,000 points randomly in this study.

Table 1 Summary of test problems

Prob.	n	Form of f	LI	NI	LE	NE	active	$\rho(\%)$
g01	13	quadratic	9	0	0	0	6	0.00022
g02	20	nonlinear	1	1	0	0	1	99.99639
g03	10	polynomial	0	0	0	1	1	0.00000
g04	5	quadratic	0	6	0	0	2	26.95954
g05	4	cubic	2	0	0	3	3	0.00000
g06	2	cubic	0	2	0	0	2	0.00655
g07	10	quadratic	3	5	0	0	6	0.00009
g08	2	nonlinear	0	2	0	0	0	0.86109
g09	7	polynomial	0	4	0	0	2	0.52759
g10	8	linear	3	3	0	0	6	0.00063
g11	2	quadratic	0	0	0	1	1	0.00000
g12	3	quadratic	0	9^3	0	0	0	4.76560
g13	5	nonlinear	0	0	1	2	3	0.00000
g14	10	nonlinear	0	0	3	0	3	0.00000
g15	3	quadratic	0	0	1	1	2	0.00000
g16	5	nonlinear	4	34	0	0	4	0.01942
g17	6	nonlinear	0	0	0	4	4	0.00000
g18	9	quadratic	0	13	0	0	4	0.00000
g19	15	nonlinear	0	5	0	0	-	33.47139
g20	24	linear	0	6	2	12	-	0.00000
g21	7	linear	0	1	0	5	6	0.00000
g22	22	linear	0	1	8	11	-	0.00000
g23	9	linear	0	2	3	1	-	0.00000
g24	2	linear	0	2	0	0	2	44.21041

5.1 Parameters Setting

The setting of algorithm parameters in the improved εDE is as follows:

- All parameters that are adjustable
  Parameters for DE are population size (N), scaling factor (F) and crossover rate (CR). Parameters for the ε constrained method are control generations (T_c) and control factor (cp). The others are gradient-based mutation rate (P_g) and the number of repeating mutation (R_g).
- Corresponding dynamic ranges
  Dynamic ranges are not studied enough, but we recommend the following ranges based on our experience to solve various benchmark problems: $N \in [2n, 20n]$, $F \in [0.6, 0.8]$, $CR \in [0.6, 0.95]$, $T_c \in [0.1T_{max}, 0.8T_{max}]$, $cp \in [2, 100]$, $P_g \in [0.01, 0.2]$, $R_g \in [1, 5]$.
- Guidelines on how to adjust the parameters
  Higher N, higher F, higher CR, higher T_c, lower cp, higher P_g and higher R_g make search process more robust, but less fast. To solve problems with many equality constraints, P_g should be a large value such as $P_g = 0.1$. To solve problems with equality constraints faster, cp should be a large value such as $cp = 100$.
- Actual parameter values used
  $N = 40$, $F = 0.7$, $CR = 0.9$, $T_c = 0.2T_{max} = 2500$ ($T_{max} = 12500$), $cp = 100$, $P_g = 0.1$, $R_g = 3$.

Table 2 Error Values Achieved When FES= 5×10^3 , FES= 5×10^4, FES= 5×10^5 for Problems 1-6

FES		g01	g02	g03	g04	g05	g06
5×10^3	Best	6.7406e-01(0)	2.1940e-01(0)	8.5332e-02(0)	1.2975e+00(0)	1.2028e-03(0)	1.2460e-04(0)
	Median	1.0597e+00(0)	2.8279e-01(0)	1.4252e-01(0)	2.9936e+00(0)	2.0853e-03(0)	1.1971e-03(0)
	Worst	1.7100e+00(0)	3.1206e-01(0)	2.9125e-01(0)	6.9507e+00(0)	2.6011e-02(0)	1.5896e-02(0)
	c	0,0,0	0,0,0	0,0,0	0,0,0	0,0,0	0,0,0
	$\bar{v}$	0	0	0	0	0	0
	Mean	1.1072e+00	2.7979e-01	1.5990e-01	3.3517e+00	4.9968e-03	3.1239e-03
	Std	2.3352e-01	2.2376e-02	5.1138e-02	1.5431e+00	5.6951e-03	3.8737e-03
5×10^4	Best	3.0198e-14(0)	1.8987e-03(0)	1.8231e-07(0)	0.0000e+00(0)	0.0000e+00(0)	1.1823e-11(0)
	Median	2.5757e-13(0)	1.0350e-02(0)	1.0915e-06(0)	3.6380e-12(0)	0.0000e+00(0)	1.1823e-11(0)
	Worst	6.4126e-13(0)	2.6937e-02(0)	9.8736e-06(0)	3.6380e-12(0)	0.0000e+00(0)	1.5461e-11(0)
	c	0,0,0	0,0,0	0,0,0	0,0,0	0,0,0	0,0,0
	$\bar{v}$	0	0	0	0	0	0
	Mean	2.9978e-13	1.0975e-02	1.6736e-06	2.7649e-12	0.0000e+00	1.1969e-11
	Std	1.8085e-13	6.0298e-03	1.9943e-06	1.5537e-12	0.0000e+00	7.1290e-13
5×10^5	Best	0.0000e+00(0)	1.0335e-09(0)	-4.4409e-16(0)	0.0000e+00(0)	0.0000e+00(0)	1.1823e-11(0)
	Median	0.0000e+00(0)	1.8181e-08(0)	-4.4409e-16(0)	0.0000e+00(0)	0.0000e+00(0)	1.1823e-11(0)
	Worst	0.0000e+00(0)	7.7944e-08(0)	8.0469e-13(0)	3.6380e-12(0)	0.0000e+00(0)	1.5461e-11(0)
	c	0,0,0	0,0,0	0,0,0	0,0,0	0,0,0	0,0,0
	$\bar{v}$	0	0	0	0	0	0
	Mean	0.0000e+00	2.2816e-08	5.1639e-14	8.7311e-13	0.0000e+00	1.1969e-11
	Std	0.0000e+00	1.8981e-08	1.7257e-13	1.5537e-12	0.0000e+00	7.1290e-13

5.2 Results Achieved

Tables 2, 3, 4 and 5 show best, worst, median, mean, and standard deviation of the difference between the best value found f^{best} and the optimal value f^*, or $f^{\text{best}} - f^*$, after 5×10^3 , 5×10^4 and 5×10^5 function evaluations (FES) for independent 25 runs. Numbers in parenthesis after objective function values show the corresponding number of violated constraints. Number of constraints, which are infeasible at a median solution by more than 1, 0.01, and 0.0001, are shown in c, respectively. Mean violation for the median solution is shown in $\bar{v}$.

The improved εDE succeeded to find feasible and optimal solutions for all problems in all runs except for g20. For g22, the improved εDE succeeded to find very good objective values between 236.37028113 and 236.37031321, which are better than the known objective value 236.430975504001 [10] that had been found by the εDE. For g20, the improved εDE could not find any feasible solutions.

Table 3 Error Values Achieved When FES=5×10^3, FES=5×10^4, FES=5×10^5 for Problems 7-12

FES		g07	g08	g09	g10	g11	g12
5×10^3	Best	1.1299e+01(0)	4.1633e-17(0)	1.0939e+00(0)	1.3994e+03(0)	4.3009e-06(0)	3.8174e-10(0)
	Median	1.7196e+01(0)	5.5511e-17(0)	2.5933e+00(0)	2.6530e+03(0)	4.8662e-05(0)	8.0516e-07(0)
	Worst	2.3544e+01(0)	5.5511e-17(0)	5.5203e+00(0)	4.4121e+03(0)	1.0740e-04(0)	2.0815e-04(0)
	c	0,0,0	0,0,0	0,0,0	0,0,0	0,0,0	0,0,0
	$\bar{v}$	0	0	0	0	0	0
	Mean	1.7041e+01	5.4956e-17	3.0236e+00	2.6088e+03	5.5876e-05	2.8718e-05
	Std	3.3444e+00	2.7195e-18	1.1468e+00	6.0973e+02	3.4898e-05	5.6363e-05
5×10^4	Best	9.1400e-04(0)	4.1633e-17(0)	3.4106e-13(0)	5.3731e-02(0)	0.0000e+00(0)	0.0000e+00(0)
	Median	1.7440e-03(0)	4.1633e-17(0)	1.5916e-12(0)	2.7823e-01(0)	0.0000e+00(0)	0.0000e+00(0)
	Worst	3.3469e-03(0)	5.5511e-17(0)	2.1828e-11(0)	7.4373e-01(0)	0.0000e+00(0)	0.0000e+00(0)
	c	0,0,0	0,0,0	0,0,0	0,0,0	0,0,0	0,0,0
	$\bar{v}$	0	0	0	0	0	0
	Mean	1.8915e-03	4.3299e-17	3.0832e-12	3.0692e-01	0.0000e+00	0.0000e+00
	Std	7.6199e-04	4.5097e-18	4.3673e-12	1.7405e-01	0.0000e+00	0.0000e+00
5×10^5	Best	-1.8474e-13(0)	4.1633e-17(0)	0.0000e+00(0)	-1.8190e-12(0)	0.0000e+00(0)	0.0000e+00(0)
	Median	-1.8119e-13(0)	4.1633e-17(0)	0.0000e+00(0)	-9.0949e-13(0)	0.0000e+00(0)	0.0000e+00(0)
	Worst	8.5212e-11(0)	4.1633e-17(0)	1.1369e-13(0)	-9.0949e-13(0)	0.0000e+00(0)	0.0000e+00(0)
	c	0,0,0	0,0,0	0,0,0	0,0,0	0,0,0	0,0,0
	$\bar{v}$	0	0	0	0	0	0
	Mean	3.3003e-12	4.1633e-17	4.0927e-14	-9.4587e-13	0.0000e+00	0.0000e+00
	Std	1.6723e-11	1.2326e-32	5.4570e-14	1.7822e-13	0.0000e+00	0.0000e+00

Table 4 Error Values Achieved When FES=5×10^3, FES=5×10^4, FES=5×10^5 for Problems 13-18

FES		g13	g14	g15	g16	g17	g18
5×10^3	Best	8.8295e-06(0)	1.0416e+00(0)	5.2736e-05(0)	4.3125e-03(0)	7.7665e-01(0)	4.1136e-01(0)
	Median	8.8614e-05(0)	2.2340e+00(0)	1.5219e-04(0)	6.2111e-03(0)	2.4712e+00(0)	5.3794e-01(0)
	Worst	1.1696e-03(0)	3.4674e+00(0)	3.8836e-04(0)	1.9258e-02(0)	6.2198e+00(0)	3.6596e-01(2)
	c	0,0,0	0,0,0	0,0,0	0,0,0	0,0,0	0,0,0
	$\bar{v}$	0	0	0	0	0	0
	Mean	1.9987e-04	2.1947e+00	1.6343e-04	7.5447e-03	2.7685e+00	5.3969e-01
	Std	2.8399e-04	5.7073e-01	6.5098e-05	3.5710e-03	1.4548e+00	8.9829e-02
5×10^4	Best	-9.7145e-17(0)	3.1397e-07(0)	0.0000e+00(0)	4.8850e-15(0)	1.8190e-12(0)	8.2423e-05(0)
	Median	-7.6328e-17(0)	1.8238e-06(0)	0.0000e+00(0)	4.8850e-15(0)	1.8190e-12(0)	3.2159e-04(0)
	Worst	-6.9389e-18(0)	2.1913e-05(0)	1.1369e-13(0)	4.8850e-15(0)	3.6380e-12(0)	1.5091e-03(0)
	c	0,0,0	0,0,0	0,0,0	0,0,0	0,0,0	0,0,0
	$\bar{v}$	0	0	0	0	0	0
	Mean	-8.1601e-17	2.9338e-06	1.3642e-14	4.8850e-15	1.9645e-12	4.0924e-04
	Std	2.1614e-17	4.1627e-06	3.6944e-14	7.8886e-31	4.9348e-13	3.2403e-04
5×10^5	Best	-9.7145e-17(0)	2.1316e-14(0)	0.0000e+00(0)	4.8850e-15(0)	1.8190e-12(0)	3.3307e-16(0)
	Median	-9.7145e-17(0)	2.1316e-14(0)	0.0000e+00(0)	4.8850e-15(0)	1.8190e-12(0)	3.3307e-16(0)
	Worst	-9.7145e-17(0)	2.7384e-11(0)	1.1369e-13(0)	4.8850e-15(0)	1.8190e-12(0)	4.4409e-16(0)
	c	0,0,0	0,0,0	0,0,0	0,0,0	0,0,0	0,0,0
	$\bar{v}$	0	0	0	0	0	0
	Mean	-9.7145e-17	1.1596e-12	4.5475e-15	4.8850e-15	1.8190e-12	3.5971e-16
	Std	0.0000e+00	5.3561e-12	2.2278e-14	7.8886e-31	1.2117e-27	4.7416e-17

Table 5 Error Values Achieved When FES=5×10^3, FES=5×10^4, FES=5×10^5 for Problems 19-24

FES		g19	g20	g21	g22	g23	g24
5×10^3	Best	6.1712e+01(0)	-3.2125e-02(3)	7.6170e-02(0)	1.4021e+04(3)	7.8657e+01(0)	1.7618e-09(0)
	Median	9.0711e+01(0)	-4.9224e-02(3)	1.6307e+00(0)	1.3511e+04(8)	1.3045e+02(0)	6.3322e-09(0)
	Worst	1.4288e+02(0)	-7.8404e-02(20)	1.0099e+01(0)	1.6054e+04(8)	2.2753e+02(0)	6.5194e-08(0)
	c	0,0,0	0,2,1	0,0,0	4,0,3	0,0,0	0,0,0
	$\bar{v}$	0	0.0227925	0	5246.23	0	0
	Mean	9.4099e+01	-1.8307e-02	2.4895e+00	1.2599e+04	1.3146e+02	1.2123e-08
	Std	2.0358e+01	8.5212e-02	2.7393e+00	5.4900e+03	3.4703e+01	1.3942e-08
5×10^4	Best	2.3584e-01(0)	5.7104e-02(6)	4.1885e-07(0)	4.3505e+03(3)	9.6017e-04(0)	5.7732e-14(0)
	Median	4.0632e-01(0)	1.9165e-03(6)	1.2294e-06(0)	3.2286e+03(4)	4.5635e-03(0)	5.7732e-14(0)
	Worst	9.1655e-01(0)	1.6405e-02(7)	4.0614e-06(0)	5.2416e+03(3)	1.5784e-02(0)	5.7732e-14(0)
	c	0,0,0	0,0,0	0,0,0	0,3,1	0,0,0	0,0,0
	$\bar{v}$	0	1.62563e-08	0	0.0219335	0	0
	Mean	4.5938e-01	9.8623e-03	1.3382e-06	3.9235e+03	6.2415e-03	5.7732e-14
	Std	1.6285e-01	2.5305e-02	7.8608e-07	4.1432e+03	4.0184e-03	2.5244e-29
5×10^5	Best	3.5527e-14(0)	2.4799e-03(6)	-1.7053e-13(0)	-6.0694e-02(0)	0.0000e+00(0)	5.7732e-14(0)
	Median	4.2633e-14(0)	2.2254e-02(6)	-2.8422e-14(0)	-6.0663e-02(0)	0.0000e+00(0)	5.7732e-14(0)
	Worst	2.3874e-12(0)	2.2755e-02(6)	1.4211e-13(0)	-6.0662e-02(0)	5.6843e-14(0)	5.7732e-14(0)
	c	0,0,0	0,0,0	0,0,0	0,0,0	0,0,0	0,0,0
	$\bar{v}$	0	2.25546e-66	0	0	0	0
	Mean	2.4130e-13	2.6661e-02	-3.2969e-14	-6.0666e-02	2.2737e-15	5.7732e-14
	Std	5.6098e-13	2.7282e-02	5.2518e-14	7.2824e-06	1.1139e-14	2.5244e-29

Table 6 Number of FES to achieve the fixed accuracy level $((f(\mathbf{x}) - f(\mathbf{x}*)) \leq 0.0001)$, Success Rate, Feasible Rate and Success Performance on Number of Constraints Evaluations, Objective Function Evaluations and Gradient Matrix Evaluations

Prob.	Best	Median	Worst	Mean	Std	Feasible Rate	Success Rate	Success Performance	
g01	18594	19502	19917	19322	760.1064	100%	100%	19322	(12755,0)
g02	108303	114347	129255	114163	8396.7284	100%	100%	114163	(61313,0)
g03	30733	35470	41716	35993	2704.9874	100%	100%	35993	(6392,6874)
g04	12771	13719	14466	13601	652.4922	100%	100%	13601	(7794,0)
g05	15402	16522	17238	16458	511.2713	100%	100%	16458	(6836,2431)
g06	5037	5733	6243	5669	355.3901	100%	100%	5669	(2915,0)
g07	60873	67946	75569	68447	3965.7744	100%	100%	68447	(19877,0)
g08	621	881	1173	876	147.7844	100%	100%	876	(396,0)
g09	19234	21080	21987	20878	851.7690	100%	100%	20878	(10046,0)
g10	87848	92807	107794	92911	6577.1924	100%	100%	92911	(16846,0)
g11	4569	4569	4569	4425	1033.8126	100%	100%	4425	(2896,378)
g12	2901	4269	5620	4048	810.8960	100%	100%	4048	(409,0)
g13	2707	4918	11759	5285	1655.9506	100%	100%	5285	(901,1293)
g14	30925	32172	32938	32090	2069.4215	100%	100%	32090	(9085,4853)
g15	4053	6805	10880	7851	2223.6332	100%	100%	7851	(3850,1089)
g16	8965	10159	11200	10106	606.1822	100%	100%	10106	(4173,0)
g17	15913	16511	16934	16354	624.8844	100%	100%	16354	(6365,2505)
g18	46856	57910	60108	57638	5335.9858	100%	100%	57638	(12382,0)
g19	147772	162947	178724	161687	10726.1824	100%	100%	161687	(31250,0)
g20	—	—	—	—	—	0%	0%	—	—
g21	31620	35293	35797	35369	1676.9561	100%	100%	35369	(9299,7075)
g22	241270	261355	288750	263655	20465.3694	100%	100%	263655	(8425,76349)
g23	70349	79059	88523	78090	5543.3487	100%	100%	78090	(12279,17649)
g24	1959	2451	2739	2402	216.5576	100%	100%	2402	(1641,0)

However, the improved εDE found solutions with smaller mean constraint violation ($\bar{v}$) $2.9095095 \times 10^{-66}$ than that of the best known solution (0.00718768) [32]. So, the improved εDE succeeded in finding more feasible solutions stably.

Table 6 shows the number of FES needed for satisfying the success condition of $f^{\text{best}} - f^* \leq 0.0001$ and $\mathbf{x}^{\text{best}}$ being feasible. The ratio of runs where feasible solutions or successful solutions can be found, and the estimated FES to find successful

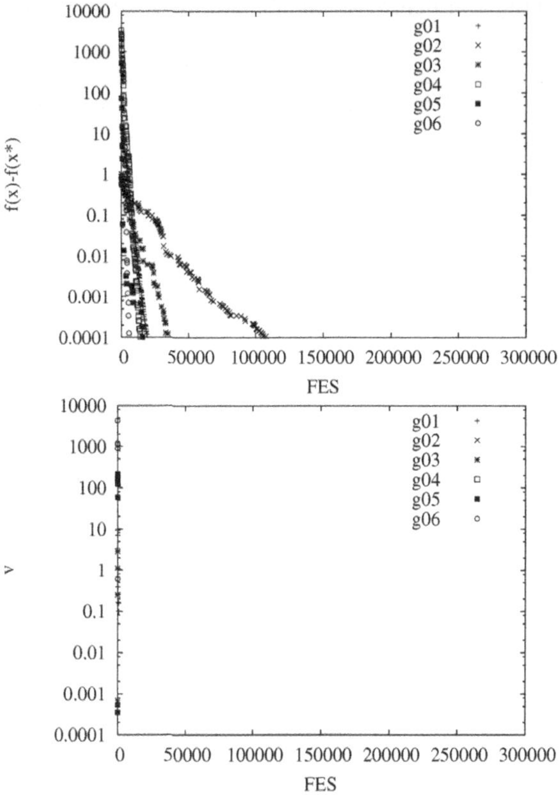

Fig. 2 Convergence Graph for Problems 1-6

solutions, or success performance which is defined by "mean FES/success rate", are shown, too. In the εDE, the objective function and the constraint violation are treated separately, and the evaluation of the objective function can often be omitted when the ε-level comparison can be calculated only by the constraint violation. Numbers in parenthesis after the estimated FES show the substantial number of evaluations for objective function and the number of evaluations for gradient matrix to find successful solutions, respectively.

It is shown that the improved εDE can find near optimal solutions very stably and efficiently; The success performance is less than 5,000 for 4 problems (g08, g11, g12 and g24), is less than 50,000 for 12 problems (g01, g03, g04, g05, g06, g09, g13, g14, g15, g16, g17 and g21), and is less than 100,000 for 4 problems (g07, g10, g18 and g23). Especially, the substantial number of evaluations for objective function is less than 5,000 for 8 problems (g06, g08, g11, g12, g13, g15, g16 and g24), is less than 50,000 for 14 problems (g01, g03, g04, g05, g07, g09, g10, g14, g17, g18, g19, g21, g22 and g23). Also, although the improved εDE needs the calculation of

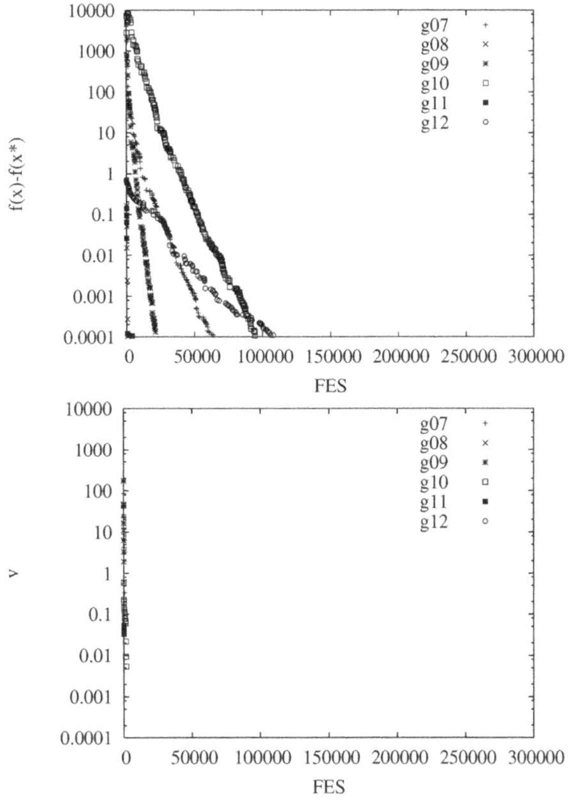

Fig. 3 Convergence Graph for Problems 7-12

gradient matrix, the number of evaluations for the matrix is fairly small compared with the number of evaluations for constraints.

Figures 2, 3, 4 and 5 show the graphs of $f^{\text{best}} - f^*$ and $\bar{v}$ over FES at the median run. In Figures 4 and 5, enough points are not plotted in some problems with equality constraints; After a feasible solution was found by gradient-based mutation, the violation $\bar{v}$ is not plotted. Also, the εDE searches points by relaxing constraints, the points often have better objective value (and worse constraint violation) than those of optimal solution, and the value $f(\mathbf{x}) - f(\mathbf{x}^*)$ becomes negative.

The figures show that the improved εDE could find near optimal solutions very efficiently except for g20. The old known objective value and the violation in the CEC2006 special session were 0.2049794002 and 0.096737, respectively. In this experiment, the solution with violation $2.2554557184 \times 10^{-66}$, which is far smaller violation than the old one, was found. As the result, the objective value became large and the value was 0.26530789.

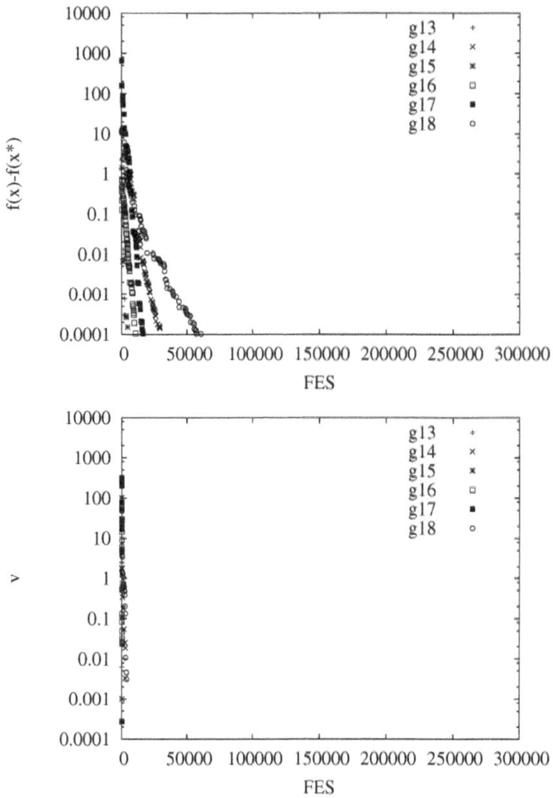

Fig. 4 Convergence Graph for Problems 13-18

5.3 Comparison

Table 7 shows the comparison between the improved εDE and the εDE [32], which is the top-ranked algorithm in CEC2006 special session, for the success rate in the column labeled "Success Rate", success performance in "Performance", the number of evaluations for objective function in "#Obj" and the number of evaluations for gradient matrix in "#Grad". The ratio of the success performance between the improved εDE and the εDE is shown in the column labeled "Performance Ratio", too. The better results are highlighted in boldface.

It is clear that the improved εDE outperformed the εDE. As for the success rate, the improved εDE can find near optimal solution stably in g22, although the εDE cannot found any near optimal solutions. As for the success performance, the improved εDE found near optimal solutions in half number of function evaluations compared with the εDE for all problems with equality constraints (g03, g05, g11, g13, g14, g15, g17, g21 and g23) except for g20. It is thought that the faster reduction of the relaxation of equality constraints, the higher gradient-based mutation

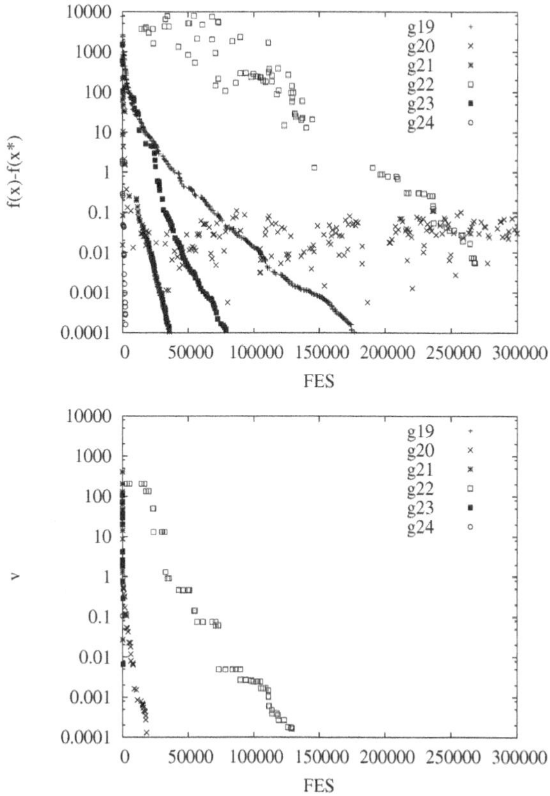

Fig. 5 Convergence Graph for Problems 19-24

rate and the cutting off operation are very effective to solve problems with equality constraints and improves the efficiency and stability of the εDE. Also, the success performance by the improved εDE is under half of that by the εDE for problems g01 and g19. The success performance by the improved εDE is better than that by the εDE in all problems without equality constraints. It is thought that the reflecting back operation is very effective to solve problems with inequality constraints and improves the efficiency of the εDE. For the number of evaluations for objective function, the improved εDE can find near optimal solutions with less number of evaluations in all problems except for g02, g12 and g18.

6 Conclusions

Differential evolution is a recently proposed variant of an evolutionary algorithm. DE is known as a simple, efficient and robust search algorithm that can solve unconstrained optimization problems. In this study, we proposed the improved εDE; The ε constrained method is applied to DE, and in order to solve problems with many

Table 7 Comparison of Success Rate, Performance, Objective Function Calls and Gradient Calls between the improved εDE and the εDE

Prob.	Improved εDE				εDE [32]				Performance
	Success Rate	Performance	#Obj	#Grad	Success Rate	Performance	#Obj	#Grad	Ratio
g01	100%	**19322**	**12755**	0	100%	59308	16979	0	32.6%
g02	100%	**114163**	61313	0	100%	149825	**59351**	0	76.2%
g03	100%	**35993**	**6392**	6874	100%	89407	40217	**188**	40.3%
g04	100%	**13601**	**7794**	0	100%	26216	10125	0	51.9%
g05	100%	**16458**	**6836**	2431	100%	97431	43998	**445**	16.9%
g06	100%	**5669**	**2915**	0	100%	7381	3544	0	76.8%
g07	100%	**68447**	**19877**	0	100%	74303	20430	0	92.1%
g08	100%	**876**	**396**	0	100%	1139	490	0	76.9%
g09	100%	**20878**	**10046**	0	100%	23121	10791	0	90.3%
g10	100%	**92911**	**16846**	0	100%	105234	17743	0	88.3%
g11	100%	**4425**	**2896**	378	100%	16420	11064	**67**	26.9%
g12	100%	**4048**	409	0	100%	4124	**366**	0	98.2%
g13	100%	**5285**	**901**	1293	100%	34738	15152	**113**	15.2%
g14	100%	**32090**	**9085**	4853	100%	113439	37864	**470**	28.3%
g15	100%	**7851**	**3850**	1089	100%	84216	40857	**423**	9.3%
g16	100%	**10106**	**4173**	0	100%	12986	4770	0	77.8%
g17	100%	**16354**	**6365**	2505	100%	98861	36786	**246**	16.5%
g18	100%	**57638**	12382	0	100%	59153	**11936**	0	97.4%
g19	100%	**161687**	**31250**	0	100%	356350	40415	0	45.4%
g20	0%	—	—	—	0%	—	—	—	—
g21	100%	**35369**	**9299**	7075	100%	135143	34633	**483**	26.2%
g22	**100%**	**263655**	**8425**	**76349**	0%	—	—	—	—
g23	100%	**78090**	**12279**	17649	100%	200765	34437	**533**	38.9%
g24	100%	**2402**	**1641**	0	100%	2952	1925	0	81.4%

equality constraints faster, which are very difficult problems for numerical optimization, the gradient-based mutation with high mutation rate and faster reduction of the relaxed constraints to the original constraints is adopted. Also, the cutting off and the reflecting back solutions outside of search space are adopted. We showed that the improved εDE could solve 23 problems out of 24 benchmark problems very efficiently. Also, by comparing the improved εDE with the εDE, it was shown that the improved εDE was a more efficient and stable algorithm than the εDE.

In the future, we will apply the improved εDE to various real world problems that have large numbers of decision variables and constraints.

Acknowledgements. This research is supported in part by Grant-in-Aid for Scientific Research (C) (2) (No. 17510139, 20500138) of Japan society for the promotion of science and Hiroshima City University Grant for Special Academic Research (General Studies) 7111.

References

1. Camponogara, E., Talukdar, S.N.: A genetic algorithm for constrained and multiobjective optimization. In: Alander, J.T. (ed.) 3rd Nordic Workshop on Genetic Algorithms and Their Applications (3NWGA), August 1997, pp. 49–62. University of Vaasa, Finland (1997)
2. Chakraborty, U.K. (ed.): Advances in Differential Evolution. Springer, Heidelberg (2008)
3. Chootinan, P., Chen, A.: Constraint handling in genetic algorithms using gradient-based repair method. Computers & Operations Research 33(8), 2263–2281 (2006)

4. Coath, G., Halgamuge, S.K.: A comparison of constraint-handling methods for the application of particle swarm optimization to constrained nonlinear optimization problems. In: Proc. of IEEE Congress on Evolutionary Computation, Canberra, Australia, pp. 2419–2425 (2003)
5. Coello, C.A.C.: Theoretical and numerical constraint-handling techniques used with evolutionary algorithms: A survey of the state of the art. Computer Methods in Applied Mechanics and Engineering 191(11-12), 1245–1287 (2002)
6. Deb, K.: An efficient constraint handling method for genetic algorithms. Computer Methods in Applied Mechanics and Engineering 186(2/4), 311–338 (2000)
7. Eiben, A.E., Smith, J.E.: Introduction to Evolutionary Computing. Springer, Heidelberg (2003)
8. Hu, X., Eberhart, R.C.: Solving constrained nonlinear optimization problems with particle swarm optimization. In: Proc. of the Sixth World Multiconference on Systemics, Cybernetics and Informatics, Orlando, Florida (2002)
9. Kukkonen, S., Lampinen, J.: Constrained real-parameter optimization with generalized differential evolution. In: Proceedings of the 2006 IEEE Congress on Evolutionary Computation, July 16-2, 2006, pp. 207–214. IEEE Press, Vancouver (2006)
10. Liang, J.J., Runarsson, T.P., Mezura-Montes, E., Clerc, M., Suganthan, P.N., Coello, C.A.C., Deb, K.: Problem definitions and evaluation criteria for the CEC 2006 special session on constrained real-parameter optimization (2006), `http://www.ntu.edu.sg/home/EPNSugan/cec2006/technical_report.pdf`
11. Michalewicz, Z.: A survey of constraint handling techniques in evolutionary computation methods. In: Proceedings of the 4th Annual Conference on Evolutionary Programming, pp. 135–155. The MIT Press, Cambridge (1995)
12. Michalewicz, Z., Schoenauer, M.: Evolutionary algorithms for constrained parameter optimization problems. Evolutionary Computation 4(1), 1–32 (1996)
13. Parsopoulos, K.E., Vrahatis, M.N.: Particle swarm optimization method for constrained optimization problems. In: Sincak, P., Vascak, J., et al. (eds.) Intelligent Technologies — Theory and Application: New Trends in Intelligent Technologies. Frontiers in Artificial Intelligence and Applications, vol. 76, pp. 214–220. IOS Press, Amsterdam (2002)
14. Price, K.V., Storn, R.M., Lampinen, J.A.: Differential Evolution: A Practical Approach to Global Optimization. Springer, Heidelberg (2005)
15. Ray, T., Liew, K.M., Saini, P.: An intelligent information sharing strategy within a swarm for unconstrained and constrained optimization problems. Soft Computing – A Fusion of Foundations, Methodologies and Applications 6(1), 38–44 (2002)
16. Runarsson, T.P., Yao, X.: Stochastic ranking for constrained evolutionary optimization. IEEE Transactions on Evolutionary Computation 4(3), 284–294 (2000)
17. Sakai, S., Takahama, T.: Constrained optimization by the ε constrained differential evolution. In: Numerical Optimization methods, theory and applications, RIMS Kokyuroku, February 2008, vol. 1584, pp. 90–101 (2008) (in Japanese)
18. Campbell, S.L., Meyer, C.J.: Generalized Inverses of Linear Transformations. Dover Publications (1979)
19. Storn, R., Price, K.: Minimizing the real functions of the ICEC 1996 contest by differential evolution. In: Proc. of the International Conference on Evolutionary Computation, pp. 842–844 (1996)
20. Storn, R., Price, K.: Differential evolution – A simple and efficient heuristic for global optimization over continuous spaces. Journal of Global Optimization 11, 341–359 (1997)

21. Surry, P.D., Radcliffe, N.J.: The COMOGA method: Constrained optimisation by multi-objective genetic algorithms. Control and Cybernetics 26(3), 391–412 (1997)
22. Takahama, T., Sakai, S.: Tuning fuzzy control rules by α constrained method which solves constrained nonlinear optimization problems. The Transactions of the Institute of Electronics, Information and Communication Engineers J82-A(5), 658–668 (1999) (in Japanese)
23. Takahama, T., Sakai, S.: Learning fuzzy control rules by α-constrained simplex method. The Transactions of the Institute of Electronics, Information and Communication Engineers J83-D-I(7), 770–779 (2000) (in Japanese)
24. Takahama, T., Sakai, S.: Tuning fuzzy control rules by the α constrained method which solves constrained nonlinear optimization problems. Electronics and Communications in Japan, Part3: Fundamental Electronic Science 83(9), 1–12 (2000)
25. Takahama, T., Sakai, S.: Constrained optimization by α constrained genetic algorithm (αGA). The Transactions of the Institute of Electronics, Information and Communication Engineers J86-D-I(4), 198–207 (2003) (in Japanese)
26. Takahama, T., Sakai, S.: Learning fuzzy control rules by α-constrained simplex method. Systems and Computers in Japan 34(6), 80–90 (2003)
27. Takahama, T., Sakai, S.: Constrained optimization by α constrained genetic algorithm (αGA). Systems and Computers in Japan 35(5), 11–22 (2004)
28. Takahama, T., Sakai, S.: Constrained optimization by combining the α constrained method with particle swarm optimization. In: Proc. of Joint 2nd International Conference on Soft Computing and Intelligent Systems and 5th International Symposium on Advanced Intelligent Systems (2004), `http://www.chi.its.hiroshima-cu.ac.jp/~takahama/eng/papers/aPSO0409c.pdf`
29. Takahama, T., Sakai, S.: Constrained optimization by applying the α constrained method to the nonlinear simplex method with mutations. IEEE Transactions on Evolutionary Computation 9(5), 437–451 (2005)
30. Takahama, T., Sakai, S.: Constrained optimization by the α constrained particle swarm optimizer. Journal of Advanced Computational Intelligence and Intelligent Informatics 9(3), 282–289 (2005)
31. Takahama, T., Sakai, S.: Constrained optimization by ε constrained particle swarm optimizer with ε-level control. In: Proc. of the 4th IEEE International Workshop on Soft Computing as Transdisciplinary Science and Technology (WSTST 2005), May 2005, pp. 1019–1029 (2005)
32. Takahama, T., Sakai, S.: Constrained optimization by the ε constrained differential evolution with gradient-based mutation and feasible elites. In: Proc. of the 2006 IEEE Congress on Evolutionary Computation, July 2006, pp. 308–315 (2006)
33. Takahama, T., Sakai, S.: Constrained optimization by the ε constrained genetic algorithm. IPSJ Journal 47(6), 1861–1871 (2006) (in Japanese)
34. Takahama, T., Sakai, S.: Solving constrained optimization problems by the ε constrained particle swarm optimizer with adaptive velocity limit control. In: Proc. of the 2nd IEEE International Conference on Cybernetics & Intelligent Systems, June 2006, pp. 683–689 (2006)
35. Takahama, T., Sakai, S., Iwane, N.: Constrained optimization by the e constrained hybrid algorithm of particle swarm optimization and genetic algorithm. In: Zhang, S., Jarvis, R. (eds.) AI 2005. LNCS, vol. 3809, pp. 389–400. Springer, Heidelberg (2005)
36. Takahama, T., Sakai, S., Iwane, N.: Solving nonlinear constrained optimization problems by the ε constrained differential evolution. In: Proc. of the 2006 IEEE Conference on Systems, Man, and Cybernetics, October 2006, pp. 2322–2327 (2006)

Constrained Real-Parameter Optimization with ε-Self-Adaptive Differential Evolution

Janez Brest

Abstract. Differential Evolution (DE) algorithms belong to Evolutionary Algorithms (EAs). They are widely used for optimizing continuous functions. In this chapter we present a self-adaptive differential evolution algorithm which uses (1) a self-adaptive mechanism on control parameters F and CR, (2) more strategies during the mutation operation, (3) a population size (NP) reduction mechanism during the evolutionary process, and (4) the ε constrained method. The performance of our algorithm is reported over the set of twenty four CEC2006 constrained benchmark functions.

Keywords: differential evolution, self-adaptation, optimization, constraints.

1 Introduction

Differential Evolution (DE) is a floating-point encoding evolutionary algorithm for global optimization over continuous spaces [12, 27, 32]. Although the original DE algorithm uses control parameters that are fixed during the optimization process, many adaptive and self-adaptive approaches have been proposed recently [1, 2, 7, 9, 17, 26, 28, 35].

Although the DE algorithm has been shown to be a simple yet powerful algorithm, many practical improvements were introduced to make the DE algorithm more powerful, and robust, especially in the following directions:

- hand-tuning of control parameters is replaced with adaptive and/or self-adaptive mechanisms,
- more mutation DE strategies are used during the optimization process, and
- hybrid combination of DE with other optimization techniques.

Janez Brest
Faculty of Electrical Engineering and Computer Science, University of Maribor, Smetanova ul. 17, 2000 Maribor, Slovenia
e-mail: janez.brest@uni-mb.si
http://marcel.uni-mb.si/janez

E. Mezura-Montes (Ed.): Constraint-Handling in Evolution. Optimi., SCI 198, pp. 73–93.
springerlink.com

The general nonlinear programming problem regarding optimization algorithm concerns finding $\mathbf{x}$ so as to optimize $f(\mathbf{x})$; $\mathbf{x} = [x_1, x_2, ..., x_D]^T$. D is the dimensionality of the search space. Domains of the variables are defined by their lower and upper bounds: $x_{j,low}$, $x_{j,upp}$; $j \in \{1, ..., D\}$. The feasible region $\mathscr{F}$ is defined by a set of m additional constraints ($m \geq 0$):

$$g_i(\mathbf{x}) \leq 0, \quad i = 1, ..., q \tag{1}$$

$$h_j(\mathbf{x}) = 0, \quad j = q+1, ..., m \tag{2}$$

If we denote the whole search space with $\mathscr{S}$, then it should be clear that $\mathscr{F} \subseteq \mathscr{S}$. Constraint-handling requires special attention in many real-world optimization problems.

The main objective of this chapter is a performance evaluation of our self-adaptive algorithm *ε-jDE* which uses a self-adaptive mechanism on the control parameters F and CR, more strategies during mutation operation, the ε-constrained method, and population size reduction during the evolutionary process. The performance of the algorithm is evaluated on a set of 24 benchmark functions provided for the CEC2006 special session on real parameter optimization [16].

This chapter is structured as follows. Section 2 gives an overview of related work. Section 3 briefly summarizes differential evolution. Section 4 describes those methods proposed for constraint-handling optimization. Section 5 revises our version of self-adaptive mechanism. Section 6 describes a new differential evolution *ε-jDE* algorithm. Section 7 presents the experimental results of our algorithm on CEC 2006 benchmark functions. Section 8 concludes the chapter with some final remarks.

2 Related Work

The DE algorithm was proposed by Storn and Price [31, 32], and since then it has been used in many practical cases. The original DE was modified, and many new versions proposed. Liu and Lampinen [18] reported that the effectiveness, efficiency, and robustness of the DE algorithm are sensitive to the settings of control parameters. The best settings for control parameters depend on the function and requirements for consumption time and accuracy.

Modifications to the original DE, in order to improve its efficiency and robustness, are introduced by using adaptation and/or self-adaptation mechanisms for DE control parameters [1, 7, 17, 19, 26, 28, 35], incorporation of more mutation strategies [6, 14, 28], or the use of other approaches [3–5, 20, 34].

Qin and Suganthan in [14, 28] proposed a Self-adaptive Differential Evolution algorithm (SaDE), where it is unnecessary to predefine the choice of learning strategy and the two control parameters F and CR. During evolution, a more suitable learning strategy and parameter settings are gradually self-adapted according to the learning experience.

J. Brest et al. [7] have proposed a jDE algorithm, which uses self-adaptive control parameters F and CR. The comparison of jDE and jDE-2 [6] algorithms showed that the jDE-2 algorithm, which uses two DE strategies, performed better than jDE algorithm, which uses only one strategy.

Differential evolution and its modified versions were widely used for constrained optimization [5, 9, 14, 20, 23, 30, 34].

An empirical comparison of some DE variants for solving global optimization problems is presented by E. Mezura Montes et al. [22].

In work [21], the authors present comparisons between four bio-inspired algorithms with similar constraint-handling technique (Deb's feasibility rules) on a set of 24 benchmark functions.

Recently, Wang et al. in [36] proposed an adaptive trade-off model (ATM) for constrained evolutionary computation. This model considers:

- the evolution of infeasible solutions when the population contains only infeasible individuals,
- balancing feasible and infeasible solutions when the population consists of a combination of feasible and infeasible individuals, and
- the selection of feasible solutions when the population is composed of feasible solutions only.

The ATM model was incorporated with evolutionary strategy (ES).

3 The Differential Evolution Algorithm

The population of the original DE algorithm [29, 31, 32] contains NP individuals. An individual is defined as a D-dimensional vector. If G denotes the generation, the population at generation G consists of:

$$\mathbf{x}_{i,G} = \{x_{i,1,G}, x_{i,2,G}, ..., x_{i,D,G}\},\ i = 1,2,...,NP. \tag{3}$$

During one generation for each vector $\mathbf{x}_{i,G}$, DE employs mutation and crossover operations to produce a trial vector:

$$\mathbf{u}_{i,G} = \{u_{i,1,G}, u_{i,2,G}, ..., u_{i,D,G}\},\ i = 1,2,...,NP. \tag{4}$$

Then a selection operation is used to choose vectors for the next generation $(G+1)$.

The initial population is usually selected uniformly randomly between the lower $(x_{j,low})$ and upper $(x_{j,upp})$ bounds defined for each variable x_j. These bounds are specified according to the nature of the problem.

3.1 Mutation Operation

Mutation for each population vector $\mathbf{x}_{i,G}$ creates a mutant vector $\mathbf{v}_{i,G}$:

$$\mathbf{v}_{i,G} = \{v_{i,1,G}, v_{i,2,G}, ..., v_{i,D,G}\},\ i = 1,2,...,NP. \tag{5}$$

A new mutant vector can be created using one of the mutation strategies. The most useful strategies are [12, 27, 32]:

$$
\begin{aligned}
&\text{'DE/rand/1'} && \mathbf{v}_{i,G} = \mathbf{x}_{r_1,G} + F \cdot (\mathbf{x}_{r_2,G} - \mathbf{x}_{r_3,G}) && (6)\\
&\text{'DE/best/1'} && \mathbf{v}_{i,G} = \mathbf{x}_{best,G} + F \cdot (\mathbf{x}_{r_1,G} - \mathbf{x}_{r_2,G}) && (7)\\
&\text{'DE/current to best/1'} && \mathbf{v}_{i,G} = \mathbf{x}_{i,G} + F \cdot (\mathbf{x}_{best,G} - \mathbf{x}_{i,G}) + F \cdot (\mathbf{x}_{r_1,G} - \mathbf{x}_{r_2,G}) && (8)\\
&\text{'DE/best/2'} && \mathbf{v}_{i,G} = \mathbf{x}_{best,G} + F \cdot (\mathbf{x}_{r_1,G} - \mathbf{x}_{r_2,G}) + F \cdot (\mathbf{x}_{r_3,G} - \mathbf{x}_{r_4,G}) && (9)\\
&\text{'DE/rand/2'} && \mathbf{v}_{i,G} = \mathbf{x}_{r_1,G} + F \cdot (\mathbf{x}_{r_2,G} - \mathbf{x}_{r_3,G}) + F \cdot (\mathbf{x}_{r_4,G} - \mathbf{x}_{r_5,G}) && (10)
\end{aligned}
$$

where the indexes $r_d, d = 1, ..., 5$ represent the random and mutually different integers generated within the range $[1, NP]$ and also different from index i. F is a mutation scale factor within the range $[0, 2]$, usually less than 1. Vector $\mathbf{x}_{best,G}$ is the best vector in generation G. Some authors use the name 'target to best' for the strategy in eq. (8).

3.2 Crossover Operation

After mutation, a 'binary' crossover operation forms the trial vector $\mathbf{u}_{i,G}$ according to the target vector $\mathbf{x}_{i,G}$ and its corresponding mutant vector $\mathbf{v}_{i,G}$.

$$
u_{i,j,G} = \begin{cases} v_{i,j,G} & \text{if } rand(0,1) \leq CR \text{ or } j = j_{rand}, \\ x_{i,j,G} & \text{otherwise} \end{cases} \tag{11}
$$

$$
i = 1, 2, ..., NP \text{ and } j = 1, 2, ..., D.
$$

CR is a crossover control parameter or factor within the range $[0, 1)$ and presents the probability of creating parameters for a trial vector from the mutant vector. Index j_{rand} is a randomly chosen integer within the range $[1, NP]$. It is responsible for the trial vector containing at least one parameter from the mutant vector. Here we have described the binary crossover operation ('bin'). The other DE crossover operation is exponential ('exp'), but this is rarely used in practical optimization.

If some components of the trial vector are out of bounds, the proposed solutions in literature [27, 29, 31, 32] are: they are reflected from bounds, set on bounds or used as they are (out of bounds).

3.3 Selection Operation

The DE algorithm uses a greedy selection. The selection operation selects, according to the fitness value of the target vector and its corresponding trial vector, which vector will survive to be a member of the next generation. For example, if we have a minimization problem, we will use the following selection rule:

$$\mathbf{x}_{i,G+1} = \begin{cases} \mathbf{u}_{i,G} & \text{if } f(\mathbf{u}_{i,G}) < f(\mathbf{x}_{i,G}), \\ \mathbf{x}_{i,G} & \text{otherwise.} \end{cases} \quad (12)$$

4 Constraint Handling

Over the last few years several methods have been proposed for handling constraints by genetic algorithms for parameter optimization problems. These methods have been grouped by Michalewicz et al. [15,25] into four categories:

1. *Methods based on preserving the feasibility of solutions.* The idea behind the method is based on specialized operators which transform feasible parents into feasible offspring. The method assumes linear constraints only and a feasible starting point or feasible initial population.
2. *Methods based on penalty functions.* Many evolutionary algorithms incorporate a constraint-handling method based on the concept of exterior penalty functions which penalize infeasible solutions. These methods differ in important details, such as how the penalty function is designed and applied to infeasible solutions.
3. *Methods which make a clear distinction between feasible and infeasible solutions.* There are a few methods which emphasize the distinction between feasible and infeasible solutions in the search space. One of those methods distinguishes between feasible and infeasible individuals: for any feasible individual $\mathbf{x}$ and any infeasible individual $\mathbf{y}$: $f(\mathbf{x}) < f(\mathbf{y})$, i.e. any feasible solution is better than any infeasible one.
4. *Other hybrid methods.* These methods combine evolutionary computation techniques with deterministic procedures for numerical optimization problems.

Most constrained problems can be handled by the penalty function method. A measure of the constraint violation is often useful when handling constraints. A solution $\mathbf{x}$ is regarded as *feasible* if

$$g_i(\mathbf{x}) \leq 0, \; i = 1, ..., q, \quad (13)$$

$$|h_j(\mathbf{x})| - \varepsilon \leq 0, \; j = q+1, ..., m, \quad (14)$$

where equality constraints are transformed into inequalities. In CEC2006 [16] ε is set to 0.0001. The mean value of all constraints' violations $\bar{v}$ is defined as:

$$\bar{v} = \frac{(\sum_{i=1}^{q} G_i(\mathbf{x}) + \sum_{j=q+1}^{m} H_j(\mathbf{x}))}{m}, \quad (15)$$

where

$$G_i(\mathbf{x}) = \begin{cases} g_i(\mathbf{x}), & g_i(\mathbf{x}) > 0, \\ 0, & g_i(\mathbf{x}) \leq 0, \end{cases} \quad (16)$$

$$H_j(\mathbf{x}) = \begin{cases} |h_j(\mathbf{x})|, & |h_j(\mathbf{x})| - \varepsilon > 0, \\ 0, & |h_j(\mathbf{x})| - \varepsilon \leq 0. \end{cases} \quad (17)$$

The sum of all constraint violations is zero for feasible solutions and positive when at least one constraint is violated. An obvious application of the constraint violation is to use it to guide the search towards feasible areas of the search space. There has been quite a lot of work on such ideas and other constraint techniques within the EA-community. A summary of these techniques can be found in [11,24], which also contains information on many other stochastic techniques.

Recently the DE algorithm has become very popular over many research areas. No special extensions of the DE algorithm are necessary to make it suitable for handling constraints.

5 The Self-adaptive Differential Evolution Algorithm

This section revises our version of a self-adaptive mechanism. It was used in algorithms [6–8] for unconstrained optimization, and in jDE-2 algorithm [9] for constrained optimization. This self-adaptive mechanism is used on the control parameters F and CR.

The self-adaptive control parameter mechanism of '*DE/rand/1/bin*' strategy was used in [7]. This strategy is the most often used in practice [13,18,32,33].

Let us describe this self-adaptive mechanism. Each individual in the population is extended with the control parameters F and CR. Better values of these control parameters lead to better individuals which, in turn, are more likely to survive and produce offspring and, hence, propagate these better parameter values.

New control parameters $F_{i,G+1}$ and $CR_{i,G+1}$ are calculated as follows:

$$F_{i,G+1} = \begin{cases} F_l + rand(0,1) \cdot F_u & \text{if } rand(0,1) < \tau_1, \\ F_{i,G} & \text{otherwise,} \end{cases} \tag{18}$$

$$CR_{i,G+1} = \begin{cases} rand(0,1) & \text{if } rand(0,1) < \tau_2, \\ CR_{i,G} & \text{otherwise.} \end{cases} \tag{19}$$

and they produce control parameters F and CR in a new trial vector. $rand(0,1)$ is uniform random value within the range $[0,1]$. Values τ_1 and τ_2 represent probabilities of adjusting control parameters F and CR, respectively. τ_1, τ_2, F_l, F_u are taken fixed values $0.1, 0.1, 0.1, 0.9$ [7], respectively. The new F takes a value from $[0.1, 1.0]$ and the new CR from $[0,1]$ in a random manner. $F_{i,G+1}$ and $CR_{i,G+1}$ are obtained before the mutation operation is performed and so influence the mutation, crossover and selection operations of the new vector $\mathbf{x}_{i,G+1}$.

Some ideas, on how to improve the jDE algorithm, are reported in [6]. Here we outline certain features, which can make our jDE algorithm more general and also improve its performance. The rest of this section outlines them.

To keep the solution of bound-constrained problems feasible, those trial parameters that violate boundary constraints are set to bound values by jDE algorithm [7]. Rönkönen, Kukkonen and Price [29] suggest that those solutions that violate boundary constraints should be reflected back from the bound by the amount of violation:

Table 1 Characteristics of the 24 benchmark problems. D is the number of decision variables, $\rho = |\mathscr{F}|/|\mathscr{S}|$ is the estimated ratio between the feasible region and the search space, LI is the number of linear inequality constraints, NI the number of nonlinear inequality constraints, LE is the number of linear equality constraints and NE is the number of nonlinear equality constraints. a is the number of active constraints at $\mathbf{x}^*$

Prob.	D	$\mathbf{f}(\mathbf{x}^*)$	Type of function	ρ	LI	NI	LE	NE	a
g01	13	-15.00000	quadratic	0.0111%	9	0	0	0	6
g02	20	-0.80361910412	nonlinear	99.9971%	0	2	0	0	1
g03	10	-1.0005001000	polynomial	0.0000%	0	0	0	1	1
g04	5	-30665.5386717834	quadratic	52.1230%	0	6	0	0	2
g05	4	5126.4967140071	cubic	0.0000%	2	0	0	3	3
g06	2	-6961.8138755802	cubic	0.0066%	0	2	0	0	2
g07	10	24.3062090681	quadratic	0.0003%	3	5	0	0	6
g08	2	-0.0958250415	nonlinear	0.8560%	0	2	0	0	0
g09	7	680.6300573745	polynomial	0.5121%	0	4	0	0	2
g10	8	7049.2480205286	linear	0.0010%	3	3	0	0	6
g11	2	0.7499	quadratic	0.0000%	0	0	0	1	1
g12	3	-1.000000	quadratic	4.7713%	0	1	0	0	0
g13	5	0.0539415140	nonlinear	0.0000%	0	0	0	3	3
g14	10	-47.7648884595	nonlinear	0.0000%	0	0	3	0	3
g15	3	961.7150222899	quadratic	0.0000%	0	0	1	1	2
g16	5	-1.9051552586	nonlinear	0.0204%	4	34	0	0	4
g17	6	8853.5396748064	nonlinear	0.0000%	0	0	0	4	4
g18	9	-0.8660254038	quadratic	0.0000%	0	13	0	0	6
g19	15	32.6555929502	nonlinear	33.4761%	0	5	0	0	0
g20	24	0.2049794002	linear	0.0000%	0	6	2	12	16
g21	7	193.7245100700	linear	0.0000%	0	1	0	5	6
g22	22	236.4309755040	linear	0.0000%	0	1	8	11	19
g23	9	-400.0551	linear	0.0000%	0	2	3	1	6
g24	2	-5.5080132716	linear	79.6556%	0	2	0	0	2

$$u_{i,j,G} = \begin{cases} 2 \cdot x_{j,low} - u_{i,j,g} & \text{if } u_{i,j,g} < x_{j,low}, \\ 2 \cdot x_{j,upp} - u_{i,j,g} & \text{if } u_{i,j,g} > x_{j,upp}. \end{cases} \tag{20}$$

The jDE-2 [6] algorithm uses both solutions for violated boundary constraints with equal probability, in a random manner:

$$t = rand(0,1), \;\; p_0 = 0.5,$$

$$u_{i,j,G} = \begin{cases} x_{j,low} & \text{if } (t \leq p_0) \wedge (u_{i,j,G} < x_{j,low}), \\ x_{j,upp} & \text{if } (t \leq p_0) \wedge (u_{i,j,G} > x_{j,upp}), \\ 2 \cdot x_{j,low} - u_{j,i,G} & \text{if } (t > p_0) \wedge (u_{i,j,G} < x_{j,low}), \\ 2 \cdot x_{j,upp} - u_{j,i,G} & \text{if } (t > p_0) \wedge (u_{i,j,G} > x_{j,upp}). \end{cases} \tag{21}$$

Table 2 Error Values Achieved When FES= 5×10^3, FES= 5×10^4, FES= 5×10^5 for Problems 1-5

FES		g01	g02	g03	g04	g05
5×10^3	Best	5.2154 (0)	0.4499 (0)	1.0005 (1)	105.1884 (0)	484.9959 (3)
	Median	7.6307 (0)	0.5163 (0)	0.9873 (1)	178.7184 (0)	774.5920 (3)
	Worst	8.6992 (0)	0.5492 (0)	0.9999 (1)	325.4142 (0)	47.6715 (4)
	c	0, 0, 0	0, 0, 0	0, 1, 1	0, 0, 0	3, 3, 3
	$\overline{v}$	0.0000	0.0000	0.0460	0.0000	3.3333
	Mean	7.4536	5.1628e-01	9.8080e-01	1.9402e+02	3.7147e+02
	Std	8.2241e-01	1.9928e-02	8.5944e-02	4.5863e+01	3.8700e+02
5×10^4	Best	0.0001 (0)	0.0416 (0)	1.0005 (1)	0.0005 (0)	9.8380e-05 (0)
	Median	0.0003 (0)	0.0698 (0)	0.9992 (1)	0.0023 (0)	0.0002 (0)
	Worst	0.0006 (0)	0.0868 (0)	0.9999 (1)	0.0047 (0)	0.0009 (0)
	c	0, 0, 0	0, 0, 0	0, 1, 1	0, 0, 0	0, 0, 0
	$\overline{v}$	0.0000	0.0000	0.0201	0.0000	0.0000
	Mean	3.3654e-04	6.7177e-02	6.5047e-01	2.2956e-03	3.2579e-04
	Std	1.1245e-04	1.3401e-02	3.8150e-01	1.2289e-03	2.0220e-04
5×10^5	Best	0 (0)	2.0003e-08 (0)	-1.0003e-11 (0)	7.2759e-11 (0)	-1.8189e-12 (0)
	Median	0 (0)	1.3415e-07 (0)	-1.0002e-11 (0)	7.2759e-11 (0)	-1.8189e-12 (0)
	Worst	0 (0)	6.4191e-07 (0)	-9.9813e-12 (0)	7.6397e-11 (0)	-1.8189e-12 (0)
	c	0, 0, 0	0, 0, 0	0, 0, 0	0, 0, 0	0, 0, 0
	$\overline{v}$	0.0000	0.0000	0.0000	0.0000	0.0000
	Mean	0.0000	1.7534e-07	-1.0002e-11	7.2905e-11	-1.8190e-12
	Std	0.0000	1.5210e-07	4.2841e-15	7.2760e-13	1.2367e-27

Strategy 'DE/rand/1/bin' is used in jDE algorithm and control parameters F and CR are encoded in each individual. In [28] the authors proposed the self-adaptive SaDE algorithm which uses two of five original DE's strategies to be applied to individuals in the current population. In [6] more DE's strategies are applied to this algorithm. Each strategy uses its own control parameters.

6 The ε-*jDE* Algorithm

In this section we describe the new version of the DE algorithm, called ε-*jDE*. The ε-*jDE* algorithm follows the jDE-2 algorithm and emphasizes constraints as follows. It compares two solutions, say i and j, during the selection operation (see section 3.3):

$$\mathbf{x}_{i,G+1} = \begin{cases} \mathbf{x}_{j,G} & \text{if } (\overline{v}_{i,G} > \overline{v}_{j,G}), \\ \mathbf{x}_{j,G} & \text{else if } (\overline{v}_{j,G} = 0) \wedge (f(\mathbf{x}_{i,G}) > f(\mathbf{x}_{j,G})), \\ \mathbf{x}_{i,G} & \text{otherwise.} \end{cases} \quad (22)$$

Table 3 Error Values Achieved When FES= 5×10^3, FES= 5×10^4, FES= 5×10^5 for Problems 6-10

FES		g06	g07	g08	g09	g10
	Best	43.6145 (0)	95.1110 (0)	2.0131e-07 (0)	54.0914 (0)	7744.8729 (0)
	Median	400.4579 (0)	199.9499 (0)	1.5087e-05 (0)	156.0776 (0)	21269.6291 (0)
	Worst	1201.7190 (0)	418.8726 (0)	6.3616e-05 (0)	323.5793 (0)	6480.4880 (2)
5×10^3	c	0, 0, 0	0, 0, 0	0, 0, 0	0, 0, 0	0, 0, 0
	$\overline{v}$	0.0000	0.0000	0.0000	0.0000	0.0000
	Mean	4.5631e+02	2.2266e+02	2.0956e-05	1.6388e+02	1.3974e+04
	Std	2.7814e+02	9.2684e+01	1.7118e-05	5.6918e+01	4.0107e+03
	Best	1.2605e-09 (0)	0.5034 (0)	8.1964e-11 (0)	0.0094 (0)	165.4944 (0)
	Median	1.0360e-08 (0)	0.7904 (0)	8.1964e-11 (0)	0.0237 (0)	247.8820 (0)
	Worst	7.9960e-08 (0)	1.3064 (0)	8.1964e-11 (0)	0.0546 (0)	395.6845 (0)
5×10^4	c	0, 0, 0	0, 0, 0	0, 0, 0	0, 0, 0	0, 0, 0
	$\overline{v}$	0.0000	0.0000	0.0000	0.0000	0.0000
	Mean	1.6667e-08	8.2681e-01	8.1964e-11	2.6217e-02	2.6657e+02
	Std	1.8548e-08	1.8677e-01	6.9389e-18	1.0998e-02	6.9506e+01
	Best	4.4565e-11 (0)	7.9761e-11 (0)	8.1964e-11 (0)	-9.8339e-11 (0)	6.1845e-11 (0)
	Median	4.4565e-11 (0)	7.9786e-11 (0)	8.1964e-11 (0)	-9.8225e-11 (0)	6.2755e-11 (0)
	Worst	4.4565e-11 (0)	8.0714e-11 (0)	8.1964e-11 (0)	-9.8111e-11 (0)	4.8826e-07 (0)
5×10^5	c	0, 0, 0	0, 0, 0	0, 0, 0	0, 0, 0	0, 0, 0
	$\overline{v}$	0.0000	0.0000	0.0000	0.0000	0.0000
	Mean	4.4565e-11	7.9878e-11	8.1964e-11	-9.8221e-11	1.9591e-08
	Std	1.3191e-26	2.3704e-13	2.6382e-26	3.9926e-14	9.7640e-08

The algorithm distinguishes between feasible ($\overline{v} = 0$) and infeasible individuals: any feasible solution being better than any infeasible one.

6.1 Controlling the ε Level

In our previous work [9], constraints have only been used to see whether an individual solution is feasible or not. Therefore the jDE-2 algorithm had difficulties when solving constrained optimization problems with equality constraints. Takahama and Sakai in [34] pointed out that for problems with equality constraints, the ε level should be controlled properly in order to obtain high quality solutions.

In this study we present ε level controlling in our *ε-jDE* algorithm. In the proposed method ε level constraint violation precedes the objective function. A method of controlling the ε level is defined according to equations (23)–(26). The ε level is updated until the number of generations G reaches the control generation G_c. After the number of generations exceeds G_c, the ε level is set to 0 to obtain solutions with minimum constraint violation

Table 4 Error Values Achieved When FES= 5×10^3, FES= 5×10^4, FES= 5×10^5 for Problems 11-15

FES		g11	g12	g13	g14	g15
	Best	0.0429 (0)	3.3932e-05 (0)	0.7096 (3)	-209.8728 (3)	3.5725 (2)
	Median	0.2501 (0)	0.0002 (0)	0.8416 (3)	-310.3281 (3)	4.3295 (2)
	Worst	0.0325 (1)	0.0085 (0)	0.6390 (3)	-359.8290 (3)	2.1052 (2)
5×10^3	c	0, 0, 0	0, 0, 0	0, 3, 3	3, 3, 3	1, 2, 2
	$\overline{v}$	0.0000	0.0000	0.2269	8.7142	1.6489
	Mean	2.1877e-01	1.3071e-03	7.0378e-01	-2.8378e+02	3.7171
	Std	7.3091e-02	2.5604e-03	3.0042e-01	4.8007e+01	3.2798
	Best	0.0001 (0)	0 (0)	0.0015 (3)	1.8430 (3)	2.4653 (2)
	Median	0.0075 (0)	0 (0)	0.0124 (3)	-42.4185 (3)	-0.3272 (2)
	Worst	0.2501 (0)	0 (0)	0.0072 (3)	-46.2233 (3)	1.2531 (2)
5×10^4	c	0, 0, 0	0, 0, 0	0, 3, 3	2, 3, 3	0, 2, 2
	$\overline{v}$	0.0000	0.0000	0.0605	1.3057	0.2829
	Mean	3.2902e-02	0.0000	1.0346e-01	-3.7731e+01	1.1544
	Std	7.0567e-02	0.0000	1.8287e-01	1.3116e+01	1.6465
	Best	0 (0)	0 (0)	4.1897e-11 (0)	8.5051e-12 (0)	6.0822e-11 (0)
	Median	0 (0)	0 (0)	4.1897e-11 (0)	8.5123e-12 (0)	6.0822e-11 (0)
	Worst	0 (0)	0 (0)	4.1897e-11 (0)	8.5833e-12 (0)	6.0822e-11 (0)
5×10^5	c	0, 0, 0	0, 0, 0	0, 0, 0	0, 0, 0	0, 0, 0
	$\overline{v}$	0.0000	0.0000	0.0000	0.0000	0.0000
	Mean	0.0000	0.0000	4.1898e-11	8.5200e-12	6.0822e-11
	Std	0.0000	0.0000	1.2098e-17	1.6527e-14	1.3191e-26

$$\varepsilon_0 = \varepsilon \tag{23}$$

$$v_0 = \overline{v}(\mathbf{x}_\theta) \tag{24}$$

$$\overline{v}_G = \begin{cases} \alpha_1 v_{G-1}, & \alpha_2 \overline{v}(\mathbf{x}_\beta) < v_{G-1},\ \varepsilon_{G-1} < \varepsilon,\ 0 < G < G_c \\ v_{G-1}, & \text{otherwise} \end{cases} \tag{25}$$

$$\varepsilon_G = \begin{cases} \max\{v_G(1 - \frac{G}{G_c})^{c_p}, \varepsilon\}, & 0 < G < G_c \\ 0, & G \geq G_c \end{cases} \tag{26}$$

where $\mathbf{x}_\theta$ is the top θ-th individual and $\theta = 0.3NP$. Note, that $\varepsilon(0) = 0$ when mean violation $\overline{v}(\mathbf{x}_\theta)$ is calculated. Similarly, $\mathbf{x}_\beta$ is the top β-th individual and $\beta = 0.7NP$. c_p is a parameter to control the speed of constraints' reducing relaxation, while parameters $\alpha_1 < 1$ and $\alpha_2 > 1$ adaptively control v_G value, which also controls the speed of the constraints' reducing relaxation. Parameters α_1 and α_2 can only decrease the v_G value by a small amount when top β individuals have mean violations $\overline{v}(\mathbf{x}_\beta)$ multiplied by α_1 less than v_G. Using this adaptation, the ε level could reach 0 before $G \geq G_c$.

In this study we use $c_p = 5$, $\alpha_1 = 0.8$, $\alpha_2 = 2.0$, and G_c=$0.2G_{max}$.

Table 5 Error Values Achieved When FES= 5×10^3, FES= 5×10^4, FES= 5×10^5 for Problems 16-20

FES		g16	g17	g18	g19	g20
	Best	0.1603 (0)	123.2327 (4)	0.8693 (6)	365.9424 (0)	3.9797 (20)
	Median	0.2759 (0)	368.3257 (4)	0.7791 (7)	753.6668 (0)	7.6478 (15)
	Worst	0.3678 (0)	15.2875 (4)	3.0180 (9)	1059.8608 (0)	6.9592 (20)
5×10^3	c	0, 0, 0	4, 4, 4	3, 7, 7	0, 0, 0	2, 14, 15
	$\bar{v}$	0.0000	18.6476	0.8755	0.0000	2.4169
	Mean	2.7076e-01	1.1013e+02	8.2458e-01	7.3005e+02	5.3422
	Std	4.9585e-02	1.2051e+02	1.1402	1.5942e+02	2.1635
	Best	8.0238e-05 (0)	79.1620 (4)	0.0149 (0)	7.0835 (0)	-0.0228 (20)
	Median	0.0001 (0)	16.1098 (4)	0.0354 (0)	10.7022 (0)	-0.0452 (20)
	Worst	0.0002 (0)	-72.5976 (4)	0.0782 (0)	15.2844 (0)	-0.0655 (20)
5×10^4	c	0, 0, 0	3, 4, 4	0, 0, 0	0, 0, 0	0, 14, 20
	$\bar{v}$	0.0000	11.8344	0.0000	0.0000	0.0320
	Mean	1.7416e-04	4.9222e+01	4.0341e-02	1.0709e+01	-2.8473e-02
	Std	5.7431e-05	5.9751e+01	1.7146e-02	2.2420	2.1696e-02
	Best	6.5213e-11 (0)	8.1854e-11 (0)	1.5561e-11 (0)	3.5858e-10 (0)	-9.7851e-06 (1)
	Median	6.5213e-11 (0)	8.1854e-11 (0)	1.5561e-11 (0)	9.9033e-08 (0)	-8.9602e-05 (1)
	Worst	6.5213e-11 (0)	8.1854e-11 (0)	1.5561e-11 (0)	1.6201e-05 (0)	0.0027 (3)
5×10^5	c	0, 0, 0	0, 0, 0	0, 0, 0	0, 0, 0	0, 1, 1
	$\bar{v}$	0.0000	0.0000	0.0000	0.0000	0.0072
	Mean	6.5214e-11	8.1855e-11	1.5561e-11	1.8493e-06	5.2432e-05
	Std	1.3191e-26	2.6382e-26	4.8393e-17	3.5810e-06	6.5814e-04

6.2 Population Size Reduction

Population size (*NP*) is the third control parameter of the DE algorithm. It plays also an important role during the optimization process, but has maybe been studied less in literature [10, 35].

The population size reduction mechanism [8] during the evolutionary process is used by the *ε-jDE* algorithm. One individual from the first half ($\mathbf{x}_{i,G}$) of the current population and a corresponding individual from the second half ($\mathbf{x}_{\frac{NP}{2}+i,G}$) are compared, based on their fitness values and the better one is placed (as a survivor) in the first half at position i of the current population, e.g. the first part of the current population is assumed to be the population which is to be the parent population in the next generation. In the proposed reduction scheme the new population size is equal to half the previous population size. Population size reduction is depicted in Fig. 5.

$NP_1 = NP_{init}$ is the initial population size and $NP_p(p = 1, 2, ..., pmax)$ is the population size after $p-1$ reductions.

In this chapter we set $pmax = 2$ and $NP_{init} = 200$. This implies that our algorithm performed one reduction after $\frac{maxFES}{2} = 250,000$ function evaluations, and final population size was 100.

Table 6 Error Values Achieved When FES= 5×10^3, FES= 5×10^4, FES= 5×10^5 for Problems 21-24

FES		g21	g22	g23	g24
	Best	390.8314 (4)	13849.5269 (17)	400.0551 (0)	0.0005 (0)
	Median	714.4626 (4)	19763.5690 (18)	-55.6814 (3)	0.0174 (0)
	Worst	789.8916 (5)	9121.8452 (16)	297.1619 (5)	0.0326 (0)
5×10^3	c	1, 4, 4	18, 18, 18	2, 3, 3	0, 0, 0
	$\bar{v}$	0.3697	2961574.2729	0.9883	0.0000
	Mean	4.3066e+02	1.1942e+04	8.9730e+01	1.7879e-02
	Std	2.1922e+02	6.5279e+03	3.4620e+02	8.9443e-03
	Best	70.4046 (3)	12571.0945 (19)	400.0551 (0)	4.7233e-12 (0)
	Median	39.5707 (4)	15339.1883 (19)	335.6886 (4)	4.8592e-12 (0)
	Worst	23.0029 (5)	9407.0760 (19)	120.8729 (3)	5.2642e-12 (0)
5×10^4	c	0, 0, 4	18, 19, 19	0, 3, 4	0, 0, 0
	$\bar{v}$	0.0010	356971.5355	0.0096	0.0000
	Mean	5.7134e+01	1.2194e+04	2.7195e+02	4.9032e-12
	Std	6.0601e+01	4.8953e+03	1.1045e+02	1.6761e-13
	Best	-3.5765e-10 (0)	6047.3689 (4)	-1.7053e-13 (0)	4.6735e-12 (0)
	Median	-3.2738e-10 (0)	19714.2478 (4)	1.7053e-13 (0)	4.6735e-12 (0)
	Worst	130.9783 (0)	18683.4258 (4)	3.2498e-07 (0)	4.6735e-12 (0)
5×10^5	c	0, 0, 0	1, 4, 4	0, 0, 0	0, 0, 0
	$\bar{v}$	0.0000	2.0466	0.0000	0.0000
	Mean	5.2391	1.2634e+04	1.3012e-08	4.6736e-12
	Std	2.6196e+01	6.6438e+03	6.4994e-08	0.0000

6.3 Using More Strategies

In this chapter our ε-*jDE* algorithm uses the following mutation strategies: 'DE/rand/1', 'DE/best/2' and a strategy which is similar to 'DE/current to best/1' (eq. 8) and also similar to the strategy used in [20]:

$$\mathbf{v}_{i,G} = \mathbf{x}_{r_3,G} + F \cdot (\mathbf{x}_{best,G} - \mathbf{x}_{i,G}) + F \cdot (\mathbf{x}_{r_1,G} - \mathbf{x}_{r_2,G}) \tag{27}$$

All strategies used are self-adaptive and each one uses its own pair of control parameters F and CR. The strategies are used according to the following rule:

$$s = \begin{cases} \text{'DE/best/2'}, & \text{if } (G+i) \bmod 5 = 3 \ \wedge \ t > 0.8T_{max}, \\ \text{strategy in eq. (27)}, & \text{if } (G+i) \bmod 5 = 1 \ \wedge \ t > 0.4T_{max}, \\ \text{'DE/rand/1'}, & \text{otherwise.} \end{cases} \tag{28}$$

Table 7 Number of FES to achieve the fixed accuracy level $((f(\mathbf{x}) - f(\mathbf{x}*)) \leq 0.0001)$, Success Rate (SR), Feasible Rate (FR) and Success Performance (SP)

Prob.	**Best**	**Median**	**Worst**	**Mean**	**Std**	**FR**	**SR**	**SP**	**SR** [9]	**SP** [9]
g01	51685	55211	57151	5.4831e+04	1.4235e+03	100%	100%	54831	100%	50386
g02	175090	226789	253197	2.2506e+05	1.8990e+04	100%	100%	225059	92%	145899
g03	184568	215694	254105	2.1533e+05	1.8628e+04	100%	100%	215333	0%	
g04	56730	62506	67383	6.2787e+04	2.8545e+03	100%	100%	62787	100%	40728
g05	49765	53773	57863	5.3576e+04	2.3524e+03	100%	100%	53576	68%	446839
g06	31410	34586	37033	3.4467e+04	1.6496e+03	100%	100%	34467	100%	29488
g07	184927	197901	221866	2.0075e+05	1.1398e+04	100%	100%	200753	100%	127744
g08	1905	4044	4777	3.9203e+03	6.2545e+02	100%	100%	3920	100%	3236
g09	79296	89372	98062	8.9262e+04	5.1839e+03	100%	100%	89262	100%	54919
g10	203851	220676	264575	2.2397e+05	1.3100e+04	100%	100%	223974	100%	146150
g11	52128	83442	105093	8.3713e+04	1.3924e+04	100%	100%	83713	96%	53928
g12	364	6899	10424	6.6678e+03	2.3895e+03	100%	100%	6668	100%	6356
g13	138630	147330	428869	1.6314e+05	6.2060e+04	100%	100%	163144	0%	
g14	223822	242265	256523	2.4345e+05	7.8180e+03	100%	100%	243455	100%	97845
g15	153943	157822	160014	1.5745e+05	1.7881e+03	100%	100%	157446	96%	241383
g16	48883	54081	57678	5.3723e+04	2.3374e+03	100%	100%	53723	100%	31695
g17	185888	205132	255333	2.0948e+05	1.7221e+04	100%	100%	209475	4%	11232650
g18	139131	169638	191345	1.6753e+05	1.2702e+04	100%	100%	167529	100%	104462
g19	322120	363456	427042	3.6971e+05	3.2417e+04	100%	100%	369705	100%	199850
g20						0%	0%		0%	
g21	131557	149672	158079	1.4264e+05	3.0443e+04	100%	96%	154770	92%	126507
g22						0%	0%		0%	
g23	260180	321118	464740	3.3100e+05	5.5302e+04	100%	100%	331004	92%	357452
g24	9359	12844	14827	1.2934e+04	1.2760e+03	100%	100%	12934	100%	10196

where 'mod' denotes the modulo operation, t denotes the iteration of the evolutionary process, and T_{max} denotes maximal number of iterations (e.g. function evaluations). The first two strategies are not used at the beginning of the optimization process, and they are used with the probability $p < 0.2$, respectively.

During the experiments in our previous work [9], the jDE-2 algorithm used three strategies 'DE/rand/1/bin', 'current to best/1/bin' and 'rand2/bin' and the population size NP was set to 200. The jDE-2 algorithm replaces k worst individuals at every l-th generation with parameter values distributed uniform randomly between lower and upper bounds without evaluating those k individuals. In [9] we set $l = 1000$ and $k = 70$.

In this chapter our ε-*jDE* algorithm does not use a technique for replacing k worst individuals at every l-th generation, but it uses the solution for violated boundary constraint, as presented in eq. (21).

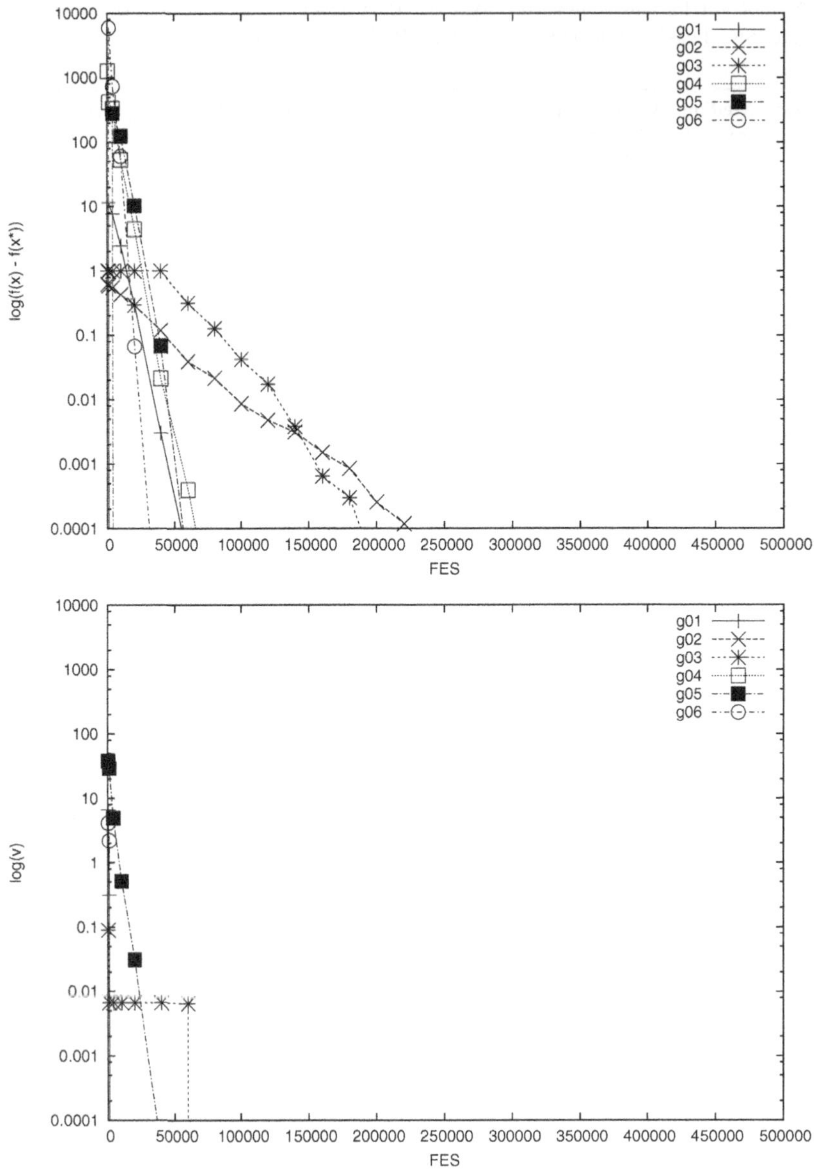

Fig. 1 Convergence Graph for Problems 1-6

7 Experimental Results

The *ε-jDE* algorithm was tested on 24 CEC2006 special-session benchmark functions [16]. The characteristics of benchmark problems are presented in Table 1. All experimental runs were terminated after 5×10^5 function evaluations (FES). The

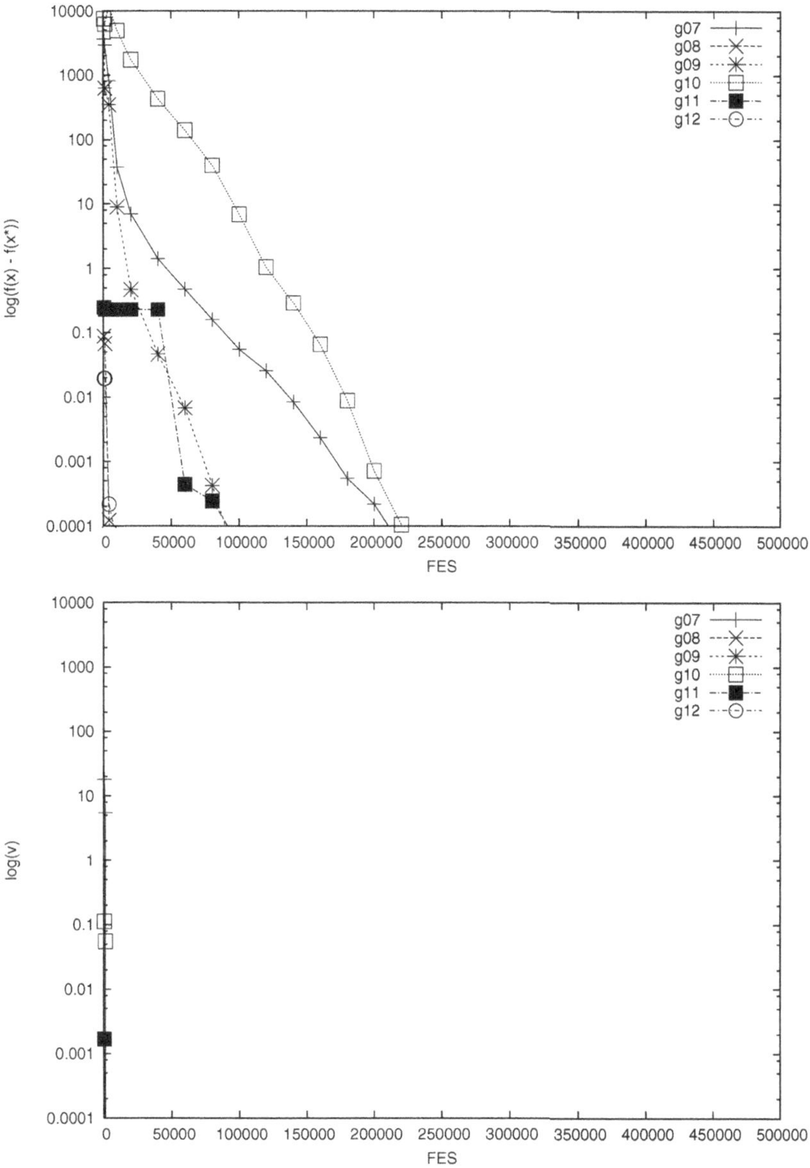

Fig. 2 Convergence Graph for Problems 7-12

obtained results are presented in Tables 2–6. 25 runs were carried-out for each test problem and the best, median, and worst results are presented along with the mean value and the standard deviation. In the tables, c represents the number of violated constraints at the median solution run. The sequence of three numbers indicates the number of constraint violations greater than 1.0, 0.01, and 0.0001, respectively. $\overline{v}$ is

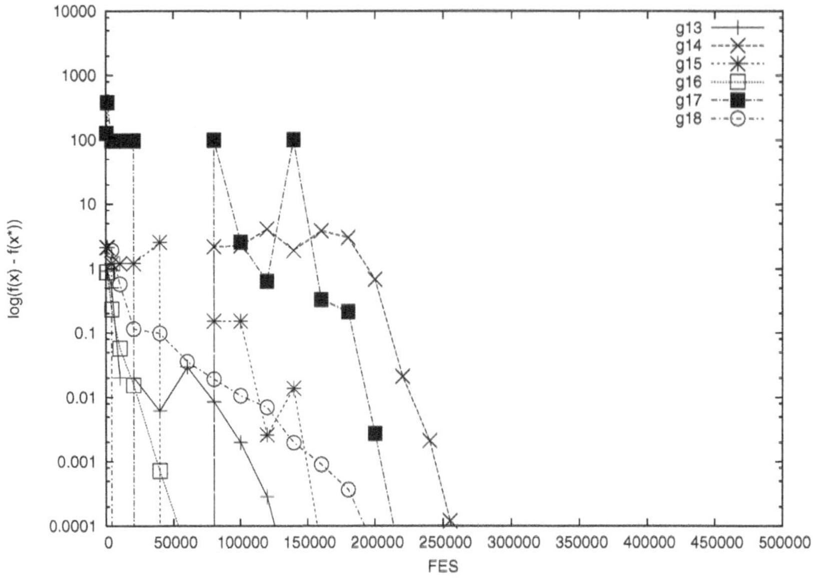

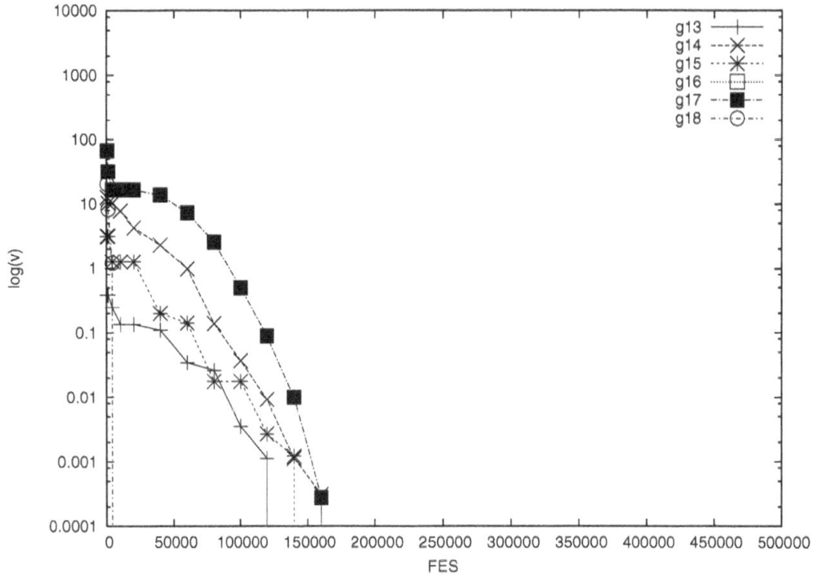

Fig. 3 Convergence Graph for Problems 13-18

the mean value of all constraints' violations at the median solution. The number in parenthesis after the error value of the best, median and worst solutions indicates the number of constraints which were unsatisfied at the corresponding location. Function $g20$ has no feasible solution [16].

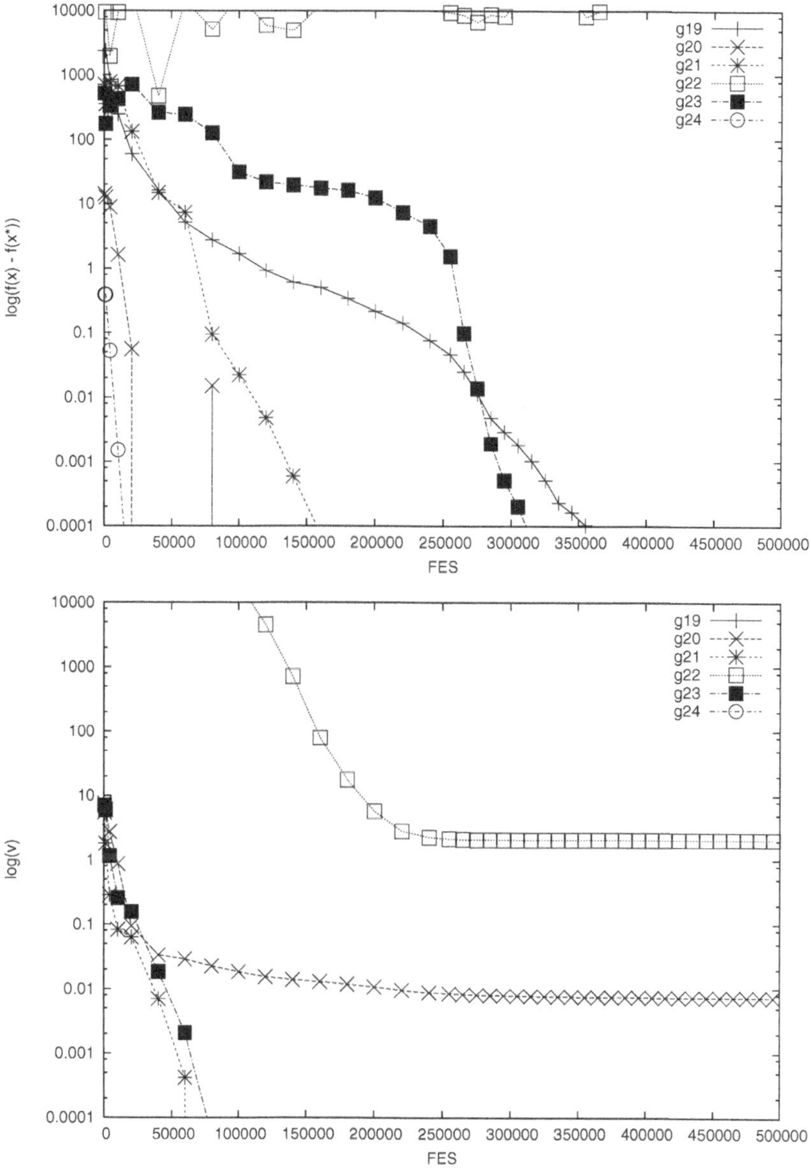

Fig. 4 Convergence Graph for Problems 19-24

For 22 functions the *ε-jDE* algorithm was successful in finding a feasible solution in all runs (FR=100%), except for functions *g*20 and *g*22. For the 22 functions with FR=100%, the success rates (SRs) were also very good, the approach reached a SR of 100% in all runs, except for function *g*21, which yielded a SR of 96%.

```
// individuals' fitness values are already stored in array named cost
// individuals' constraint violation values are already stored in array named v
for (i=0; i < NP/2; i++) {
    if (x_i is worse than x_{NP/2+i}) {   // comparison of two individuals using eq. (22)
        swap(x[i], x[NP/2+i]);            // swap individuals
        swap(cost[i], cost[NP/2+i])       // and swap their fitness values,
        swap(v[i], v[NP/2+i])             // and swap their violation values, etc.
    }
}
NP = (NP+1)/2;   // new population size
```

Fig. 5 The population size reduction scheme

Table 7 shows the number of FES required to achieve fixed accuracy level $((f(\mathbf{x}) - f(\mathbf{x}*)) \leq 0.0001)$, success rate, feasible rate and success performance performed by *ε-jDE* algorithm. The last two columns in Table 7 show the results obtained by the jDE-2 algorithm [9]. If function $g20$ has no feasible solution, the overall success rate for all remaining 23 benchmark functions was 95.48% for the *ε-jDE* algorithm, and 80.0% for the jDE-2 algorithm. The jDE-2 algorithm performed better than the *ε-jDE* algorithm when comparing those algorithms based on the success performance, but the *ε-jDE* algorithm is better when comparing the success rate and the overall success rate. Therefore, we can conclude that the *ε-jDE* algorithm is more robust compared to the jDE-2 algorithm.

Convergence graphs are presented in Figures 1–4. Logarithmic scale is used for y-axis. The figures show the development of objective function values, as well as mean violations of constraints for the median solutions. We can observe from the convergence graphs that no feasible solution was found for functions $g20$ and $g22$.

We implemented the *ε-jDE* algorithm on the GNU/Linux operating system using C/C++ language.

The population reduction mechanism is simple. It was inspired by the selection operation of the DE algorithm. However, someone could reduce population size e.g. by 20%. We did not conduct experiments in this way. Here we used a mechanism based on our previous experiences (see [8]). We can conclude based on the results of additional experiments, however ommited here, that, in our algorithm, the controlling ε level mechanism plays a more important role in obtaining better results than the population size reduction mechanism.

8 Conclusions

The performance of the self-adaptive differential evolution *ε-jDE* algorithm was evaluated on a set of 24 well-known benchmark functions.

The best settings for the control parameters depend highly on the benchmark function. A self-adaptive control mechanism is used by *ε-jDE* algorithm to change the (DE strategy) control parameters F and CR during the run.

The results in this chapter provide evidence that the *ε-jDE* algorithm is competitive for non-linear, non-separable, constrained, continuous global optimization.

Constrained optimization problems with mixed (continuous and discrete) decision variables are a challenge for future work.

Acknowledgements. Simulation studies for the differential evolution algorithm were performed on the C code downloaded from http://www.icsi.berkeley.edu/~storn/code.html. The authors would like to thank the editor and the referees for their valuable comments. This work was supported in part by the Slovenian Research Agency under program P2-0041 – Computer Systems, Methodologies, and Intelligent Services.

References

1. Abbass, H.A.: The self-adaptive pareto differential evolution algorithm. In: Proceedings of the 2002 Congress on Evolutionary Computation, 2002 (CEC 2002), vol. 1, pp. 831–836 (2002)
2. Al-Anzia, F.S., Allahverdi, A.: A self-adaptive differential evolution heuristic for two-stage assembly scheduling problem to minimize maximum lateness with setup times. European Journal of Operational Research 182(1), 80–94 (2007)
3. Ali, M.M.: Differential evolution with preferential crossover. European Journal of Operational Research 181(3), 1137–1147 (2007)
4. Ali, M.M., Törn, A.: Population Set-Based Global Optimization Algorithms: Some Modifications and Numerical Studies. Computers & Operations Research 31(10), 1703–1725 (2004)
5. Becerra, R.L., Coello, C.A.C.: Cultured differential evolution for constrained optimization. Computer Methods in Applied Mechanics and Engineering 195(33-36), 4303–4322 (2006)
6. Brest, J., Bošković, B., Greiner, S., Žumer, V., Maučec, M.S.: Performance comparison of self-adaptive and adaptive differential evolution algorithms. Soft Computing - A Fusion of Foundations, Methodologies and Applications 11(7), 617–629 (2007)
7. Brest, J., Greiner, S., Bošković, B., Mernik, M., Žumer, V.: Self-Adapting Control Parameters in Differential Evolution: A Comparative Study on Numerical Benchmark Problems. IEEE Transactions on Evolutionary Computation 10(6), 646–657 (2006)
8. Brest, J., Maučec, M.S.: Population Size Reduction for the Differential Evolution Algorithm. Applied Intelligence 29(3), 228–247 (2008)
9. Brest, J., Žumer, V., Maučec, M.S.: Self-adaptive Differential Evolution Algorithm in Constrained Real-Parameter Optimization. In: The 2006 IEEE Congress on Evolutionary Computation CEC 2006, pp. 919–926. IEEE Press, Los Alamitos (2006)
10. Brest, J., Žumer, V., Maučec, M.S.: Population size in differential evolution algorithm. Electrotechnical Review 74(1-2), 55–60 (2007) (in Slovene)
11. Coello Coello, C.: Theoretical and numerical constraint-handling techniques used with evolutionary algorithms: a survey of the state of the art. Computer Methods in Applied Mechanics and Engineering 191(11-12), 1245–1287 (2002)
12. Feoktistov, V.: Differential Evolution: In Search of Solutions (Springer Optimization and Its Applications). Springer, New York (2006)

13. Gämperle, R., Müller, S.D., Koumoutsakos, P.: A Parameter Study for Differential Evolution. In: WSEAS NNA-FSFS-EC 2002. WSEAS, Interlaken, Switzerland (2002), http://www.worldses.org/online/
14. Huang, V.L., Qin, A.K., Suganthan, P.N.: Self-adaptive Differential Evolution Algorithm for Constrained Real-Parameter Optimization. In: The 2006 IEEE Congress on Evolutionary Computation CEC 2006, pp. 17–24. IEEE Press, Los Alamitos (2006)
15. Koziel, S., Michalewicz, Z.: Evolutionary Algorithms, Homomorphous Mappings, and Constrained Parameter Optimization. Evolutionary Computation 7(1), 19–44 (1999)
16. Liang, J.J., Runarsson, T.P., Mezura-Montes, E., Clerc, M., Suganthan, N., Coello, C.A.C., Deb, K.: Problem Definitions and Evaluation Criteria for the CEC 2006 Special Session on Constrained Real-Parameter Optimization. Tech. Rep. Report #2006005, Nanyang Technological University, Singapore, et al. (December 2005), http://www.ntu.edu.sg/home/EPNSugan
17. Liu, J., Lampinen, J.: Adaptive Parameter Control of Differential Evolution. In: Proceedings of the 8th International Conference on Soft Computing (MENDEL 2002), pp. 19–26 (2002)
18. Liu, J., Lampinen, J.: On Setting the Control Parameter of the Differential Evolution Method. In: Proceedings of the 8th International Conference on Soft Computing (MENDEL 2002), pp. 11–18 (2002)
19. Liu, J., Lampinen, J.: A Fuzzy Adaptive Differential Evolution Algorithm. Soft Computing - A Fusion of Foundations, Methodologies and Applications 9(6), 448–462 (2005), http://springerlink.metapress.com/index/10.1007/s00500-004-0363-x
20. Mezura-Montes, E., Veláquez-Reyes, J., Coello, C.A.C.: Modified Differential Evolution for Constrained Optimization. In: The 2006 IEEE Congress on Evolutionary Computation CEC 2006, pp. 25–31. IEEE Press, Los Alamitos (2006)
21. Mezura-Montes, E., López-Ramírez, B.C.: Comparing Bio-Inspired Algorithms in Constrained Optimization Problems. In: Proceedings of the IEEE Congress on Evolutionary Computation (CEC 2007), pp. 662–666. IEEE Press, Los Alamitos (2007)
22. Mezura-Montes, E., Velázquez-Reyes, J., Coello, C.A.C.: Promising infeasibility and multiple offspring incorporated to differential evolution for constrained optimization. In: GECCO 2005: Proceedings of the 2005 conference on Genetic and evolutionary computation, pp. 225–232. ACM, New York (2005), http://doi.acm.org/10.1145/1068009.1068043
23. Mezura-Montes, E., Velázquez-Reyes, J., Coello, C.A.C., Muñoz Dávila, L.M.: Multiple Trial Vectors in Differential Evolution for Engineering Design. Engineering Optimization 39(5), 567–589 (2007)
24. Michalewicz, Z., Fogel, D.B.: How to Solve It: Modern Heuristics. Springer, Berlin (2000)
25. Michalewicz, Z., Schoenauer, M.: Evolutionary Algorithms for Constrained Parameter Optimization Problems. Evolutionary Computation 4(1), 1–32 (1996)
26. Nobakhti, A., Wang, H.: A simple self-adaptive Differential Evolution algorithm with application on the ALSTOM gasifier. Applied Soft Computing 8(1), 350–370 (2008), http://dx.doi.org/10.1016/j.asoc.2006.12.005
27. Price, K.V., Storn, R.M., Lampinen, J.A.: Differential Evolution, A Practical Approach to Global Optimization. Springer, Heidelberg (2005)
28. Qin, A.K., Suganthan, P.N.: Self-adaptive Differential Evolution Algorithm for Numerical Optimization. In: The 2005 IEEE Congress on Evolutionary Computation CEC 2005, pp. 1785–1791. IEEE Press, Los Alamitos (2005)

29. Rönkkönen, J., Kukkonen, S., Price, K.V.: Real-Parameter Optimization with Differential Evolution. In: The 2005 IEEE Congress on Evolutionary Computation CEC 2005, pp. 506–513. IEEE Press, Los Alamitos (2005)
30. Storn, R.: System Design by Constraint Adaptation and Differential Evolution. IEEE Transactions on Evolutionary Computation 3(1), 22–34 (1999)
31. Storn, R., Price, K.: Differential Evolution - a simple and efficient adaptive scheme for global optimization over continuous spaces. Tech. Rep. TR-95-012, Berkeley, CA (1995), `citeseer.ist.psu.edu/article/storn95differential.html`
32. Storn, R., Price, K.: Differential Evolution – A Simple and Efficient Heuristic for Global Optimization over Continuous Spaces. Journal of Global Optimization 11, 341–359 (1997)
33. Sun, J., Zhang, Q., Tsang, E.P.K.: DE/EDA: A New Evolutionary Algorithm for Global Optimization. Information Sciences 169, 249–262 (2004)
34. Takahama, T., Sakai, S.: Constrained Optimization by the ε Constrained Differential Evolution with Gradient-Based Mutation Feasible Elites. In: The 2006 IEEE Congress on Evolutionary Computation CEC 2006, pp. 17–24. IEEE Press, Los Alamitos (2006)
35. Teo, J.: Exploring dynamic self-adaptive populations in differential evolution. Soft Computing - A Fusion of Foundations, Methodologies and Applications 10(8), 673–686 (2006)
36. Wang, Y., Cai, Z., Zhou, Y., Zeng, W.: An Adaptive tradeoff Model for Constrained Evolutionary Optimization. IEEE Transactions on Evolutionary Computation 12(1), 80–92 (2008)

Self-adaptive and Deterministic Parameter Control in Differential Evolution for Constrained Optimization

Efrén Mezura-Montes and Ana Gabriela Palomeque-Ortiz

Abstract. In this Chapter we present the modification of a Differential Evolution algorithm to solve constrained optimization problems. The changes include a deterministic and a self-adaptive parameter control in two of the Differential Evolution parameters and also in two parameters related with the constraint-handling mechanism. The proposed approach is extensively tested by using a set of well-known test problems and performance measures found in the specialized literature. Besides analyzing the final results obtained by the algorithm with respect to its original version, some interesting findings regarding the behavior found in the approach and in the values observed on each of the parameters controlled are also discussed.

Keywords: Parameter Control, Constrained Optimization, Differential Evolution, Self-Adaptation.

1 Introduction

Evolutionary computing (EC) comprises a set of algorithms based on simulating the natural evolution and the survival of the fittest. These algorithms are known as Evolutionary Algorithms (EAs).

Three original EAs were proposed in the 1960's: (1) Genetic Algorithms (GAs) [10], Evolution Strategies (ES) [28] and Evolutionary Programming (EP) [9]. Despite the fact that they arose from different motivations, all of them have been used to solve complex search tasks [12] providing competitive results [1, 7, 23].

In the 1990's, Storn and Price proposed a novel EA called Differential Evolution (DE) [27]. DE shares similarities with original EAs e.g. DE uses a population of solutions called vectors to sample the search space; DE also uses a recombination

Efrén Mezura-Montes and Ana Gabriela Palomeque-Ortiz
Laboratorio Nacional de Informática Avanzada (LANIA A.C.), Rébsamen 80, Centro, Xalapa, Veracruz, 91000, Mexico
e-mail: emezura@lania.mx, apalomeque@lania.edu.mx

E. Mezura-Montes (Ed.): Constraint-Handling in Evolution. Optimi., SCI 198, pp. 95–120.
springerlink.com

and mutation operators to generate new vectors from the current population and, finally, DE has a replacement process to discard the less fit vectors. Like ES, DE uses real-value vectors to represent solutions (no decoding process is necessary as in traditional GAs with binary encoding). Unlike Gaussian distribution used in ES, DE does not use a pre-defined probability distribution for its mutation operator. Instead, DE uses the current distribution of vectors in the population to define the behavior of the mutation operator, and this seems to be one of its main advantages. Furthermore DE, in its original version, does not perform a self-adaptive process to its parameters as ES does with its mutation operator.

The optimization problem in discrete, continuous or even mixed search spaces has been solved by using EAs. However, two shortcomings can be identified in this process: (1) A set of parameter values must be defined by the user and the behavior of the algorithm in the search depends on these values and (2) in presence of constraints, a constraint-handling mechanism must be added to the EA in order to incorporate feasibility information in the selection and replacement processes, and this mechanism may involve additional parameters to be fine-tuned by the user.

Eiben and Schut [8] proposed a classification of parameter setting techniques: (1) Parameter tuning and (2) parameter control. Besides, parameter control is divided into deterministic, adaptive and self-adaptive. Parameter tuning consists on defining good values for the parameters before the run of an algorithm and then running it with these values. On the other hand, deterministic parameter control aims to modify the parameter values by a deterministic rule e.g. a fixed schedule. Adaptive parameter control aims to modify the parameter values based on some feedback from the search behavior e.g. diversity measure to update the mutation rate. Finally, self-adaptive parameter control encodes the parameter values into the chromosome of solutions and they are modified by variation operators. The expected behavior is that the search process will be able to evolve the solutions of the problems as well as to find the optimal values for the parameters of the algorithm. Eiben and Schut [8] mention that most of the work related to parameter setting is focused on variation operators (mostly on mutation) and population size.

DE, as the remaining EAs, lacks a mechanism to incorporate feasibility information into the fitness value of a given solution. Hence, the selection of an adequate constraint-handling technique for a given EA (DE in this case) is an open problem. Coello [4] proposed a taxonomy of mechanisms: (1) Penalty functions, (2) special representations and operators, (3) repair algorithms, (4) separation of objectives and constraints and (5) hybrid methods. Penalty functions [25] decrease the fitness of infeasible solutions as to prefer feasible solution in the selection process. Special representations and operators are designed to represent only feasible solutions and the operators are able to preserve the feasibility of the offspring generated. Repair algorithms aim to transform an infeasible solution into a feasible one. The separation of objectives and constraints consists on using these values as separated criteria in the selection process of an EA [19]; this is opposed to penalty functions, where the values of the objective function and the constraints are mixed into one single value. Finally, hybrid methods are a combination of different algorithms and/or mechanisms e.g. fuzzy-logic with EAs, cultural algorithms [15] and immune systems [5].

Research in parameter control for constrained optimization is scarce compared to unconstrained optimization. Furthermore, the research efforts do not usually consider, with the exception of penalty-function-based approaches, the parameters added with the constraint-handling mechanism.

Based on the aforementioned, three main motivations originated this work: (1) to propose parameter control mechanisms in a competitive EA for constrained optimization by considering parameters of the constraint-handling mechanism (2) to analyze the behavior of these controlled parameter values when solving constrained optimization problems and (3) to know the on-line behavior of the proposed approach by measuring the evaluations required to reach the feasible region and the improvement within it.

We use a very competitive approach for constrained optimization known as Diversity Differential Evolution (DDE) [21] where some of the DE parameter values and also those parameter values of its constraint-handling mechanism are controlled by deterministic and self-adaptive mechanisms. Furthermore, we analyze the behavior of each parameter during the evolutionary process in order to provide some insights about the values they use. A set of 24 test functions [16,22] and two performance measures [18] found in the specialized literature are used in the experimental design, where the aims are: (1) to compare the performance of the proposed DDE algorithm with respect to its original version and with respect to state-of-the-art approaches (2) to analyze the behavior of each parameter during the process and (3) to know the on-line behavior of the proposed approach compared with the original DDE version.

The Chapter is organized as follows: In Section 2 we formally present the problem of our interest. Section 3 offers a brief introduction to DE. After that, Section 4 presents a review of DE and parameter control in constrained optimization. In Section 5 we detail DDE, the approach which is the base of our study. Then, Section 6 introduces our parameter control proposal. The experimental design, the obtained results and their corresponding discussions are given in Section 7. Finally, Section 8 provides some conclusions and the future work.

2 Statement of the Problem

We are interested in the general nonlinear programming problem in which, without loss of generality, we want to:

$$\text{Find } \mathbf{x} \text{ which minimizes } f(\mathbf{x}) \tag{1}$$

subject to:

$$g_i(\mathbf{x}) \leq 0, \;\; i = 1, \ldots, m \tag{2}$$

$$h_j(\mathbf{x}) = 0, \;\; j = 1, \ldots, p \tag{3}$$

where $\mathbf{x}$ is the vector of solutions $\mathbf{x} = [x_1, x_2, \ldots, x_n]^T$, and each $x_i \in \mathbf{R}$, $i = 1, \ldots, n$ is bounded by lower and upper limits $L_i \leq x_i \leq U_i$. These limits define the search space of the problem; m is the number of inequality constraints and p is the number of equality constraints (in both cases, constraints could be linear or nonlinear). If we denote with $\mathscr{F}$ to the feasible region and with $\mathscr{S}$ to the whole search space, then it should be clear that $\mathscr{F} \subseteq \mathscr{S}$. For an inequality constraint that satisfies $g_i(\mathbf{x}) = 0$, then we will say that it is active at $\mathbf{x}$. All equality constraints h_j (regardless of the value of $\mathbf{x}$ used) are considered active at all points of $\mathscr{F}$. Most constraint-handling approaches used with EAs tend to deal only with inequality constraints. However, in those cases, equality constraints are transformed into inequality constraints of the form:

$$|h_j(\mathbf{x})| - \varepsilon \leq 0 \tag{4}$$

where ε is the tolerance allowed (a very small value).

3 Differential Evolution

DE is a simple, but powerful search engine that simulates natural evolution combined with a mechanism to generate multiple search directions based on the distribution of solutions in the current population. Each vector i in the population at generation G, ($\mathbf{x}_{i,G}$), called at the moment of reproduction as the target vector, will be able to generate one offspring, called trial vector ($\mathbf{u}_{i,G}$). This trial vector is generated as follows: First of all, a search direction is defined by calculating the difference between a pair of vectors r_1 and r_2, called "*differential vectors*", both of them chosen at random from the population. This difference vector is also scaled by using a user-defined parameter called $F \geq 0$ [27]. This scaled difference vector is then added to a third vector r_3, called "*base vector*". As a result, a new vector is obtained, known as the mutation vector. After that, this mutation vector is recombined with the target vector (also called parent vector) by using discrete recombination (usually binomial crossover) controlled by a crossover parameter $0 \leq CR \leq 1$ whose value determines how similar the trial vector will be with respect to the target vector. There are several DE variants [27]. However, the most known and used is DE/rand/1/bin, where the base vector is chosen at random, there is only a pair of differential vectors and a binomial crossover is used. The detailed pseudocode of this variant is presented in Figure 1.

4 Related Work

There are previous works on DE for constrained optimization. Lampinen used DE/rand/1/bin variant to tackle constrained problems [14] by using Pareto dominance in the space of constraints, Mezura et al. [20] proposed to add Deb's feasibility rules [6] into DE to deal with constraints. Kukkonen & Lampinen [13] improved

```
Begin
  G=0
  Create a random initial population x_{i,G} ∀i, i = 1,...,NP
  Evaluate f(x_{i,G}) ∀i, i = 1,...,NP
  For G=1 to MAX_GEN Do
    For i=1 to NP Do
      Select randomly r1 ≠ r2 ≠ r3 ≠ i
      j_rand = randint(1,D)
      For j=1 to n Do
        If (rand_j[0,1) < CR or j = j_rand) Then
          u_{i,j,G+1} = x_{r3,j,G} + F(x_{r1,j,G} − x_{r2,j,G})
        Else
          u_{i,j,G+1} = x_{i,j,G}
        End If
      End For
      If (f(u_{i,G+1}) ≤ f(x_{i,G})) Then
        x_{i,G+1} = u_{i,G+1}
      Else
        x_{i,G+1} = x_{i,G}
      End If
    End For
    G = G+1
  End For
End
```

Fig. 1 "DE/rand/1/bin" pseudocode. rand$[0,1)$ is a function that returns a real number between 0 and 1. randint(min,max) is a function that returns an integer number between min and max. *NP*, *MAX_GEN*, *CR* and *F* are user-defined parameters. *n* is the dimensionality of the problem

its DE-based approach now to solve constrained multiobjective optimization problems. Zielinsky & Laur also used Deb's rules [6] in DE to solve some constrained optimization problems.

Other search techniques have been combined with DE. A gradient-based mutation with DE by Takahama & Sakai was recently proposed [32]. A combination of Particle Swarm Optimization and DE (called PESO+), where the DE mutation operator is considered as a turbulence operator, was proposed by Muñoz-Zavala et al. [26]. Other authors have proposed novel DE variants for constrained optimization [22] or multi-population DE approaches [33].

On the other hand, there are some studies regarding parameter control in DE for constrained optimization. Brest et al. [2] proposed an adaptive parameter control to two DE parameters (*F* and *CR*). Huang et al. [11] presented an adaptive approach to choose the most suitable DE variant to generate new trail vectors in constrained search spaces. In this proposal, DE parameters (*F*, *K* and *CR*) were also adapted. Liu & Lampinen [17] proposed to adapt DE parameters by means of Fuzzy Logic.

Besides controlling DE parameters, in this chapter two parameters related with the constraint-handling mechanism are controlled and analyzed. Furthermore, two performance measures help to understand the impact of the control process in the performance and behavior of the approach.

5 Diversity Differential Evolution

Based on the very competitive performance shown by DE in global optimization problems [3], an adapted version to solve numerical optimization problems in presence of constraints, called Diversity Differential Evolution (DDE) was proposed in [21]. Three simple modifications were made to the original DE/rand/1/bin (detailed in Figure 1):

1. The probability of a target vector to generate a better trial vector is increased by allowing it to generate NO offspring in the same generation.
2. A simple constraint-handling mechanism based on Deb's feasibility rules [6] is added to bias the search to the feasible region of the search space.
    a. Between 2 feasible vectors, the one with the highest fitness value is preferred.
    b. If one vector is feasible and the other one is infeasible, the feasible vector is preferred.
    c. If both vectors are infeasible, the one with the lowest sum of constraint violation $\left(\sum_{i=1}^{m+p} max(0, g_i(\mathbf{x}))\right)$ is preferred.
3. A selection ratio parameter $0 \leq S_r \leq 1$ is added to control the way vectors will be selected. Based on the S_r value the selection will be made based only in the value of the objective function $f(\mathbf{x})$ regardless of feasibility. Otherwise, the selection will be made based on the feasibility rules described before.

The detailed pseudocode of DDE is presented in Figure 2

6 Self Adaptive Diversity Differential Evolution

As it can be noted in the pseudocode presented in Figure 2, DDE adds two parameters (NO and S_r) to the original four parameters used in DE (NP, MAX_GEN, CR and F). Therefore, in this work, two parameter control mechanisms are proposed as to keep the user from defining the values of four (out of six) parameters. Three parameters are self-adapted and another one uses a deterministic control. Furthermore, the behavior of these parameters and the online performance of the new approach are analyzed.

Fig. 2 DDE pseudocode. $randint(min, max)$ returns an integer value between min and max. rand$[0,1)$ returns a real number between 0 and 1. Both functions adopt a uniform probability distribution. flip(W) returns 1 with probability W. NP, MAX_GEN, CR, F, NO and S_r are user-defined parameters. n is the dimensionality of the problem

```
Begin
  G=0
  Create a random initial population x_{i,G} ∀i, i = 1,...,NP
  Evaluate f(x_{i,G}) ∀i, i = 1,...,NP
  For G=1 to MAX_GENERATIONS Do
    F=rand[0.3,0.9]
    For i=1 to NP Do
      For k=1 to NO Do
        Select randomly r1 ≠ r2 ≠ r3 ≠ i
        j_rand = randint(1,D)
        For j=1 to n Do
          If (rand_j[0,1) < CR or j = j_rand) Then
            child_j = x_{r3,j,G} + F(x_{r1,j,G} − x_{r2,j,G})
          Else
            child_j = x_{i,j,G}
          End If
        End For
        If k > 1 Then
          If (child is better than u_{i,G+1}
           based on the three selection criteria) Then
            u_{i,G+1}=child
          End If
        Else
          u_{i,G+1}=child
      End For
      If flip(S_r) Then
        If (f(u_{i,G+1}) ≤ f(x_{i,G})) Then
          x_{i,G+1} = u_{i,G+1}
        Else
          x_{i,G+1} = x_{i,G}
        End If
      Else
        If (u_{i,G+1} is better than x_{i,G}
         based on the three selection criteria) Then
          x_{i,G+1} = u_{i,G+1}
        Else
          x_{i,G+1} = x_{i,G}
        End If
      End If
    End For
    G = G+1
  End For
End
```

6.1 Self-adaptive Parameter Control

In order to get a self-adaptive parameter control, the parameters must be encoded within the solution of the problem. Motivated on the way Evolution Strategies work [30], three parameters are encoded in each solution: F, CR and NO as shown in Figure 3.

Now, each solution has its own F, CR and NO values and these values are subject to differential mutation and crossover. The process is explained in Figure 4, where the trial vector in Diversity Differential Evolution will inherit the three parameter values from the target vector if the last decision variable was taken from it . On the other hand, the values for each parameter will be calculated by using the differential mutation operator i.e. they will be inherited from the mutation vector. The decision variables are handled as in traditional DE, however, the CR parameter value used in the process is that of the target vector.

$X_{1,1,G}$	…	$X_{1,n,G}$	$F_{1,G}$	$CR_{1,G}$	$NO_{1,G}$
$X_{2,1,G}$	…	$X_{2,n,G}$	$F_{2,G}$	$CR_{2,G}$	$NO_{2,G}$
⋮	⋮	⋮	⋮	⋮	⋮
$X_{NP,1,G}$	…	$X_{NP,n,G}$	$F_{NP,G}$	$CR_{NP,G}$	$NO_{NP,G}$

Fig. 3 Encoded solutions including three parameters to be self-adapted

If (the last decision variable was inherited from the target vector) **Then**
  $\text{child}_j F = \text{F}_{i,G}$
  $\text{child}_j CR = \text{CR}_{i,G}$
  $\text{child}_j NO = \text{NO}_{i,G}$
Else
  $\text{child}_j F = \text{F}_{r_3,G} + \text{F}_{i,G}(\text{F}_{r_1,G} - \text{F}_{r_2,G})$
  $\text{child}_j CR = \text{CR}_{r_3,G} + \text{F}_{i,G}(\text{CR}_{r_1,G} - \text{CR}_{r_2,G})$
  $\text{child}_j NO = \text{NO}_{r_3,G} + \text{F}_{i,G}(\text{NO}_{r_1,G} - \text{NO}_{r_2,G})$
End If

Fig. 4 Differential mutation applied to the self-adapted parameters. Note that the F value for the target vector $F_{i,G}$ is used in all cases where differential mutation is used

6.2 Deterministic Parameter Control

Recalling the S_r parameter explanation in Section 5, this parameter controls the percentage of comparisons made between pairs of vectors by only considering the objective function value, regardless of feasibility information. Therefore, it affects the bias in the search [29]. Higher S_r values keep infeasible solutions located in promising areas of the search space, whereas lower S_r values help to reach the feasible region by using Deb's rules [6].

Based on this behavior, the S_r parameter is controlled by a fixed schedule. A simple function is used to decrease the value for this parameter in such a way that initial higher values allow DDE to focus on searching promising regions of the search space, regardless of feasibility, with the aim to approach the feasible region from a more convenient area. Later in the process, the S_r values will be lower, assuming the feasible region has been reached and that it is more important to keep good feasible solutions. The interval within S_r initial values are considered is the following: $[0.45, 0.65]$. At each generation, the S_r value will be decreased based on the expression in Equation 5:

$$S_r^{(t+1)} = S_r^t - \Delta_{S_r} \tag{5}$$

where $S_r^{(t+1)}$ is the new value for this parameter, S_r^t is the current S_r value, Δ_{S_r} is the amount decreased from this value at each generation and calculated as indicated in Equation 6:

$$\Delta_{S_r} = \left(\frac{S_r^0 - S_r^{G_{max}}}{G_{max}} \right) \tag{6}$$

where S_r^0 represents the initial value for S_r, and $S_r^{G_{max}}$ its last value in a given run.

The detailed pseudocode of Diversity Differential Evolution with the parameter control techniques, called Adaptive-DDE (A-DDE) is shown in Figure 5.

7 Experiments and Results

In order to extensively test the A-DDE algorithm, six experiments are conducted. The first experiment compares the final results of A-DDE with respect to the original DDE (called Static DDE). In order to verify that the self-adaptive mechanism is not equivalent to just generating random values for them within suggested intervals, the second experiment compares A-DDE with respect a special DDE version, called Static2 DDE, where the values for the four controlled parameters in A-DDE are just generated at random (by using a uniform distribution) within the same intervals used in A-DDE. The third experiment analyzes the convergence graphs for Static DDE, Static2 DDE and A-DDE. The fourth experiment includes the graphs for the three self-adapted parameters (F, CR and NO) in order to know which values are they taking to provide competitive results. The fifth experiment compares Static DDE, Static2 DDE and A-DDE by using two performance measures for constrained optimization in order to know (1) how fast the feasible region is reached and (2) the ability of each DE algorithm to improve inside the feasible region (difficult for most EAs as analyzed in [18]). The two measures utilized are the following:

1. Evals [14]: The number of evaluations of solutions (objective function and constraints) required to generate the first feasible solution are counted. A lower value is preferred because it means a faster approach to the feasible region.
2. Progress Ratio [18]: Originally proposed by Bäck for unconstrained optimization [1]. It is a measure of improvement inside the feasible region by using the objective function values of the first and the best feasible solution reached so far at the end of the process. The formula is as follows: $\mathrm{Pr} = \left| \ln \sqrt{\frac{f_{\min}(G_{ff})}{f_{\min}(T)}} \right|$, where $f_{\min}(G_{ff})$ is the objective function value of the first feasible solution found and $f_{\min}(T)$ is the objective function value of the best feasible solution found in all the search so far. A higher value means a better improvement inside the feasible region.

Finally, the sixth and last experiment compares the final results obtained by A-DDE with those reported by some state-of-the-art algorithms. In the first five experiments 24 well-known minimization test problems were used. These problems are used to test EAs in constrained search spaces. Details of the problems can be found in [16]. A summary of their features can be found in Table 1.

For all the experiments the results are based on 30 independent runs for each DE algorithm for each test problem. The number of evaluations performed by each

```
   Begin
     G=0
⇒    Create a random initial population X_{i,G} ∀i, i = 1,...,NP
     Evaluate f(X_{i,G}) ∀i, i = 1,...,NP
⇒    Select randomly S_r ∈ [0.45,0.65]
⇒    Select randomly S_r^{Gmax} ∈ (0,0.45)
     For G=1 to G_max Do
       For i=1 to NP Do
⇒        For k=1 to NO_{i,G} Do
           Select randomly r_1 ≠ r_2 ≠ r_3 ≠ i
           j_rand = randint(1,D)
           For j=1 to D Do
⇒            If (rand_j[0,1) < CR_{i,G} or j = j_rand) Then
               child_{j,G} = x_{r3,j,G} + F_{i,G}(x_{r1,j,G} − x_{r2,j,G})
               ban=0
             Else
               child_{j,G} = x_{ri,j,G}
               ban=1
             End If
           End For
⇒          If (ban==1) Then
             child_{F,G} = F_{i,G}
             child_{CR,G} = CR_{i,G}
             child_{NO,G} = NO_{i,G}
           Else
             child_{F,G} = F_{r3,G} + F_{i,G}(F_{r1,G} − F_{r2,G})
             child_{CR,G} = CR_{r3,G} + F_{i,G}(CR_{r1,G} − CR_{r2,G})
             child_{NO,G} = NO_{r3,G} + F_{i,G}(NO_{r1,G} − NO_{r2,G})
           End If
           If k > 1 Then
             If (child is better than u_{i,G+1}
             (Based on three selection criteria))Then
               u_{i,G+1} = child
             End If
           Else
             u_{i,G+1} = child
           End If
         End For
         If flip(S_r)
           If (f(u_{i,G+1}) ≤ f(x_{i,G})) Then
             x_{i,G+1} = u_{i,G+1}
           Else
             x_{i,G+1} = x_{i,G}
           End If
         Else
           If (u_{i,G+1} ≤ x_{i,G}(Based on three selection criteria)) Then
             x_{i,G+1} = u_{i,G+1}
           Else
             x_{i,G+1} = x_{i,G}
           End If
         End If
       End For
       G = G+1
⇒      S_r = S_r − Δ_{S_r}
     End For
   End
```

Fig. 5 A-DDE pseudocode. Arrows indicate steps where the parameter control mechanisms are involved

DDE version is 180,000 in order to promote a fair comparison (except experiment 6, where the results by the state-of-the-art algorithms were taken directly from the specialized literature). A tolerance value for equality constraints $\varepsilon = 1\text{E-}4$ was used.

Table 1 Details of the 24 test problems. "n" is the number of decision variables, $\rho = |\mathscr{F}|/|\mathscr{S}|$ is the estimated ratio between the feasible region and the search space [24], LI is the number of linear inequality constraints, NI the number of nonlinear inequality constraints, LE is the number of linear equality constraints and NE is the number of nonlinear equality constraints. a is the number of active constraints at the optimum

Prob.	n	Type of function	ρ	LI	NI	LE	NE	a
g01	13	quadratic	0.0111%	9	0	0	0	6
g02	20	nonlinear	99.9971%	0	2	0	0	1
g03	10	polynomial	0.0000%	0	0	0	1	1
g04	5	quadratic	52.1230%	0	6	0	0	2
g05	4	cubic	0.0000%	2	0	0	3	3
g06	2	cubic	0.0066%	0	2	0	0	2
g07	10	quadratic	0.0003%	3	5	0	0	6
g08	2	nonlinear	0.8560%	0	2	0	0	0
g09	7	polynomial	0.5121%	0	4	0	0	2
g10	8	linear	0.0010%	3	3	0	0	6
g11	2	quadratic	0.0000%	0	0	0	1	1
g12	3	quadratic	4.7713%	0	1	0	0	0
g13	5	nonlinear	0.0000%	0	0	0	3	3
g14	10	nonlinear	0.0000%	0	0	3	0	3
g15	3	quadratic	0.0000%	0	0	1	1	2
g16	5	nonlinear	0.0204%	4	34	0	0	4
g17	6	nonlinear	0.0000%	0	0	0	4	4
g18	9	quadratic	0.0000%	0	12	0	0	6
g19	15	nonlinear	33.4761%	0	5	0	0	0
g20	24	linear	0.0000%	0	6	2	12	16
g21	7	linear	0.0000%	0	1	0	5	6
g22	22	linear	0.0000%	0	1	8	11	19
g23	9	linear	0.0000%	0	2	3	1	6
g24	2	linear	79.6556%	0	2	0	0	2

7.1 Experiment 1

In the first experiment Static DDE and A-DDE are compared. The parameters used for each algorithm were the following:

1. Static DDE
   - $NP = 60$ y $GMAX = 600$
   - $S_r = 0.45$
   - $NO = 5$, $CR = 0.9$ and $F \in [0.3, 0.9]$ generated at random.
2. A-DDE.
   - $NP = 60$ y $GMAX = 600$
   - $S_r \in [0.45, 0.65]$, $S_r^{G_{max}} \in (0, 0.45)$, randomly generated on each independent run and then handled by the deterministic control.

- $NO \in [3,7]$, $CR \in [0.9, 1.0]$ and $F \in [0.3, 0.9]$ initially generated at random per each vector in the population and then handled with the self-adaptive control.

The statistical results (best, mean and worst values from a set of 30 independent runs) are summarized in Table 2.

From those results in Table 2, A-DDE was able to maintain the good performance of the original DDE in fourteen test problems (g01, g03, g04, g05, g06, g07, g08, g09, g11, g12, g15, g16, g18, g24). Furthermore, the statistical values were improved in six problems (g10, g14, g17, g19, g21 and g23). Finally, in problems g02 and g13 the results, mostly in the average and worst values, were not better than those provided by Static DDE. Problems g20 and g22 remained unsolved by A-DDE.

Table 2 Comparison of results obtained with the original DDE version with static parameter values (named Static DDE) and the proposed Adaptive-DDE (deterministic and self-adaptive parameter control). "-", means no feasible solution found. Values in **boldface** mean that the global optimum or best know solution was reached, values in *italic* mean that the obtained result is better (but not the optimal or best known) with respect to the approach compared

	Best	**Best**		**Mean**		**Worst**	
Test problem	known solution	Adaptive DDE	Static DDE	Adaptive DDE	Static DDE	Adaptive DDE	Static DDE
g01	**-15.000**	**-15.000**	**-15.000**	**-15.000**	**-15.000**	**-15.000**	**-15.000**
g02	-0.803619	-0.803605	*-0.803618*	-0.771090	*-0.789132*	-0.609853	*-0.747876*
g03	-1.000	**-1.000**	**-1.000**	**-1.000**	**-1.000**	**-1.000**	**-1.000**
g04	-30665.539	**-30665.539**	**-30665.539**	**-30665.539**	**-30665.539**	**-30665.539**	**-30665.539**
g05	5126.497	**5126.497**	**5126.497**	**5126.497**	**5126.497**	**5126.497**	**5126.497**
g06	-6961.814	**-6961.814**	**-6961.814**	**-6961.814**	**-6961.814**	**-6961.814**	**-6961.814**
g07	24.306	**24.306**	**24.306**	**24.306**	**24.306**	**24.306**	**24.306**
g08	-0.095825	**-0.095825**	**-0.095825**	**-0.095825**	**-0.095825**	**-0.095825**	**-0.095825**
g09	**680.63**	**680.63**	**680.63**	**680.63**	**680.63**	**680.63**	**680.63**
g10	7049.248	**7049.248**	**7049.248**	**7049.248**	7049.262	**7049.248**	7049.503
g11	0.75	**0.75**	**0.75**	**0.75**	**0.75**	**0.75**	**0.75**
g12	-1.000	**-1.000**	**-1.000**	**-1.000**	**-1.000**	**-1.000**	**-1.000**
g13	0.053942	**0.053942**	**0.053942**	0.079627	**0.053942**	0.438803	*0.053961*
g14	-47.765	**-47.765**	-	**-47.765**	-	**-47.765**	-
g15	961.715	**961.715**	**961.715**	**961.715**	**961.715**	**961.715**	**961.715**
g16	-1.905	**-1.905**	**-1.905**	**-1.905**	**-1.905**	**-1.905**	**-1.905**
g17	8853.540	**8853.540**	**8853.540**	8854.664	*8854.655*	*8858.874*	8859.413
g18	-0.866025	**-0.866025**	**-0.866025**	**-0.866025**	**-0.866025**	**-0.866025**	**-0.866025**
g19	32.656	**32.656**	**32.656**	*32.658*	32.666	*32.665*	32.802
g20	0.096700	-	-	-	-	-	-
g21	193.725	**193.725**	**193.725**	**193.725**	193.733	*193.726*	193.782
g22	236.431	-	-	-	-	-	-
g23	-400.055	**-400.055**	-	-391.415	-	-367.452	-
g24	-5.508	**-5.508**	**-5.508**	**-5.508**	**-5.508**	**-5.508**	**-5.508**

7.2 Experiment 2

A-DDE is now compared with respect to Static2 DDE, where the four parameters to be controlled in A-DDE are just generated at random between the same intervals in Static2 DDE. The parameter values used in A-DDE were the same reported in experiment 1 in Section 7.1. The parameters used in Static2 DDE were the following:

- Static2 DDE
  - $NP = 60$ y $GMAX = 600$
  - $S_r \in [0.45, 0.65]$ generated at random instead of using the deterministic parameter control.
  - $NO \in [3,7]$, $CR \in [0.9, 1.0]$ and $F \in [0.3, 0.9]$ also generated at random instead of using the self-adaptive parameter control.

Table 3 Comparison of results obtained with the DDE version with randomly-generated parameter values (named Static2 DDE) and the proposed Adaptive-DDE (deterministic and self-adaptive parameter control). "-", means no feasible solution found. Values in **boldface** mean that the global optimum or best know solution was reached, values in *italic* mean that the obtained result is better (but not the optimal or best known) with respect to the approach compared

Test problem	Best known solution	**Best** Adaptive DDE	Static2 DDE	**Mean** Adaptive DDE	Static2 DDE	**Worst** Adaptive DDE	Static2 DDE
g01	-15.000	**-15.000**	**-15.000**	**-15.000**	-14.937	**-15.000**	-13.917
g02	-0.803619	-0.803605	*-0.803610*	*-0.771090*	-0.706674	*-0.609853*	-0.483550
g03	-1.000	**-1.000**	**-1.000**	**-1.000**	**-1.000**	**-1.000**	**-1.000**
g04	-30665.539	**-30665.539**	**-30665.539**	**-30665.539**	-30660.237	**-30665.539**	-30591.889
g05	5126.497	**5126.497**	**5126.497**	**5126.497**	**5126.497**	**5126.497**	**5126.497**
g06	-6961.814	**-6961.814**	**-6961.814**	**-6961.814**	-6959.015	**-6961.814**	-6877.840
g07	24.306	**24.306**	**24.306**	**24.306**	24.945	**24.306**	38.903
g08	-0.095825	**-0.095825**	**-0.095825**	**-0.095825**	**-0.095825**	**-0.095825**	**-0.095825**
g09	680.63	**680.63**	**680.63**	**680.63**	**680.63**	**680.63**	**680.63**
g10	7049.248	**7049.248**	**7049.248**	**7049.248**	7073.779	**7049.248**	7308.826
g11	0.75	**0.75**	**0.75**	**0.75**	**0.75**	**0.75**	**0.75**
g12	-1.000	**-1.000**	**-1.000**	**-1.000**	**-1.000**	**-1.000**	**-1.000**
g13	0.053942	**0.053942**	**0.053942**	*0.079627*	0.131458	0.438803	0.438803
g14	-47.765	**-47.765**	-	**-47.765**	-	**-47.765**	-
g15	961.715	**961.715**	**961.715**	**961.715**	**961.715**	**961.715**	**961.715**
g16	-1.905	**-1.905**	**-1.905**	**-1.905**	**-1.905**	**-1.905**	-1.900
g17	8853.540	**8853.540**	**8853.540**	*8854.664*	8855.884	*8858.874*	8859.693
g18	-0.866025	**-0.866025**	**-0.866025**	**-0.866025**	**-0.866025**	**-0.866025**	**-0.866025**
g19	32.656	**32.656**	**32.656**	*32.658*	37.785	*32.665*	65.993
g20	-	-	-	-	-	-	-
g21	193.725	**193.725**	**193.725**	**193.725**	198.009	*193.726*	263.444
g22	-	-	-	-	-	-	-
g23	-400.055	**-400.055**	-	*-391.415*	-	*-367.452*	-
g24	-5.508	**-5.508**	**-5.508**	**-5.508**	**-5.508**	**-5.508**	**-5.508**

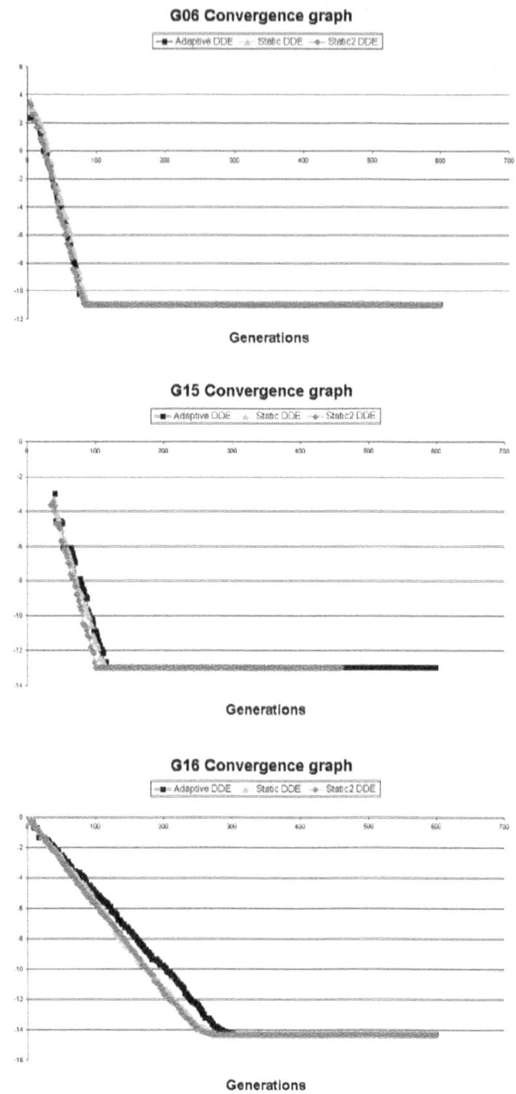

Fig. 6 Convergence graphs for problems g06, g15 and g16 where the behavior was similar in the three compared DE algorithms.

The summary of statistical values from a set of 30 independent runs is shown in Table 3.

The results in Table 3 suggest that the effect of the self-adaptive mechanism is not equivalent to just generating random values within convenient limits. A-DDE provided better statistical results (mostly in the mean and worst values from a set of 30 independent runs) in thirteen test problems (g01, g02, g04, g06, g07, g10, g13, g14, g16, g17, g19, g21 and g23). In nine problems the performance was similar

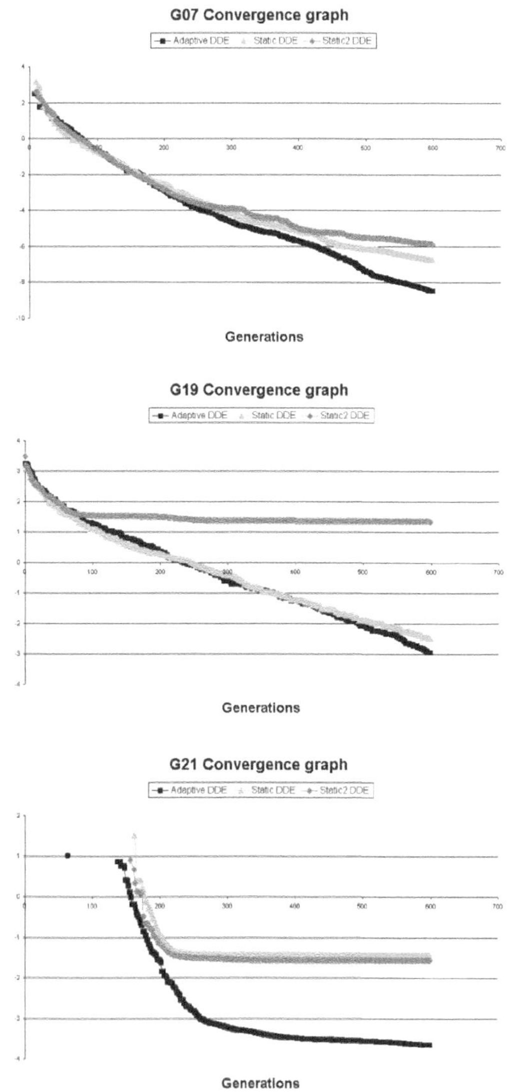

Fig. 7 Convergence graphs for problems g07, g19 and g21 where the behavior of A-DDE was better

between A-DDE and Static2 DDE (g03, g05, g08, g09, g11, g12, g15, g18 and g24). Finally, Static2 DDE was unable to provide better results in any problem.

7.3 Experiment 3

In order to analyze the convergence behavior of each DDE algorithm compared, the convergence graph of the run located in the median value of the 30 independent runs was plotted for each test problem. The graph starts when the first feasible solution is

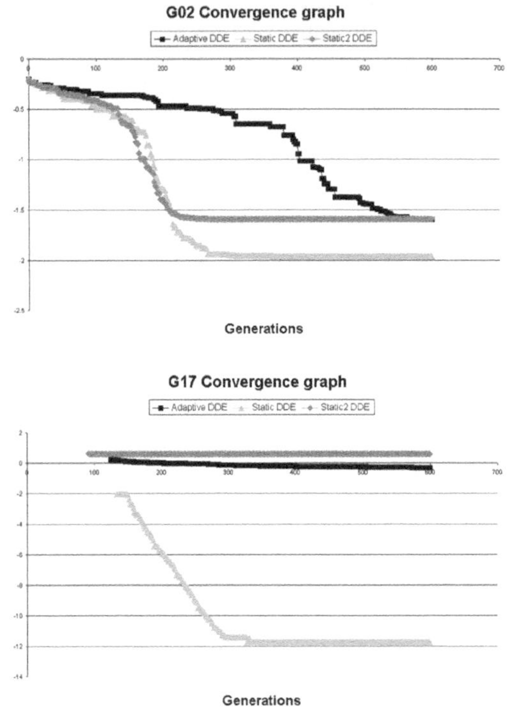

Fig. 8 Convergence graphs for problems g02 and g17 where the behavior of A-DDE was not better than those provided by the compared approaches

generated. The x-axis represents the generation number and the y-axis is calculated as follows: $log10(f(\mathbf{x}) - f(\mathbf{x}^*))$, where $f(\mathbf{x})$ is the best feasible solution found in the current generation and $f(\mathbf{x}^*)$ is the optimal solution or best known solution for the problem being solved (see Table 1).

For sake of clarity representative graphs were grouped based on the behavior observed: (1) Test problems where the convergence was similar among Static DDE, Static2 DDE and A-DDE in Figure 6, (2) problems where A-DDE converged to better solutions faster than the other two approaches in Figure 7 and (3) problems where A-DDE got trapped in a local optimum solution in Figure 8.

Regarding Figure 6, besides problems g06, g15 and g16, the convergence behavior was similar in other nine problems (g01, g03, g04, g05, g08, g09, g11, g12 and g24). A-DDE was able to present a better convergence behavior, besides problems g07, g19 and g21 in Figure 7, in other five problems (g10, g13, g14, g18 and g23). Finally, the two problems presented in Figure 8 (g02 and g17) are those where A-DDE got trapped in local optima solutions compared to the other two DDE algorithms.

There is not a clear pattern between convergence behavior and features of a test problem. However, the results indeed show that A-DDE mostly maintained the original DDE competitive convergence behavior and even was able to skip local optimum solutions and providing better results in some problems, most of them in presence of equality constraints (g13, g14, g21 and g23).

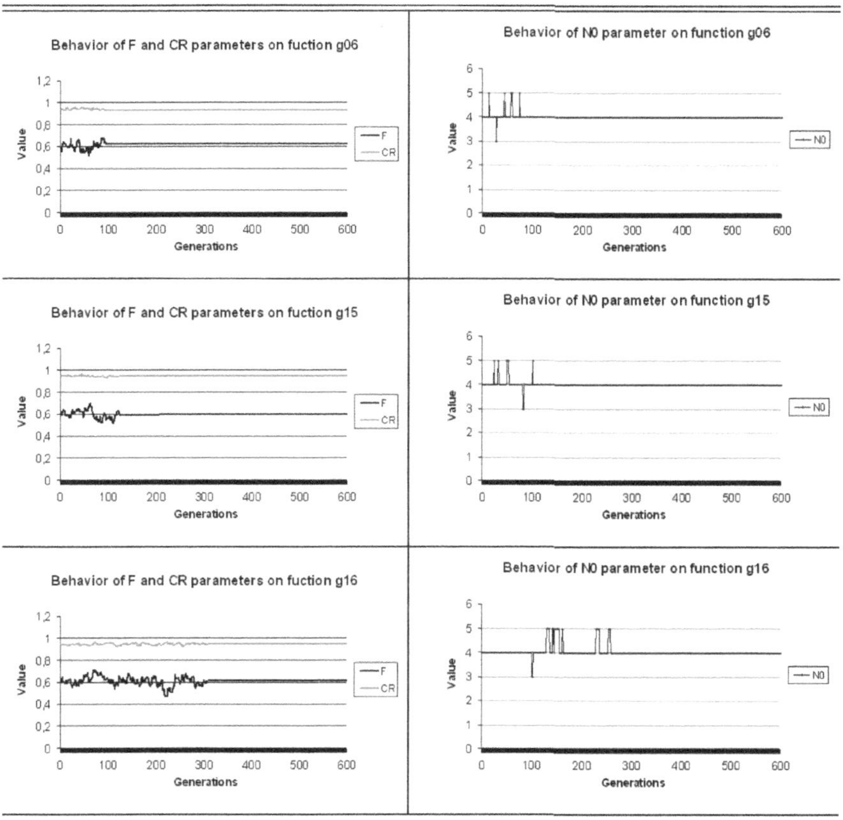

Fig. 9 Average values for F, CR y NO parameters in each generation on the run located in the median value out of 30 independent runs for problems g06, g15 and g16. The parameter values converged to a single value

7.4 Experiment 4

The results of the experiment to analyze the values taken for the controlled parameters are reported as follows: Two graphs for representative test problems are presented. One graph includes the average value for the F and CR parameters in each generation of the run located in the median value out of 30 independent runs. The other graph presents the average value of the NO parameter of the same run. As in the previous experiment, the graphs are grouped based on the behavior found: (1) Those where the parameter values converged to a specific value in Figure 9, (2) problems where the parameter values oscillated and the final results were better with respect to the compared approaches in Figure 10 and, finally, (3) test problems where the parameter values oscillated and the final results where not better than those obtained with the other two DDE versions in Figure 11.

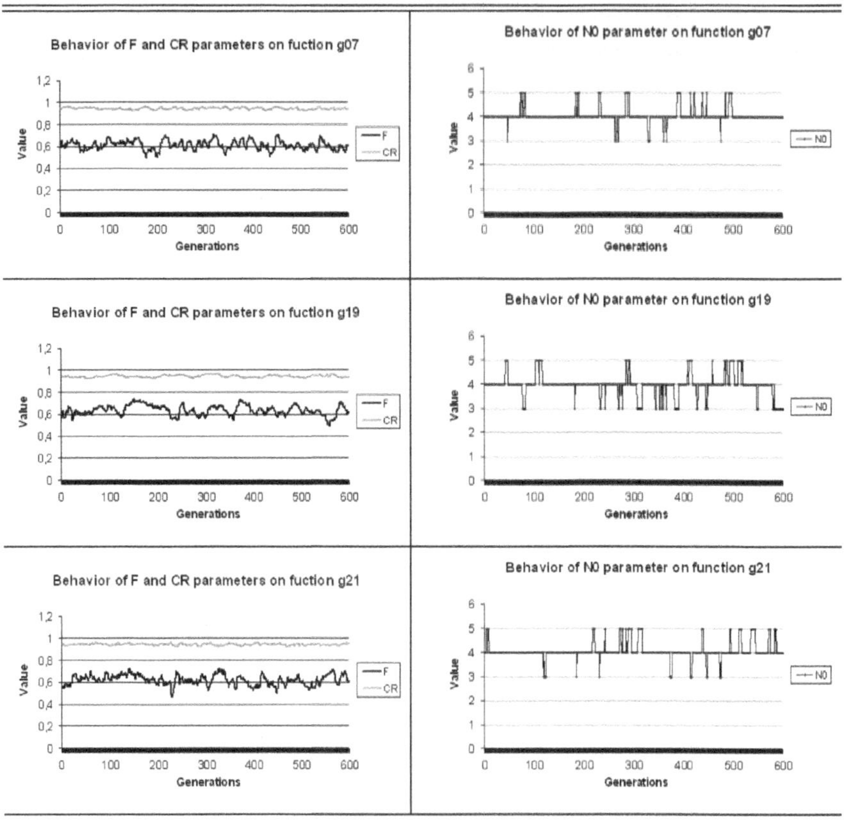

Fig. 10 Average values for F, CR y NO parameters in each generation on the run located in the median value out of 30 independent runs for problems g07, g19 and g21. The parameter values keep oscillating during all the run and the final results were better with respect to both Static DDE versions

It is clear from Figures 9, 10 and 11 that an oscillating behavior was obtained in all cases, This effect was more remarked in F and NO, whereas in CR it is barely noted. Based on these results, the self-adaptive mechanism is able to find that $CR \approx 0.9$, which means trial vectors more similar to the mutation vector and less similar to the target vector, is a suitable value on this set of test functions. On the other hand, $F \in [0.5, 0.7]$ and $NO \in [3, 5]$ are adequate boundaries for the set of constrained problems used in the experiments.

Figure 9 includes representative graphs for test problems where the parameter values reached a single value after converging to an optimum (See Figure 6). Other test problems where the behavior was similar were g04, g08, g09, g11, g12 and g24.

Figure 10 contains graphs where A-DDE provided a better final result (See Figure 7), but required more time to converge. In the same way, the parameter values kept oscillating, helping the search by varying the values, mostly for F and $N0$. Other

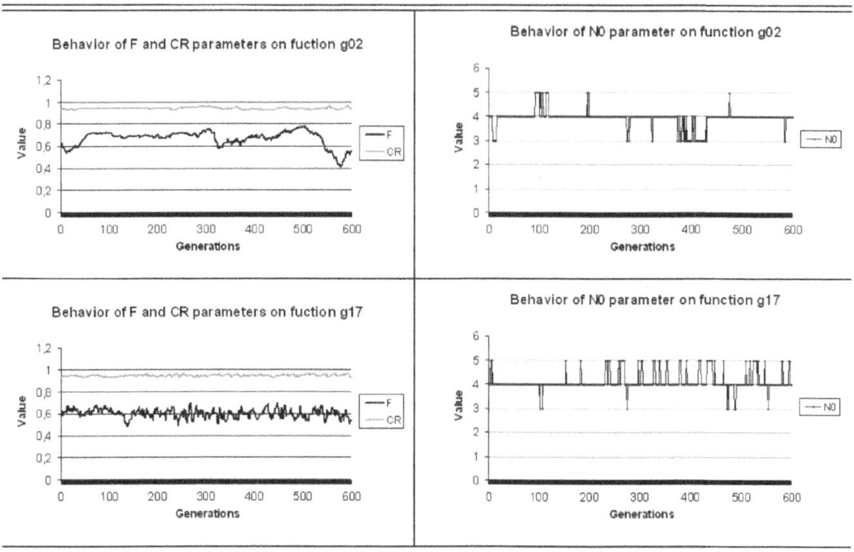

Fig. 11 Average values for *F*, *CR* y *NO* parameters in each generation on the median value out of 30 independent runs for problems g02 and g17. The parameter values keep oscillating during all the run and the final results were not better with respect to both Static DDE versions

test functions with the same type of results were g01, g03, g05, g10, g14, g17, g18 and g23.

Finally, Figure 11 shows those graphs where the parameter values kept varying while A-DDE got trapped in a local optimum solution (See Figure 8).

7.5 Experiment 5

The 95%-confidence intervals for the mean value, out of 30 independent runs are presented for both performance measures: Evals in Table 4 and Progress Ratio in Table 5. The aim is to analyze the average performance of the three DDE versions as to establish the effect of the deterministic and self-adaptive control mechanisms.

Regarding the Evals results reported in Table 4, some test problems were not considered in the discussion because the feasible region was reached in the initial population or even in the first generation. These problems are g02, g04, g08, g09, g12, g19 and g24. Problems g20 and g22 are also excluded because no feasible solutions were found by any algorithm. The confidence intervals for Evals indicate that Static2 DDE reached the feasible region faster in eight problems: g03, g05, g06, g11, g13, g15, g17 and g21. Static DDE generated feasible solutions faster in four problems: g01, g07, g16 and g18. A-DDE only found feasible solutions faster in three problems: g10, g14 and g23. These results point out that the deterministic and self-adaptive mechanisms, despite maintaining or improving the quality and

Table 4 95%-Confidence intervals for the Evals performance measure in the three DE algorithms compared. The best values are remarked in **boldface**

Problem	Evals		
	Adaptive DDE	Static	Static2
g01	[5134,5158]	**[4349,4375]**	[4852,4875]
g02	[62,62]	**[61,61]**	[61,61]
g03	[5396,5440]	[5038,5084]	**[4989,5023]**
g04	**[63,63]**	[62,63]	[65,65]
g05	[18926,18948]	[20825,20850]	**[18820,18858]**
g06	[1237,1244]	[1282,1291]	**[1221,1231]**
g07	[2723,2734]	**[2432,2443]**	[2590,2605]
g08	**[143,146]**	[152,154]	[163,166]
g09	**[163,165]**	[188,192]	[165,167]
g10	**[4137,4158]**	[4236,4259]	[4486,4509]
g11	[2681,2715]	[2393,2419]	**[2161,2190]**
g12	[79,79]	**[78,78]**	[85,85]
g13	[30866,30973]	[34440,34551]	**[29012,29164]**
g14	**[106738,106899]**	-	-
g15	[11332,11354]	[11528,11545]	**[10362,10386]**
g16	[1312,1323]	**[1159,1168]**	[1199,1210]
g17	[33804,33865]	[37488,37553]	**[30929,31013]**
g18	[10785,10818]	**[10282,10310]**	[10434,10468]
g19	[64,64]	**[62,62]**	[64,64]
g20	-	-	-
g21	[38214,38289]	[47745,47856]	**[34957,35127]**
g22	-	-	-
g23	**[59992,60107]**	-	-
g24	**[62,62]**	[62,63]	**[62,62]**

consistency of the final results, delayed the arrival to the feasible region with respect to the other two DDE variants.

The results for the Progress Ration in Table 5, where again problems g20 and g22 are excluded from discussion because no feasible solutions were found, show that A-DDE obtained a better improvement inside the feasible region in ten problems: g02, g04, g09, g10, g14, g15, g16, g17, g21 and g23. Static DDE obtained a better Progress Ratio interval in eight problems g01, g06, g07, g08, g11, g12, g18 and g19, while Static2 DDE was better in four problems: g03, g05, g13 and g24.

After taking more evaluations to reach the feasible region, A-DDE was able to improve the fist feasible solution in more problems with respect to Static and Static2 DDE. This behavior suggests that A-DDE enters the feasible region from a more promising region, based on a better exploration of the search space due to the suitable parameter values. However, this issue requires further and more detailed research.

Table 5 95%-Confidence intervals for the Progress Ratio performance measure in the three DE algorithms compared. The best values are remarked in **boldface**

Problem	Progress Ratio		
	Adaptive DDE	Static	Static2
g01	[1.004,1.013]	**[1.064,1.073]**	[0.900,0.906]
g02	**[1.071,1.074]**	[1.060,1.062]	[1.005,1.008]
g03	[1.199,1.216]	[1.255,1.273]	**[1.411,1.427]**
g04	**[0.77E-01,0.78E-01]**	[0.63E-01,0.63E-01]	[0.62E-01,0.63E-01]
g05	[0.83E-06,0.85E-06]	[0.17E-05,0.18E-05]	**[0.20E-05,0.21E-05]**
g06	[0.375,0.379]	**[0.446,0.450]**	[0.396,0.400]
g07	[1.689,1.699]	**[1.849,1.857]**	[1.750,1.761]
g08	[1.465,1.479]	**[1.691,1.701]**	[1.479,1.492]
g09	**[2.622,2.647]**	[2.522,2.558]	[2.492,2.528]
g10	**[0.468,0.470]**	[0.481,0.483]	[0.484,0.487]
g11	[0.52E-01,0.53E-01]	**[0.58E-01,0.59E-01]**	[0.30E-01,0.31E-01]
g12	[0.102,0.104]	**[0.123,0.124]**	[0.100,0.102]
g13	[0.11E-02,0.11E-02]	[0.89E-03,0.97E-03]	**[0.34E-02,0.36E-02]**
g14	**[0.47E-01,0.47E-01]**	-	-
g15	**[0.11E-05,0.12E-05]**	[0.10E-05,0.11E-05]	[0.92E-06,0.96E-06]
g16	**[0.215,0.216]**	[0.205,0.207]	[0.207,0.209]
g17	**[0.12E-02,0.12E-02]**	[0.41E-03,0.44E-03]	[0.61E-03,0.64E-03]
g18	[0.731,0.738]	**[0.763,0.771]**	[0.710,0.714]
g19	[3.124,3.129]	**[3.150,3.155]**	[2.990,2.997]
g20	-	-	-
g21	**[0.54E-01,0.55E-01]**	[0.31E-01,0.31E-01]	[0.35E-01,0.35E-01]
g22	-	-	-
g23	**[0.290,0.294]**	-	-
g24	[0.397,0.401]	[0.310,0.313]	**[0.506,0.517]**

7.6 Experiment 6

In order to compare the final results obtained with A-DDE with respect to state-of-the-art approaches, a summary of statistical values on the first 13 test problems (the most used for comparison in the specialized literature) are presented in Table 6. The approaches used for comparison are: (1) The Generic Framework for constrained optimization by Venkatraman & Yen [35], where the search is divided in two phases, one where only the feasibility of solutions is considered and another one where the feasibility and the optimization of the objective function are taken into account, (2) the self-adaptive penalty function by Tessema & Yen [34], where a parameter-free penalty function is used to deal with the constraints of the problem and (3) a mathematical programming approach combined with a mutation operator by Takahama & Sakai [31]. The comparison shows that A-DDE is indeed very competitive with other evolutionary approaches for constrained optimization based on the quality (best result obtained so far) and consistency (better mean and standard deviation values) of the final results.

Table 6 Statistical results obtained by A-DDE with respect to those provided by state-of-the-art approaches on 13 benchmark problems. Values in **boldface** mean that the global optimum or best know solution was reached, values in *italic* mean that the obtained result is better (but not the optimal or best known) with respect to the approaches compared. No results reported for problems $g12$ and $g13$ were found in [35]

Problem/BKS	Statistic	Venkatraman & Yen [35]	Tessema & Yen [34]	Takahama & Sakai [31]	A-DDE
g01	Best	**-15.000**	**-15.000**	**-15.000**	**-15.000**
-15.000	Median	**-15.000**	-14.966	**-15.000**	**-15.000**
	Worst	-12.000	-13.097	**-15.000**	**-15.000**
	St. Dev.	8.51E-01	7.00E-01	**6.40E-06**	7.00E-06
g02	Best	-0.803190	-0.803202	**-0.803619**	-0.803605
-0.803619	Median	-0.755332	*-0.789398*	-0.785163	-0.777368
	Worst	-0.672169	-0.745712	*-0.754259*	-0.609853
	St. Dev.	3.27E-02	1.33E-01	**1.30E-02**	3.66E-02
g03	Best	**-1.000**	**-1.000**	**-1.000**	**-1.000**
-1.000	Median	-0.949	-0.971	**-1.000**	**-1.000**
	Worst	**-1.000**	-0.887	**-1.000**	**-1.000**
	St. Dev.	4.89E-02	3.01E-01	**8.50E-14**	9.30E-12
g04	Best	-30665.531	-30665.401	**-30665.539**	**-30665.539**
-30665.539	Median	-30663.364	-30663.921	**-30665.539**	**-30665.539**
	Worst	-30651.960	-30656.471	**-30665.539**	**-30665.539**
	St. Dev.	3.31E+00	2.04E+00	4.20E-11	**3.20E-13**
g05	Best	5126.510	5126.907	**5126.497**	**5126.497**
5126.497	Median	5170.529	5208.897	**5126.497**	**5126.497**
	Worst	6112.223	5564.642	**5126.497**	**5126.497**
	St. Dev.	3.41E+02	2.47E+02	3.50E-11	**2.10E-11**
g06	Best	-6961.179	-6961.046	**-6961.814**	**-6961.814**
-6961.814	Median	-6959.568	-6953.823	**-6961.814**	**-6961.814**
	Worst	-6954.319	-6943.304	**-6961.814**	**-6961.814**
	St. Dev.	1.27E+00	5.88E+00	1.30E-10	**2.11E-12**
g07	Best	24.411	24.838	**24.306**	**24.306**
24.306	Median	26.736	25.415	**24.306**	**24.306**
	Worst	35.882	33.095	24.307	**24.306**
	St. Dev.	2.61E+00	2.17E+00	1.30E-04	**4.20E-05**
g08	Best	**-0.095825**	**-0.095825**	**-0.095825**	**-0.095825**
-0.095825	Median	**-0.095825**	**-0.095825**	**-0.095825**	**-0.095825**
	Worst	**-0.095825**	-0.092697	**-0.095825**	**-0.095825**
	St. Dev.	**0**	1.06E-03	3.80E-13	9.10E-10
g09	Best	680.76	680.77	**680.63**	**680.63**
680.63	Median	681.71	681.24	**680.63**	**680.63**
	Worst	684.13	682.08	**680.63**	**680.63**
	St. Dev.	7.44E-01	3.22E-01	2.90E-10	**1.15E-10**
g10	Best	7060.553	7069.981	**7049.248**	**7049.248**
7049.248	Median	7723.167	7201.017	**7049.248**	**7049.248**
	Worst	12097.408	7489.406	**7049.248**	**7049.248**
	St. Dev.	7.99E+02	1.38E+02	**4.70E-06**	3.23E-4
g11	Best	**0.75**	**0.75**	**0.75**	**0.75**
0.75	Median	**0.75**	**0.75**	**0.75**	**0.75**
	Worst	0.81	0.76	**0.75**	**0.75**
	St. Dev.	9.30E-03	2.00E-03	**4.90E-16**	5.35E-15
g12	Best	NA	**-1.000**	**-1.000**	**-1.000**
-1.000	Median	NA	**-1.000**	**-1.000**	**-1.000**
	Worst	NA	**-1.000**	**-1.000**	**-1.000**
	St. Dev.	NA	1.41E-04	**3.90E-10**	4.10E-9
g13	Best	NA	**0.053941**	0.053942	0.053942
0.053942	Median	NA	0.054713	**0.053942**	**0.053942**
	Worst	NA	0.885276	0.438803	0.438803
	St. Dev.	NA	2.75E-01	**6.90E-02**	9.60E-02

8 Conclusions and Future Work

In this chapter, a deterministic and self-adaptive parameter control mechanisms were added to a competitive DE-based approach, called Diversity Differential Evolution (DDE) to solve constrained optimization problems. The proposed approach, called Adaptive-DDE (A-DDE) considered the encoding of three parameters per each vector in the population, two from the original DE (F and CR) and one related with the constraint-handling mechanism (NO, number of trial vectors generated per target vector). Traditional mutation and crossover DE operators were used to self-adapt these three values per each vector. Furthermore, other parameter which controls the bias in the search to keep infeasible solutions located in promising areas of the search space (based on the objective function value, regardless of feasibility), called S_r was deterministically controlled by a decreasing function, focusing first on keeping good infeasible solutions and, after that, maintaining mostly good feasible solutions and discarding those infeasible ones.

A-DDE was extensively compared with respect to the original DDE and also with respect to other state-of-the-art approaches. Six experiments were conducted and the following findings were obtained:

- A-DDE maintained the very competitive performance of the original DDE and it also was able to improve the final results in some test problems.
- A-DDE's performance was clearly superior with respect to a DDE version where just random values were generated per each parameter within adequate limits. The self-adaptive mechanism seems to be effective in most of the test problems.
- A-DDE convergence behavior was similar in most cases with respect to the original DDE. However, in some problems with equality constraints A-DDE was able to avoid local optimum solutions.
- An oscillating behavior dominated the self-adaptive mechanism on the three parameters encoded on each vector in the population. The effect was more remarked in $F \in [0.5, 0.7]$ and in $NO \in [3, 5]$, whereas $CR \approx 0.9$ was almost a constant in all the test problems. These results indicate that DDE requires (1) different scale values for the search directions generated in the process, (2) to allow each target vector to generate at least 3 trail vectors and (3) to let them be more similar to the mutation vector instead of being similar to the trial vector.
- The results obtained in the two performance measures showed that A-DDE requires more evaluations to reach the feasible region with respect to the original DDE. However, A-DDE is capable of generating a better improvement inside it.
- A-DDE provided very competitive results with respect to some state-of-the-art approaches.

Part of the future work derived from the present research is to analyze more in depth how A-DDE approaches the feasible region with respect to the original DDE as to get more knowledge on the effects of the parameter control mechanisms added. Moreover, we will use A-DDE to solve complex engineering design problems.

Finally, we will test other type of special operators in order to self-adapt the parameters encoded in each vector of the population.

Acknowledgements. The first author acknowledges support from CONACyT through project No. 79809-Y. The second author acknowledges support from CONACyT through a scholarship to pursue graduate studies at LANIA.

References

1. Bäck, T.: Evolutionary Algorithms in Theory and Practice. Oxford University Press, New York (1996)
2. Brest, J., Žumer, V., Maučecc, M.S.: Control Parameters in Self-Adaptive Differential Evolution. In: Filipič, B., Šilc, J. (eds.) Bioinspired Optimization Methods and Their Applications, Ljubljana, Slovenia, October 2006, pp. 35–44. Jožef Stefan Institute (2006)
3. Chakraborty, U.K. (ed.): Advances in Differential Evolution. Studies in Computational Intelligence Series. Springer, Heidelberg (2008)
4. Coello Coello, C.A.: Theoretical and Numerical Constraint Handling Techniques used with Evolutionary Algorithms: A Survey of the State of the Art. Computer Methods in Applied Mechanics and Engineering 191(11-12), 1245–1287 (2002)
5. Cruz-Cortés, N., Trejo-Pérez, D., Coello Coello, C.A.: Handling Constraints in Global Optimization using an Artificial Immune System. In: Jacob, C., Pilat, M.L., Bentley, P.J., Timmis, J.I. (eds.) ICARIS 2005. LNCS, vol. 3627, pp. 234–247. Springer, Heidelberg (2005)
6. Deb, K.: An Efficient Constraint Handling Method for Genetic Algorithms. Computer Methods in Applied Mechanics and Engineering 186(2/4), 311–338 (2000)
7. Eiben, A., Smith, J.E.: Introduction to Evolutionary Computing. Natural Computing Series. Springer, Heidelberg (2003)
8. Eiben, G., Schut, M.: New Ways to Calibrate Evolutionary Algorithms. In: Siarry, P., Michalewicz, Z. (eds.) Advances in Metaheuristics for Hard Optimization, Natural Computing, pp. 153–177. Springer, Heidelberg (2008)
9. Fogel, L.J.: Intelligence Through Simulated Evolution. Forty years of Evolutionary Programming. John Wiley & Sons, New York (1999)
10. Holland, J.H.: Adaptation in Natural and Artificial Systems. An Introductory Analysis with Applications to Biology, Control and Artificial Intelligence. University of Michigan Press, Ann Arbor (1975)
11. Huang, V.L., Qin, A.K., Suganthan, P.N.: Self-adaptative Differential Evolution Algorithm for Constrained Real-Parameter Optimization. In: 2006 IEEE Congress on Evolutionary Computation (CEC 2006), Vancouver, BC, Canada, July 2006, pp. 324–331. IEEE, Los Alamitos (2006)
12. Jong, K.A.D.: Evolutionary Computation. A Unified Approach. MIT Press, Cambridge (2006)
13. Kukkonen, S., Lampinen, J.: Constrained Real-Parameter Optimization with Generalized Differential Evolution. In: 2006 IEEE Congress on Evolutionary Computation (CEC 2006), Vancouver, BC, Canada, July 2006, pp. 911–918. IEEE, Los Alamitos (2006)
14. Lampinen, J.: A Constraint Handling Approach for the Differential Evolution Algorithm. In: Proceedings of the Congress on Evolutionary Computation 2002 (CEC 2002), Piscataway, New Jersey, May 2002, pp. 1468–1473. IEEE Service Center (2002)

15. Landa-Becerra, R., Coello Coello, C.A.: Optimization with Constraints using a Cultured Differential Evolution Approach. In: Beyer, H.G., et al. (eds.) Proceedings of the Genetic and Evolutionary Computation Conference (GECCO 2005), Washington DC, USA, June 2005, vol. 1, pp. 27–34. ACM Press, New York (2005)
16. Liang, J., Runarsson, T.P., Mezura-Montes, E., Clerc, M., Suganthan, P., Coello Coello, C., Deb, K.: Problem Definitions and Evaluation Criteria for the CEC 2006 Special Session on Constrained Real-Parameter Optimization. Technical report, CEC (March 2006), `http://www.lania.mx/~emezura/documentos/tr_cec06.pdf`
17. Liu, J., Lampinen, J.: A fuzzy adaptive differential evolution algorithm. Soft Comput 9(6), 448–462 (2005)
18. Mezura-Montes, E., Coello Coello, C.A.: Identifying On-line Behavior and Some Sources of Difficulty in Two Competitive Approaches for Constrained Optimization. In: 2005 IEEE Congress on Evolutionary Computation (CEC 2005), Edinburgh, Scotland, September 2005, vol. 2, pp. 1477–1484. IEEE Press, Los Alamitos (2005)
19. Mezura-Montes, E., Coello Coello, C.A.: Constrained Optimization via Multiobjective Evolutionary Algorithms. In: Knowles, J., Corne, D., Deb, K. (eds.) Multiobjective Problem Solving from Nature, pp. 53–75. Springer, Heidelberg (2008)
20. Mezura-Montes, E., Coello Coello, C.A., Tun-Morales, E.I.: Simple feasibility rules and differential evolution for constrained optimization. In: Monroy, R., Arroyo-Figueroa, G., Sucar, L.E., Sossa, H. (eds.) MICAI 2004. LNCS (LNAI), vol. 2972, pp. 707–716. Springer, Heidelberg (2004)
21. Mezura-Montes, E., Velázquez-Reyes, J., Coello Coello, C.A.: Promising Infeasibility and Multiple Offspring Incorporated to Differential Evolution for Constrained Optimization. In: Beyer, H., et al. (eds.) Proceedings of the Genetic and Evolutionary Computation Conference (GECCO 2005), New York, Washington DC, USA, June 2005, vol. 1, pp. 225–232. ACM Press, New York (2005)
22. Mezura-Montes, E., Velázquez-Reyes, J., Coello Coello, C.A.: Modified Differential Evolution for Constrained Optimization. In: 2006 IEEE Congress on Evolutionary Computation (CEC 2006), Vancouver, BC, Canada, July 2006, pp. 332–339. IEEE, Los Alamitos (2006)
23. Michalewicz, Z., Fogel, D.B.: How to Solve It: Modern Heuristics, 2nd edn. Springer, Heidelberg (2004)
24. Michalewicz, Z., Schoenauer, M.: Evolutionary Algorithms for Constrained Parameter Optimization Problems. Evolutionary Computation 4(1), 1–32 (1996)
25. Miettinen, K., Makela, M., Toivanen, J.: Numerical comparison of some penalty-based constraint handling techniques in genetic algorithms. Journal of Global Optimization 27(4), 427–446 (2003)
26. Muñoz-Zavala, A.E., Hernández-Aguirre, A., Villa-Diharce, E.R., Botello-Rionda, S.: PESO+ for Constrained Optimization. In: 2006 IEEE Congress on Evolutionary Computation (CEC 2006), Vancouver, BC, Canada, July 2006, pp. 935–942. IEEE, Los Alamitos (2006)
27. Price, K., Storn, R., Lampinen, J.: Differential Evolution: A Practical Approach to Global Optimization. Natural Computing Series. Springer, Heidelberg (2005)
28. Rechenberg, I.: Evolutionsstrategie: Optimierung technischer Systeme nach Prinzipien der biologischen Evolution. Frommann-Holzboog, Stuttgart (1973)
29. Runarsson, T.P., Yao, X.: Stochastic Ranking for Constrained Evolutionary Optimization. IEEE Transactions on Evolutionary Computation 4(3), 284–294 (2000)
30. Schwefel, H.-P. (ed.): Evolution and Optimization Seeking. Wiley, New York (1995)

31. Takahama, T., Sakai, S.: Constrained Optimization by Applying the α Constrained Method to the Nonlinear Simplex Method with Mutations. IEEE Transactions on Evolutionary Computation 9(5), 437–451 (2005)
32. Takahama, T., Sakai, S.: Constrained Optimization by the ε Constrained Differential Evolution with Gradient-Based Mutation and Feasible Elites. In: 2006 IEEE Congress on Evolutionary Computation (CEC 2006), Vancouver, BC, Canada, July 2006, pp. 308–315. IEEE, Los Alamitos (2006)
33. Tasgetiren, M.F., Suganthan, P.N.: A Multi-Populated Differential Evolution Algorithm for Solving Constrained Optimization Problem. In: 2006 IEEE Congress on Evolutionary Computation (CEC 2006), Vancouver, BC, Canada, July 2006, pp. 340–354. IEEE, Los Alamitos (2006)
34. Tessema, B., Yen, G.G.: A Self Adaptative Penalty Function Based Algorithm for Constrained Optimization. In: 2006 IEEE Congress on Evolutionary Computation (CEC 2006), Vancouver, BC, Canada, July 2006, pp. 950–957. IEEE, Los Alamitos (2006)
35. Venkatraman, S., Yen, G.G.: A Generic Framework for Constrained Optimization Using Genetic Algorithms. IEEE Transactions on Evolutionary Computation 9(4), 424, 435 (2005)

An Adaptive Penalty Function for Handling Constraint in Multi-objective Evolutionary Optimization

Gary G. Yen

Abstract. This chapter proposes a constraint handling technique for multi-objective evolutionary algorithms based on an adaptive penalty function and a distance measure. These two functions vary dependent upon the objective function value and the sum of constraint violations of an individual. Through this design, the objective space is modified to account for the performance and constraint violation of each individual. The modified objective functions are used in the non-dominance sorting to facilitate the search of optimal solutions not only in the feasible space but also in the infeasible regions. The search in the infeasible space is designed to exploit those individuals with better objective values and lower constraint violations. The number of feasible individuals in the population is used to guide the search process either toward finding more feasible solutions or favor in search for optimal solutions. The proposed method is simple to implement and does not need any parameter tuning. The constraint handling technique is tested on several constrained multi-objective optimization problems and has shown superior results compared to some chosen state-of-the-art designs.

Keywords: Multiobjective evolutionary algorithm, constraint handling, adaptive penalty.

1 Introduction

Evolutionary algorithms (EAs) have been successfully applied to solve optimization problems in the fields of science and engineering [15]. EAs were originally designed for solving unconstrained optimization problems, but in recent years, researchers have been able to tailor constraint handling techniques into these algorithms. The great challenges in constrained optimization problems arise from the

Gary G. Yen
Oklahoma State University, School of Electrical and Computer Engineering, Stillwater, OK 74078, USA
e-mail: gyen@okstate.edu

E. Mezura-Montes (Ed.): Constraint-Handling in Evolution. Optimi., SCI 198, pp. 121–143.
springerlink.com

various limits on the decision variables, the types of constraints involved, the interference among constraints, and the interrelationship between the constraints and the objective function [34]. In the mean time, researchers were also developing evolutionary approaches for solving multi-objective optimization problems (MOPs). These multi-objective evolutionary algorithms (MOEAs) are capable of simultaneously optimizing a set of competing objectives [10]. Nevertheless, limited research was conducted in the area of constrained multi-objective optimization. Such problems involve multiple conflicting objectives that are subject to various equality and inequality constraints [24,25].

A constrained multi-objective optimization problem (CMOP) can be mathematically formulated as: Minimize / Maximize

$$f_i(\mathbf{x}) = f_i(x_1, x_2, ..., x_n), i = 1, ..., p \tag{1}$$

subject to

$$g_j(\mathbf{x}) = g_j(x_1, x_2, ..., x_n) \leq 0, j = 1, ..., q$$
$$h_j(\mathbf{x}) = h_j(x_1, x_2, ..., x_n) = 0, j = q+1, ..., m$$
$$x_k^{\min} \leq x_k \leq x_k^{\max}, k = 1, ..., n$$

There are p objective functions that are required to be simultaneously optimized. Each objective function $f_i(\mathbf{x})$ is defined on the search space $S \subseteq \Re^n$. Usually the search space is an n-dimensional hyperbox in $\Re^n$. Each dimension of the search space is bounded by its upper ($x_k^{\max}$) and lower ($x_k^{\min}$) limits.$g_j(\mathbf{x})$ is the jth-inequality constraint, while $h_j(\mathbf{x})$ is the jth-equality constraint. There are a total of m constraints, q inequality and $m-q$ equality, which are required to be satisfied by the optimum solution. The presence of equality and inequality constraints will restrict the search space to a feasible region $F \subseteq S$, where a usable solution can be found.

This chapter extends the single-objective constrained optimization algorithm proposed by Tessema and Yen [33] to CMOPs. The proposed algorithm basically modifies the objective function value of an individual using its distance measure and penalty value. These modified objective function values are ranked through the nondominance sorting of the multi-objective optimization. Distance measures are found for each dimension of the objective space by incorporating the effect of an individual's constraint violation into its objective function. The penalty function , on the other hand, introduces additional penalty for infeasible individuals based on their objective values and constraint violations. The balance between two components, one based on objective function and the other on constraint violation, is controlled by the number of feasible individuals currently present in the population. If few feasible individuals are present, then those infeasible individuals with higher constraint violations are penalized more than those with lower constraint violations. On the other hand, if a sufficient number of feasible individuals exists, then those infeasible individuals with worse objective values are penalized more than those with better objective values. However, if the number of feasible individuals is in the middle of the two extremes, then the individual with lower constraint violation and better objective function is less penalized. The two components of the penalty function allow the algorithm to switch between feasibility and optimality at anytime during the evolutionary process. Furthermore, since priority is initially given to finding feasible individuals before searching for optimal solutions, the algorithm is capable of finding feasible solutions when the feasible space is very small compared to the search space.

This chapter is structured as follows: Section 2 provides a brief overview of the various evolutionary approaches developed for constrained multi-objective optimization problems. Then, in Section 3, the proposed CMOP is presented and analyzed in detail. Next, in Section 4, various CMOP test problems are adopted to evaluate the proposed algorithm as opposed to two chosen state-of-the-art designs in literature. Finally, the results of the experiments and conclusion with a summary of this chapter and ideas for future work are discussed.

2 Literature Survey

This section presents a brief review of evolutionary approaches developed for constrained multi-objective optimization problems. The review begins with a discussion about the MOEA for unconstrained MOPs and then extends into limited work in the MOEA designs for CMOPs.

Over the last decade, several MOEAs have been developed to solve multi-objective optimization problems. The earlier MOEAs are non-elitism based methods that assign fitness to population members based on non-dominated sorting. In addition, they exploited different techniques to preserve diversity among solutions of the same non-dominated front. Of these types, the Multi-Objective Genetic Algorithm (MOGA) [13] by Fonseca and Fleming and the Non-dominated Sorting Genetic Algorithm (NSGA) [29] by Srinivas and Deb were very popular. MOGA uses the niche-formation technique to preserve diversity over the Pareto optimal region, and sharing is performed on the objective function values. On the other hand, sharing is performed on the decision variable space for NSGA .

More recently, elitism based algorithms have been suggested to enhance the convergence properties of MOEAs. The Pareto Archived Evolution Strategy (PAES) [21] by Knowles and Corne uses a (1+1) evolution strategy with a historical archive that records all the non-dominated solutions found until the current generation. It also designs a novel approach to maintain diversity which consists of a crowding procedure that divides objective space in a recursive manner into several grids. This procedure is adaptive and has lower computational complexity than the traditional niching based approaches. Zitzler and Thiele introduce the Strength Pareto Evolutionary Algorithm (SPEA) [36] that uses an external archive to preserve non-dominated solutions. In each generation, the non-dominated solutions in the external set are given a strength value that is proportional to the number of individuals they dominate. Fitness of individuals in the main population are computed according to the strengths of all external non-dominated solutions that dominate it. In addition, a clustering technique is used to preserve diversity. Today, many advanced versions of MOEAs have been constantly made available in literature in continue pursuit of the performance frontier [10].

On the other hand, constraint handling for single objective optimization problems has also been actively researched over the past two decades. Penalty functions are the simplest and most commonly used methods for handling constraints using EAs. In death penalty function methods such as [2], individuals that violate any one of the constraints are rejected and no information is extracted from infeasible individuals. If the added penalties do not depend on the current generation number and remain constant during the entire evolutionary process, then the penalty function is called static penalty function. In static penalty function methods [17], the penalties are the weighted sum of the constraint violations. If, alternatively, the current generation number is considered in determining the penalties, then the method is called dynamic penalty function method [19]. In adaptive penalty function methods [3, 12], information gathered from the search process will be used to control the amount of penalty added to infeasible individuals.

In [9, 19], methods based on preference of feasible solutions over infeasible solutions are employed. In these types of techniques, feasible solutions are always considered better than infeasible ones. Therefore, when population fitness ranking is performed, feasible individuals will come first followed by infeasible individuals with low constraint violation. In [27,28], Runarsson and Yao introduce the stochastic ranking method to achieve a balance between objective and penalty functions stochastically. A probability factor is used to determine whether the objective function value or the constraint violation value determines the rank of each individual. In [30,31], similar algorithms are proposed where constraint violation and objective function are optimized separately. A satisfaction level for the constraints was introduced to indicate how well a search point meets the constraints, which was then used for dominance comparison.

More recently, multi-objective optimization techniques have been used to solve constrained optimization problems [22, 24]. In [32], a multi-objective optimization technique that uses population-based algorithm generator and infeasible solutions archiving and replacement mechanism is introduced. In [33], a two-phase algorithm that is based on multi-objective optimization technique is proposed. In the first phase of the algorithm, the objective function is completely disregarded and the constraint optimization problem is treated as a constraint satisfaction problem. In the second phase, both constraint satisfaction and objective optimization are treated as a bi-objective optimization problem. An algorithm that combines penalty function approach and multi-objective optimization technique is also suggested in [1]. The algorithm has a similar structure as the penalty-based approach but borrows the ranking scheme from multi-objective optimization techniques.

Although multi-objective optimization and constraint handling have received a lot of attention individually, very little effort has been devoted in solving constrained multi-objective optimization problems [5]. Coello Coello and Christiansen [8] propose a naï¿½ve approach to solve CMOPs by ignoring any solution that violates any of the assigned constraints. This method is easy to implement but it often experiences difficulty in searching for even a single feasible solution.

In [4], Binh and Korn propose the Multi-objective Evolution Strategy (MOBES), which takes into account the objective function vector as well as the degree of constraint violation of infeasible solutions in order to evaluate their fitness. Infeasible individuals are divided into different classes according to their "nearness" to the feasible region, and ranking is performed based on the class. In addition, a mechanism to maintain a feasible Pareto optimal set is employed.

In [11], Deb *et al.* propose a constrained multi-objective algorithm based on constrained dominance of individuals. According to their algorithm, a solution i is said to constrained-dominate a solution j if 1) i is feasible while j is infeasible; 2) both are infeasible and i has less constraint violation; or 3) both are feasible and i dominates j. Feasible solutions constrained-dominate all infeasible solutions. However, when two feasible individuals are compared, the usual dominance relationship is used. The level of constraint violation is used to compare two infeasible individuals.

In [18], Jimenez *et al.* propose the Evolutionary algorithm of Non-dominated Sorting with Radial Slots (ENORA), which employs the min-max formulation for constraint handling. Feasible individuals evolve toward optimality, while infeasible individuals evolve toward feasibility. In addition, a diversity technique based on partitioning of the search space in a set of radial slots along which the successive populations generated by the algorithm are positioned is introduced.

In [26], Ray *et al.* suggest using three different non-dominated rankings of the population. The first ranking is performed using the objective function values; the second is performed using different constraints; and the last ranking is based on the combination of all objective functions and constraints. Depending on these rankings, the algorithm performs according to the predefined rules. In [6], Chafekar *et al.* propose two novel approaches for solving constrained multi-objective optimization problems. One method, called Objective Exchange Genetic Algorithm of Design Optimization (OEGADO), runs several GAs concurrently with each GA optimizing one objective and exchanging information about its objective with others. The other method, called Objective Switching Genetic Algorithm for Design Optimization (OSGADO), runs each objective sequentially with a common population for all objectives.

In [35], Young proposes a constrained multi-objective evolutionary algorithm called Blended Space EA (BSEA). The algorithm checks dominance by using a rank obtained by blending an individual's rank in objective space with its rank in constraint space. A similar approach is proposed by Angantyr *et al.* [1] that uses the weighted average rank of the ranks in the constraint and objective space. Although their algorithm was examined only for testing problems with one objective function and several constraints, a simple adjustment in their formulation will provide a constrained multi-objective optimization tool.

Fonseca and Fleming [14] propose a unified approach for multi-objective optimization and multiple constraint handling. Their algorithm handles constraints by assigning high priority to constraints and low priority to objective functions, which allows search of feasible solutions followed by search of optimal solutions.

Harada *et al.* [16] propose Pareto Descent Repair (PDR) operator that searches for feasible solutions out of infeasible individuals in the constraint function space. This operator involves gradients that are usually unavailable in functions justified to be optimized using EAs.

In light of superior performance achieved in [33] for the single objective constraint optimization, a similar idea is extended in this study into the uses of multi-objective constraint optimization. In the next section, we introduce the proposed constrained multi-objective evolutionary algorithm (CMOEA).

3 Proposed Constrained MOEA

The proposed algorithm extends the single-objective constrained evolutionary algorithm proposed by Tessema and Yen [33] into a multi-objective framework. The major difference in various constraint handling techniques used in multi-objective optimization arises from the variations in the involvement of infeasible individuals in the evolutionary process. The main purpose of involving infeasible individuals in the search process is to exploit the information they carry. Since EAs are stochastic search techniques, discarding infeasible individuals might lead to the EA being stuck in local optima, especially in problems with discontinuous feasible regions. In addition, in some highly constrained optimization problems, finding a single feasible individual by itself might be a daunting challenge when the algorithm must be able to extract information from the previous infeasible individuals.

The proposed algorithm uses modified objective function values for checking dominance in the population. The modification is based on the constraint violation of the individual and its objective performance. The modified objective value has two components: distance measure and adaptive penalty. The two components are discussed below in detail.

3.1 Distance Values

Distance measure is found for each dimension of the objective space by including the effect of an individual's constraint violation into its objective function. The major steps in calculating the distance measure starts with obtaining the minimum and maximum values of each objective function in the current population, $P(t)$, as:

$$f^i_{\min} = \min_{\mathbf{x} \in P(t)} f_i(\mathbf{x}), \text{ and} \tag{2}$$

$$f^i_{\max} = \max_{\mathbf{x} \in P(t)} f_i(\mathbf{x}). \tag{3}$$

Using these values, normalize each objective function i for every individual $\mathbf{x}$,

$$\tilde{f}_i(\mathbf{x}) = \frac{f_i(\mathbf{x}) - f^i_{\min}}{f^i_{\max} - f^i_{\min}}, \tag{4}$$

where $\tilde{f}_i(\mathbf{x})$ is the normalized i^{th}-objective value of individual $\mathbf{x}$.

Constraint violation , $v(\mathbf{x})$, of individual $\mathbf{x}$ is then calculated as the summation of the normalized violations of each constraint divided by the total number of constraints,

$$v(\mathbf{x}) = \frac{1}{m}\sum_{j=1}^{m}\frac{c_j(\mathbf{x})}{c^j_{\max}}, \tag{5}$$

where

$$c_j(\mathbf{x}) = \begin{cases} \max(0, g_j(\mathbf{x})), & j = 1, \cdots, q \\ \max(0, |h_j(\mathbf{x})| - \delta), & j = q+1, \cdots, m \end{cases} \tag{6}$$

$$c^j_{\max} = \max_{\mathbf{x}\in P(t)} c_j(\mathbf{x}), \tag{7}$$

δ is the tolerance value for equality constraints (usually 0.001 or 0.0001). q is the number of inequality constraints, and $m-q$ is the number of equality constraints. If the constraint violation $c_j(\mathbf{x})$ is greater than zero, then the individual $\mathbf{x}$ violates the jth-constraint. On the other hand, if the constraint violation $c_j(\mathbf{x})$ is equal to zero, then the individual $\mathbf{x}$ satisfies the jth-constraint and the constraint violation $c_j(\mathbf{x})$ is set to zero. Then the "distance" value of individual $\mathbf{x}$ in each objective function dimension i is formulated as follows:

$$d_i(\mathbf{x}) = \begin{cases} v(\mathbf{x}), & \text{if } r_f = 0 \\ \sqrt{\tilde{f}_i(\mathbf{x})^2 + v(\mathbf{x})^2}, & \text{otherwise} \end{cases}, \tag{8}$$

where

$$r_f = \frac{\text{number of feasible individuals in current population}}{\text{population size}} \tag{9}$$

From Equation 8, we observe that if there is no feasible individual in the current population, then the distance values are equal to the constraint violation of the individual. In this case, according to the distance values, an infeasible individual with smaller constraint violation will dominate another infeasible individual with higher constraint violation regardless of their objective function values. This is the best way to compare infeasible individuals in the absence of feasible individuals, and it will help us approach the feasible regions quickly.

On the other hand, if there is more than one feasible solution in the population, then the distance values will have the properties summarized below:

1. For a feasible individual $\mathbf{x}$, the distance value in a given objective function dimension i is equal to $\tilde{f}_i(\mathbf{x})$. Hence, those feasible individuals with smaller objective function values will have smaller distance values in that given dimension.
2. For infeasible individuals, the distance value has two components: the objective function value and the constraint violation. Hence, individuals closer to the origin in the $\tilde{f}_i(\mathbf{x}) - v(\mathbf{x})$ two-dimensional space would have lower distance

value in that objective function dimension than those farther away from the origin.

3. If we compare the distance values of infeasible and feasible individuals, then either one may have a smaller value. But if the two individuals have similar objective function values, then the feasible individual will have a smaller distance value in the corresponding objective function dimension.

3.2 Two Penalties

In addition to the penalty imposed upon infeasible individuals by the distance measure, two other penalty functions are also added. These functions introduce additional penalty for infeasible individuals based on their corresponding objective value and constraint violation . The first penalty function is based on the objective functions, and the second is based on the constraint violation. The balance between the two components is controlled by the number of feasible individuals currently present in the population. These penalties have two major purposes:

1. To further reduce the fitness of infeasible individuals as the penalty imposed by the distance formulation alone is small.
2. To identify the best infeasible individuals in the population by adding different amount of penalty to each infeasible individual's fitness.

The two penalties are formulated for individual $\mathbf{x}$ in the ith-objective function dimension as follows:

$$p_i(\mathbf{x}) = (1 - r_f)X_i(\mathbf{x}) + r_f Y_i(\mathbf{x}), \tag{10}$$

where

$$X_i(\mathbf{x}) = \begin{cases} 0, & \text{if } r_f = 0 \\ v(\mathbf{x}), & \text{otherwise} \end{cases} \tag{11}$$

$$Y_i(\mathbf{x}) = \begin{cases} 0, & \text{if x is a feasible individual} \\ \tilde{f}_i(\mathbf{x}), & \text{if x is an infeasible individual} \end{cases} \tag{12}$$

From the penalty function definition in Equations (10,11-12), we observe that if the feasibility ratio of the population is small (but not zero), then the first penalty ($X_i(\mathbf{x})$) will have more impact than the second penalty ($Y_i(\mathbf{x})$). The first penalty is formulated to have large value for individuals with large amount of constraint violation. Hence, in the case when there are few feasible individuals present in the population (r_f is small), infeasible individuals with higher constraint violation will be more penalized than those with lower constraint violation. On the other hand, if there are many feasible solutions in the population (r_f is large), the second penalty will have more effect than the first one. In this case, infeasible individuals with larger objective function value will be more penalized than infeasible individuals with smaller objective function value. Additionally, if there are no feasible individuals in the population ($r_f = 0$), both penalties will be zero.

The two components of the penalty function allow the algorithm to switch between finding more feasible solutions and finding better solutions at anytime during the evolutionary process. Furthermore, because priority is initially given to the search for feasible individuals, the algorithm is capable of finding feasible solutions in cases where the feasible space is small or discontinuous compared to the search space.

3.3 Final Modified Objective Value Formulation

The final modified objective value of individual **x**, using which non-dominance sorting is performed, is formulated as the sum of the distance measure and penalty function in the ith-objective function dimension,

$$F_i(\mathbf{x}) = d_i(\mathbf{x}) + p_i(\mathbf{x}). \tag{13}$$

This modified objective value formulation is flexible and will allow us to utilize infeasible individuals efficiently and effectively. Most constraint optimization algorithms in literature are "rigid" in a sense that they always prefer certain types of infeasible individuals throughout the entire evolutionary process. For example, they might always give priority to those individuals with small constraint violation only or those individuals with low objective value only. According to our new fitness formulation, the infeasible individuals that are considered valuable are not always similar. Here are some of the interesting properties of this modified objective value formulation:

1. If there is no feasible individual in the current population, each $d_i(\mathbf{x})$ will be equal to the constraint violation ($v(\mathbf{x})$), and each $p_i(\mathbf{x})$ term will be zero. In this case, the objective values of the individuals will be totally disregarded, and all individuals will be compared based on their constraint violation only. This will help us find feasible individuals before looking for optimal solutions.
2. If there are feasible individuals in the population, then individuals with both low objective function values and low constraint violation values will dominate individuals with high objective function values or high constraint violation or both.
3. If two individuals have equal or very close distance values, then the penalty term ($p_i(\mathbf{x})$) determines the dominant individual. According to our penalty formulation, if the feasibility ratio (r_f) in the population is small, then the individual closer to the feasible space will be dominant. On the other hand, the individual with smaller objective function values will be dominant. Otherwise, the two individuals will be non-dominant solutions.
4. If there is no infeasible individual in the population ($r_f = 1$), then individuals will be compared based on their objective function values alone.

After the computation of the modified objective values, the standard features of NSGA-II, such as non-dominant ranking and diversity through crowding distances, will be used based on these modified values. During the archiving process, the best feasible individuals are given priority over infeasible individuals, as the goal of constrained multi-objective optimization is eventually to find feasible optimal solutions. The proposed constraint handling technique is very generic, involving only the modification of fitness function through adaptive penalty measure. It can be easily extended to other MOEAs.

4 Experimental Results and Observations

4.1 Experimental Setup

The algorithm is tested on several constrained multi-objective benchmark problems available from literature. The simulations are conducted with a population size of 100, crossover rate of 0.8, mutation rate of 0.2, and maximum generation number of 100 for all implementations. In addition, we use SBX crossover and mutation. Tournament selection is adopted in a recombination and replacement scheme. These parameters are chosen to be consistent with what were used in other constrained MOEAs for a fair comparison.

Fourteen benchmark problems have been used to test the performance of the proposed algorithm. These problems are all minimization problems and are denoted as BNH [4], SRN [7, 29], OSY [23], TNK [32], CTP1 [10], CTP2 [10], CTP3 [10], CTP4 [10], CTP5 [10], CTP6 [10], CTP7 [10], CTP8 [10], CONSTR [10], and Welded Beam Problem [6]. Each benchmark problem is run 50 times, and the performance metrics are measured statistically. Both quantitative and qualitative comparisons are made to validate the proposed algorithm. For qualitative comparison, the plots of final non-dominated fronts that were obtained from the same initial population are presented. The quantitative comparison is performed using hypervolume indicator and additive epsilon indicator. These two Pareto compliant performance metrics are able to measure the performance of algorithms with respect to their dominance relations and diversity preservation. A detailed discussion about these measures can be found in [20, 37]. The quantitative comparisons are illustrated by statistical box plots, and a Mann-Whitney rank-sum test is implemented to evaluate whether the difference in performance between two independent samples is significant [20].

4.2 Comparative Study

The performance metric for hypervolume indicator (I_H value) is computed for each CMOEA over 50 independent runs. Figures 1, 2 and 3 presents the box plots of I_H indicator found in all CMOEAs, in which 1 is denoted by the proposed algorithm, 2 as the NSGA-II and 3 as the Ray-Tai-Seow's. Higher I_H value indicates the ability of the algorithm to dominate a larger region in the objective space. The

Table 1 The distribution of I_H values tested using Mann-Whitney rank-sum Test [20]. The table presents the p-values with respect to the alternative hypothesis (i.e., p-value $< \alpha$=0.05) for each pair of the proposed algorithm and a selected CMOEA. The distribution of the proposed algorithm has significant differences than those selected CMOEA unless stated. (*) No diference. (**) No feasible solution

Test Functions	I_H (Proposed, NSGA-II)	I_H (Proposed, Ray-Tai-Seow)
BNH	0.0315	4.11E-05
SRN	0.0244	4.11E-05
OSY	0.1081>0.05	4.11E-05
TNK	0.0315	1.65E-04
CTP1	0.0078	0.0476
CTP2	0.04	0.3284>0.05 (*)
CTP3	0.2224>0.05 (*)	4.11E-05
CTP4	0.0051	4.11E-05
CTP5	0.0315	0.0027
CTP6	(**)	0.0228
CTP7	0.04	0.0175
CTP8	0.0625>0.05 (*)	0.7785>0.05 (**)
CONSTR	0.0056	4.11E-05
Welded Beam	0.0078	5.35E-04

figures show that the proposed algorithm has the highest I_H values for the test functions BNH, SRN, TNK, CTP1, CTP4, CTP5, CTP6, CTP7, CONSTR, and Welded Beam. The proposed algorithm and NSGA-II showed comparable I_H values for test problems OSY and CTP3. Similarly, Ray-Tai-Seow's algorithm showed comparable I_H values for CTP2 test problem. All algorithms performed well for CTP8 test problem. NSGA-II showed a higher I_H value than Ray-Tai-Seow for test functions SRN, CTP3, CTP5, CTP7, and welded Beam. On the contrary, Ray-Tai-Seow's showed better I_H value than NSGA-II for test functions OSY, CTP1, CTP2, and CTP8. NSGA-II was not able to find a feasible solution for CTP6 test problem due to the extent of constraints imposed. In some of the problems shown in Figures 1, 2 and 3, it is hard to determine whether the proposed algorithm is significantly better than the other CMOEAs since they attain close I_H values. Hence, the Mann-Whitney rank-sum test is used to examine the distribution of the I_H values. The tested results are presented in Table 1, and they indicate that the proposed algorithm's performance has a significant advantage compared to the distribution in NSGA-II and Ray-Tai-Seow's in most test functions except OSY, CTP2, CTP3, and CTP8. In addition, Figures 1, 2 and 3 shows that the standard deviations for the proposed algorithm are consistently lower, which indicates that the proposed algorithm is more reliable in producing better solutions than those selected CMOEAs.

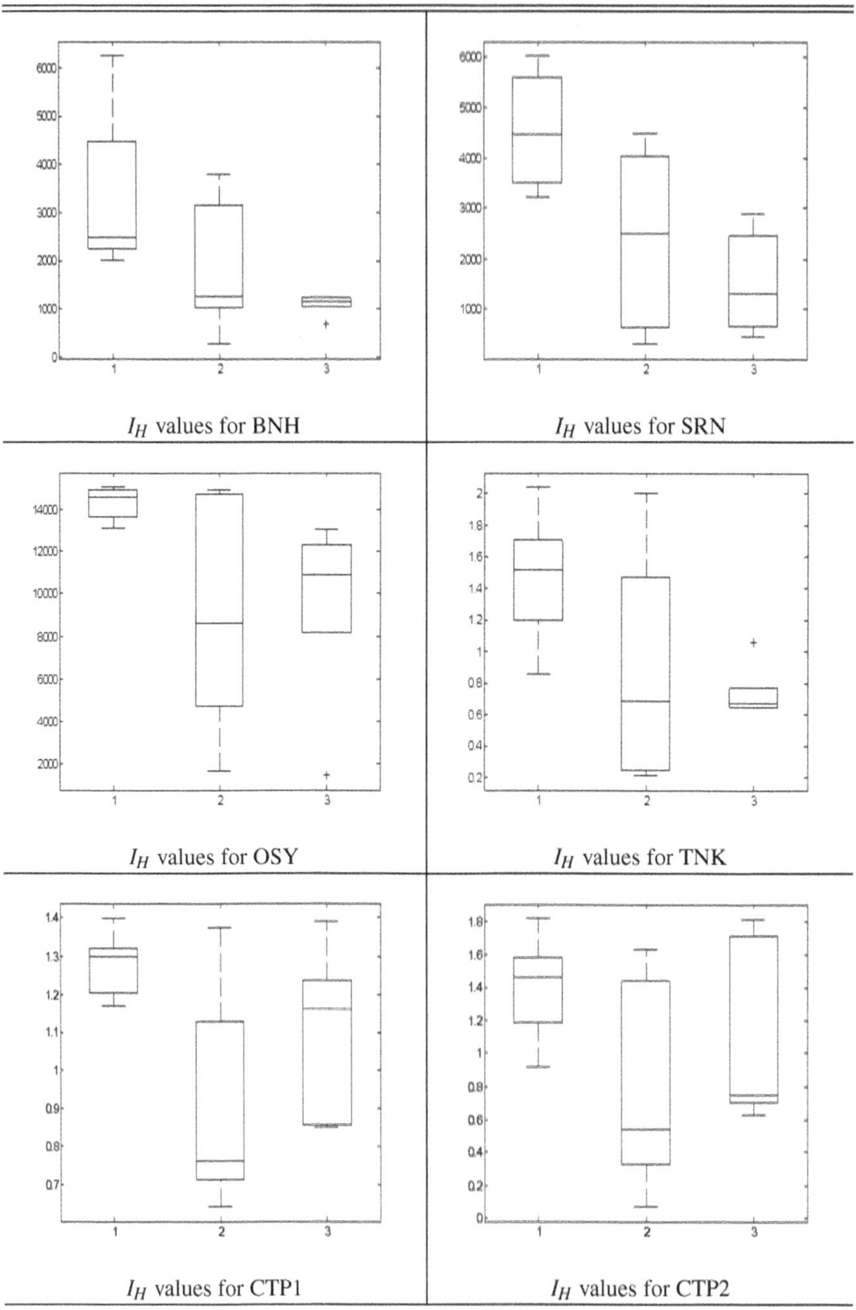

Fig. 1 Box plot of hypervolume indicator (I_H values) for all test functions by algorithms 1-3 represented (in order): the Proposed algorithm, NSGA-II, and Ray-Tai-Seow's (Part I)

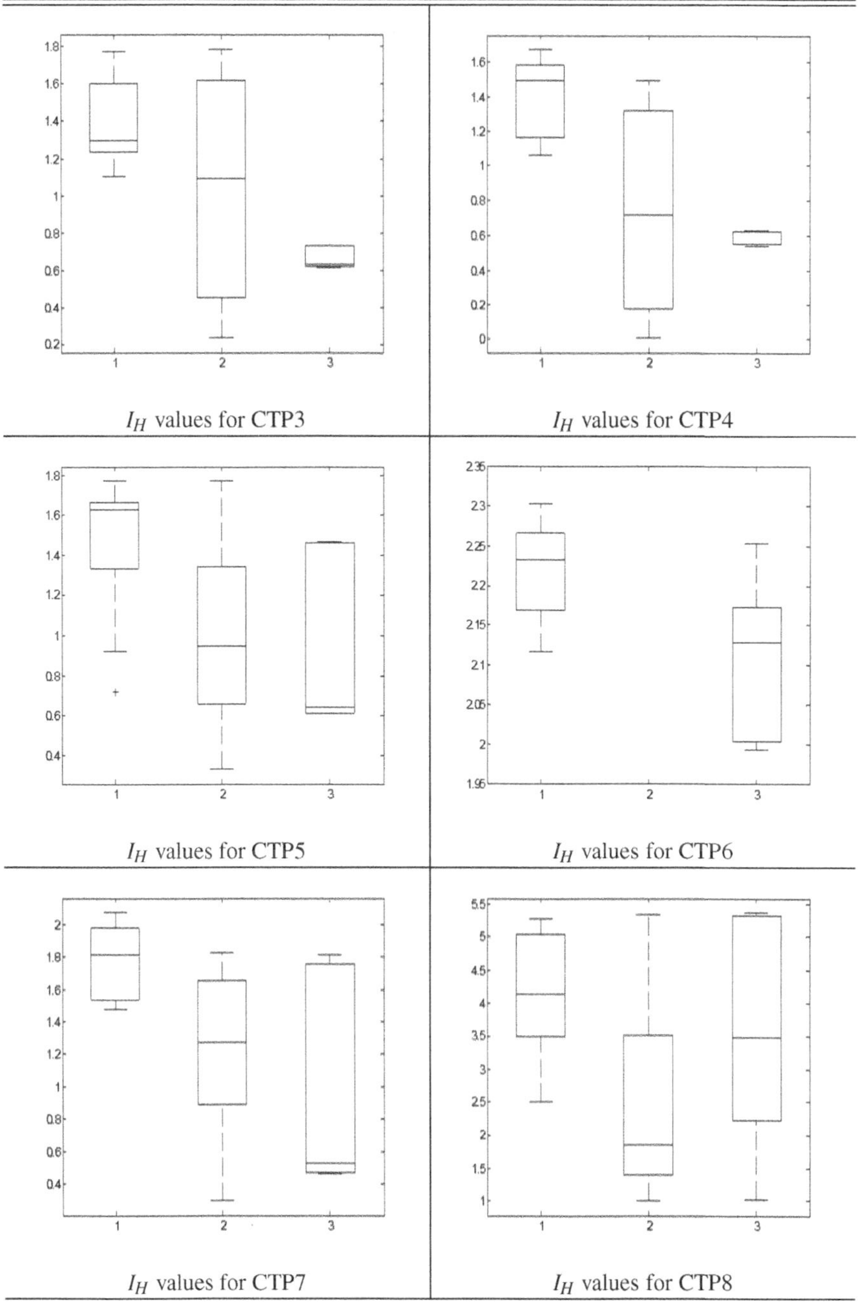

I_H values for CTP3

I_H values for CTP4

I_H values for CTP5

I_H values for CTP6

I_H values for CTP7

I_H values for CTP8

Fig. 2 Box plot of hypervolume indicator (I_H values) for all test functions by algorithms 1-3 represented (in order): the Proposed algorithm, NSGA-II, and Ray-Tai-Seow's (Part II)

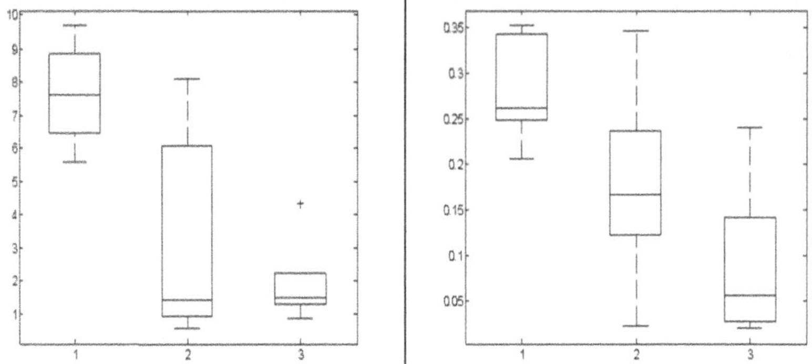

I_H values for CONSTR | I_H values for Welded Beam

Fig. 3 Box plot of hypervolume indicator (I_H values) for all test functions by algorithms 1-3 represented (in order): the Proposed algorithm, NSGA-II, and Ray-Tai-Seow's (Part III)

Table 2 The distribution of Iε+ values tested using Mann-Whitney rank-sum Test [20]. The table presents the p-values with respect to the alternative hypothesis (i.e., p-value $< \alpha$=0.05) for each pair of the proposed and a selected CMOEA. The distribution of the proposed algorithm has significant differences than those selected CMOEA unless stated. (∗) No diference. (∗∗) No feasible solution

Test Functions	Ie+ (Proposed,NSGA-II)	Ie+ (Proposed,Ray-Tai-Seow)
BNH	4.11E-05	8.23E-05
SRN	4.11E-05	4.11E-05
OSY	4.11E-05	4.11E-05
TNK	2.00E-03	1.42E-02
CTP1	0.1359>0.05 (*)	0.3734>0.05 (*)
CTP2	4.11E-05	4.11E-05
CTP3	4.11E-05	1.65E-04
CTP4	0.4755>0.05 (*)	1.65E-04
CTP5	7.82E-04	0.0181
CTP6	(**)	8.23E-05
CTP7	1.65E-04	2.88E-04
CTP8	0.5457>0.05 (*)	0.1903>0.05 (*)
CONSTR	5.60E-03	2.60E-03
Welded Beam	8.23E-05	1.65E-04

Figures 4, 5, 6, 7 and 8 illustrate the results of additive epsilon indicator using statistical box plots. There are two box plots for each test problem, i.e., $I_{\varepsilon+}(A,X_{1,2})$ and $I_{\varepsilon+}(X_{1,2},A)$, in which algorithm A is referred to as the proposed design, while

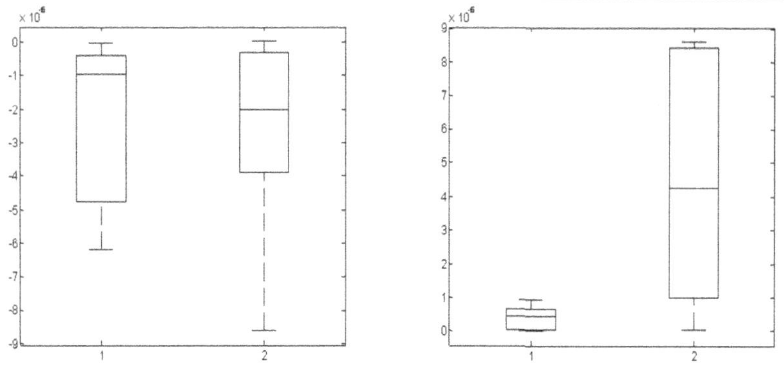

$I_{\varepsilon+}(A, X_{1,2})$ and $I_{\varepsilon+}(X_{1,2}, A)$ for BNH

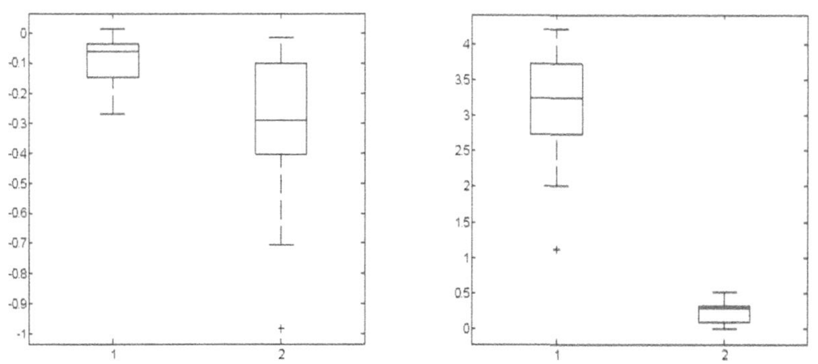

$I_{\varepsilon+}(A, X_{1,2})$ and $I_{\varepsilon+}(X_{1,2}, A)$ for SRN

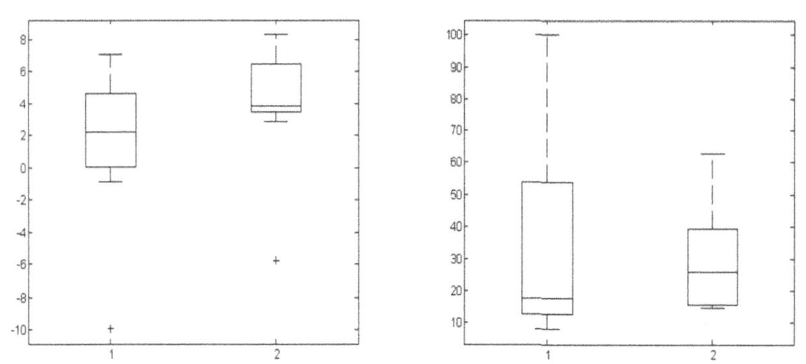

$I_{\varepsilon+}(A, X_{1,2})$ and $I_{\varepsilon+}(X_{1,2}, A)$ for OSY

Fig. 4 Box plots of additive epsilon indicator (Iε+ values) ('A' corresponds to the proposed algorithm, while 'X1, 2 ' refers to NSGA-II and Ray-Tai-Seow's, respectively). Part I

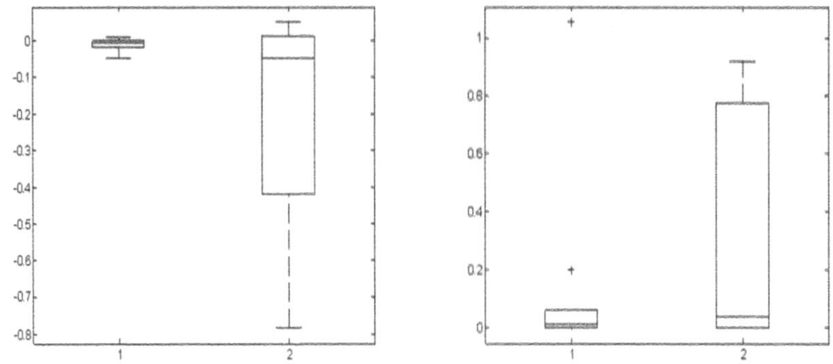

$I_{\varepsilon+}(A, X_{1,2})$ and $I_{\varepsilon+}(X_{1,2}, A)$ for TNK

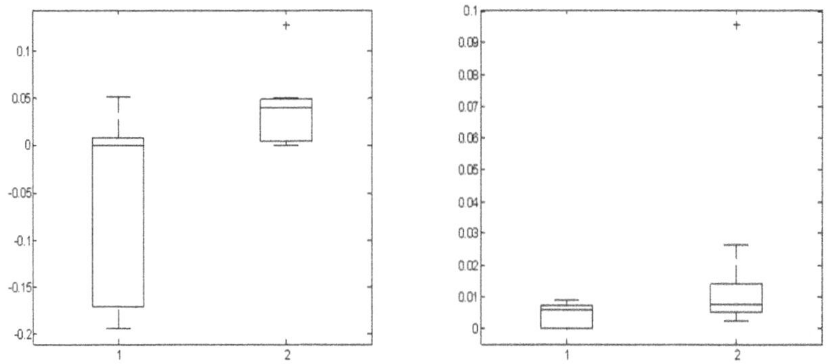

$I_{\varepsilon+}(A, X_{1,2})$ and $I_{\varepsilon+}(X_{1,2}, A)$ for CTP1

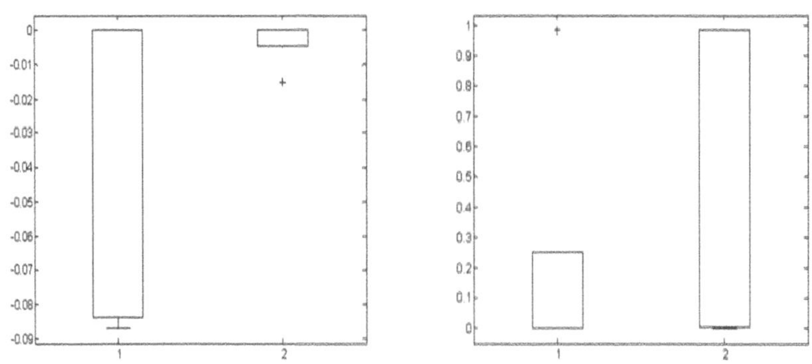

$I_{\varepsilon+}(A, X_{1,2})$ and $I_{\varepsilon+}(X_{1,2}, A)$ for CTP2

Fig. 5 Box plots of additive epsilon indicator (Iε+ values) ('A' corresponds to the proposed algorithm, while 'X1, 2 ' refers to NSGA-II and Ray-Tai-Seow's, respectively). Part II

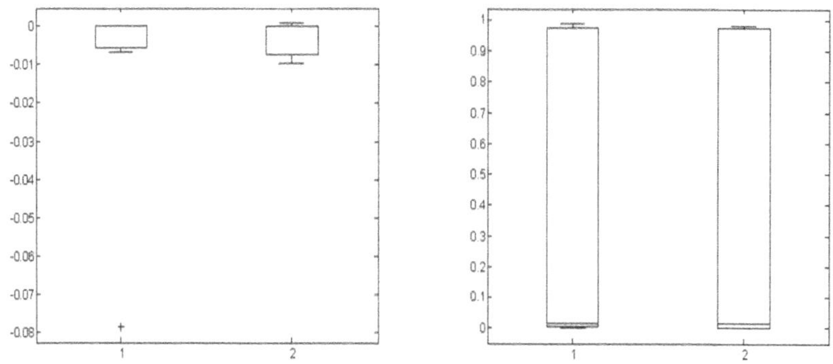

$I_{\varepsilon+}(A, X_{1,2})$ and $I_{\varepsilon+}(X_{1,2}, A)$ for CTP3

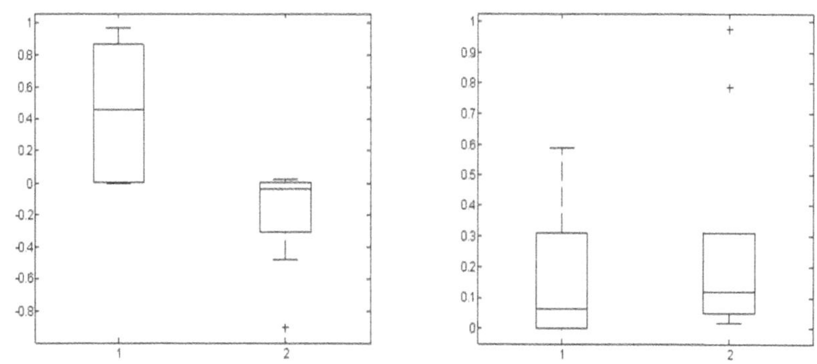

$I_{\varepsilon+}(A, X_{1,2})$ and $I_{\varepsilon+}(X_{1,2}, A)$ for CTP4

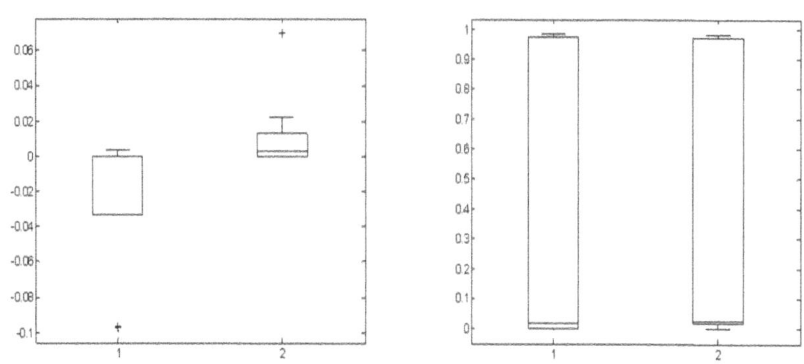

$I_{\varepsilon+}(A, X_{1,2})$ and $I_{\varepsilon+}(X_{1,2}, A)$ for CTP5

Fig. 6 Box plots of additive epsilon indicator (Iε+ values) ('A' corresponds to the proposed algorithm, while 'X1, 2 ' refers to NSGA-II and Ray-Tai-Seow's, respectively). Part III

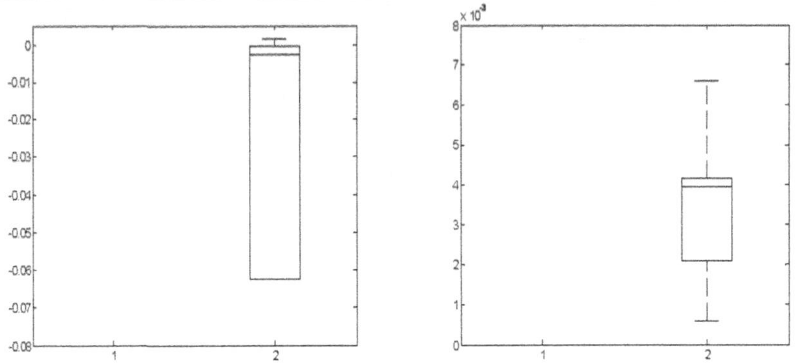

$I_{\varepsilon+}(A, X_{1,2})$ and $I_{\varepsilon+}(X_{1,2}, A)$ for CTP6

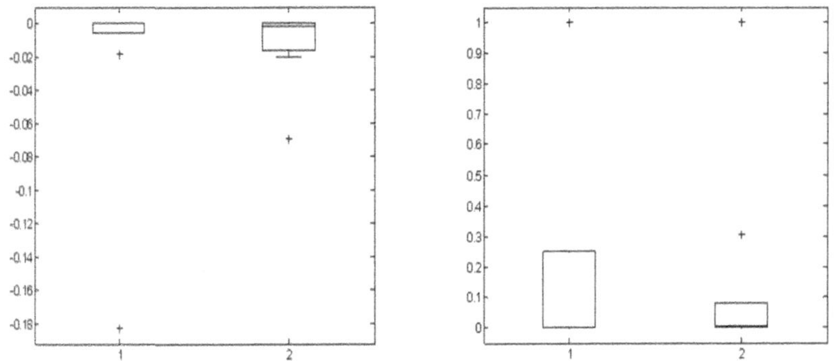

$I_{\varepsilon+}(A, X_{1,2})$ and $I_{\varepsilon+}(X_{1,2}, A)$ for CTP7

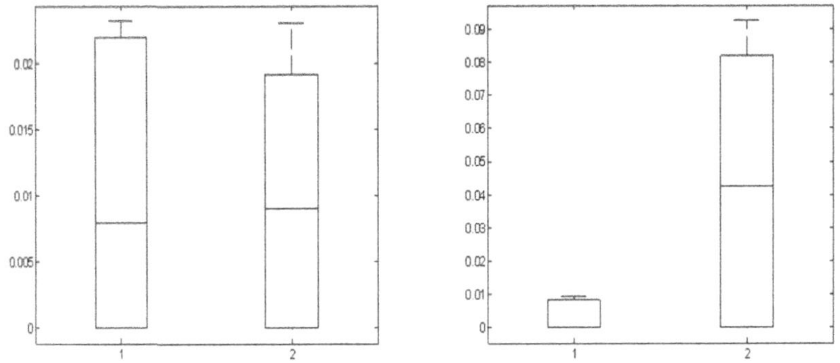

$I_{\varepsilon+}(A, X_{1,2})$ and $I_{\varepsilon+}(X_{1,2}, A)$ for CTP8

Fig. 7 Box plots of additive epsilon indicator (Iε+ values) ('A' corresponds to the proposed algorithm, while 'X1, 2 ' refers to NSGA-II and Ray-Tai-Seow's, respectively). Part IV

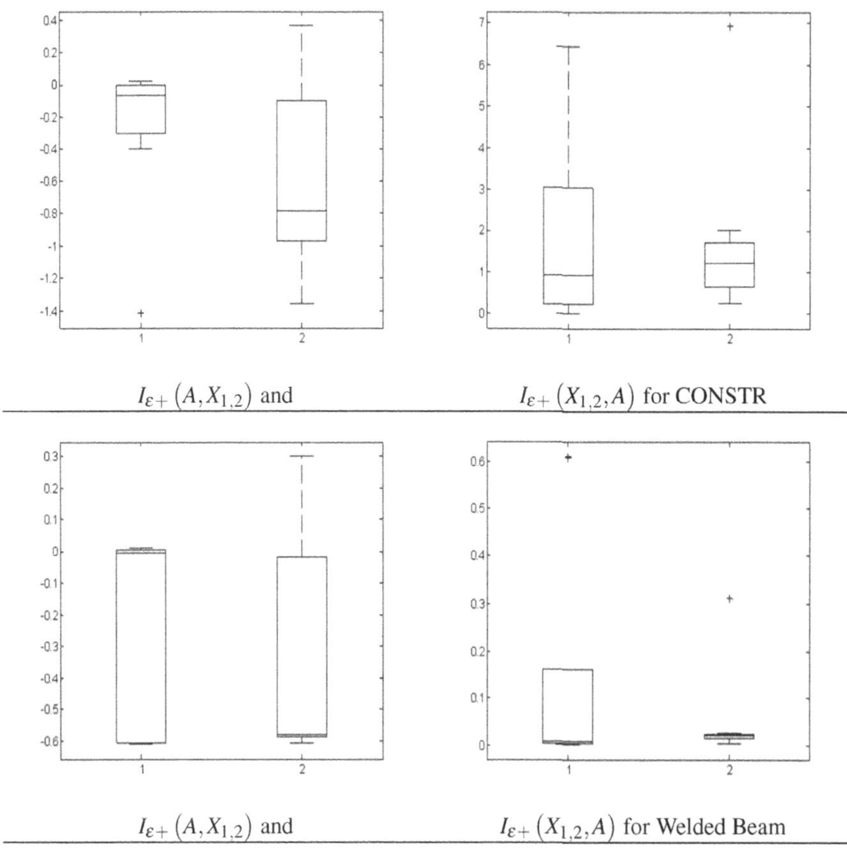

Fig. 8 Box plots of additive epsilon indicator (Iε+ values) ('A' corresponds to the proposed algorithm, while 'X1, 2 ' refers to NSGA-II and Ray-Tai-Seow's, respectively). Part V

algorithms X_1 and X_2 represent NSGA-II and Ray-Tai-Seow's, respectively. It seems that the proposed algorithm performs relatively better with respect to dominance relation than most of the CMOEAs for all functions except OSY, CTP2, CTP4, CTP5, and CTP8. For example, Figures 4, 5, 6, 7 and 8 show that the proposed algorithm strictly dominates NSGA-II on Welded Beam problem because $I_{\varepsilon+}(A, X_1) \leq 0$ and $I_{\varepsilon+}(X_1, A) > 0$. On the other hand, the box plot on CTP8 in Figures 4, 5, 6, 7 and 8 may indicate that the proposed algorithm does not strictly dominate NSGA-II because $I_{\varepsilon+}(A, X_1) > 0$ and $I_{\varepsilon+}(X_1, A) > 0$.

In summary, NSGA-II and the proposed algorithm showed no difference for CTP4 test function; Ray-Tai-Seow's and the proposed algorithm showed comparable results for CTP2 and CTP5 test functions; and finally the proposed algorithm seems to perform as well as NSGA-II and Ray-Tai-Seow's for functions OSY and CTP8. For the rest of the test functions, the proposed algorithm showed better performance compared to the other CMOEAs. Moreover, we can observe that the proposed algorithm has relatively lower standard deviations, which are consistent with those shown in Figures 1, 2 and 3. For further analysis, the distributions of $I_{\varepsilon+}$ values are analyzed via the Mann-Whitney rank-sum test, which are presented in Table 2. In general, results in Table 2 and Figures 4, 5, 6, 7 and 8 confirm that the proposed algorithm is significantly better than most or even all of the CMOEAs on all benchmark test problems in terms of the chosen performance metrics.

The success of the proposed algorithm is mainly due to the exploitation of the evolutionary information contained in infeasible individuals in addition to that contained in feasible individuals. The constraint handling normally used in NSGA-II [11] compares infeasible individuals solely based on their constraint violation. This way of non-dominance ranking ignores how well each individual performed in the objective space and may result in the inefficient use of some evolutionary materials. The proposed algorithm, on the other hand, uses a combined measure of constraint violation and objective performance to arrive at the fitness of individuals that will govern the evolutionary process. The number of feasible individuals available in the current population is used to control the relative emphasis given to either constraint violation or objective performance in the final fitness calculation. As can be observed from the test results of the proposed algorithm, this way of fitness formulation provides better solutions compared to other constrained multiobjective evolutionary algorithms.

5 Conclusions

In this chapter, we propose an adaptive constraint handling technique for solving constrained multi-objective optimization problems. Besides the search for optimal solutions in the feasible region, the algorithm also exploits the information hidden in infeasible individuals with better objective values and lower constraint violation. This is achieved by using the modified objective values in the non-dominance ranking of the multi-objective evolutionary algorithm. The modified objective values incorporate the effects of the individuals' constraint violation. They are composed of distance measures and penalty functions. These values are associated with how well an individual performs and how much it violates the constraints. They are obtained for every objective function dimension. For feasible individuals, the distance values are just the normalized objective function values. For infeasible individuals, the distance values are obtained from the normalized objective function values and their constraint violation. The penalty function, on the other hand, will be applied to infeasible individuals in order to further decrease their fitness compared to feasible individuals. The number of feasible individuals in the population adaptively controls the emphasis given to objective values or constraint violation in the modified objective function formulation. If there is no feasible individual in the population, the algorithm uses the constraint violations as the primary means to rank the individuals. This adaptive formulation allows further exploitation of the evolutionary information possessed by infeasible individuals with low objective values and low constraint violation. Involving the infeasible individuals in the evolutionary process helps the algorithm to find additional feasible individuals, even in cases where the feasible space is very small and discontinuous. Furthermore, since there is no parameter tuning, this makes the algorithm easy to implement. Moreover, the additional evaluations are simple arithmetic operations and do not impose any significant increase in the computational cost.

The proposed constraint handling technique is implemented on NSGA-II simply due to its popularity as an MOEA. The proposed constraint handling technique can be easily extended to other multi-objective evolutionary algorithms. The performance of the algorithm was tested on fourteen constrained multi-objective test problems. From the simulation results, it is observed that the algorithm is capable of finding better-fit feasible solutions that are well spread over the Pareto front in all the runs of all test problems. In addition, the results of the algorithm are compared with some of the constrained multi-objective algorithms suggested so far. The comparison results indicate that the proposed algorithm performs better than the other algorithms in that it is able to provide a well distributed Pareto front that has optimal individuals. For future work, the authors recommend applying the proposed constraint handling technique using modified objective function formulation to other multi-objective evolutionary approaches, such as SPEA2.

References

1. Angantyr, A., Andersson, J., Aidanpaa, J.: Constrained Optimization based on a Multiobjective Evolutionary Algorithms. In: Proceedings of the Congress on Evolutionary Computation 2003 (CEC 2003), Canberra, Australia, December 2003, pp. 1560–1567. IEEE Service Center (2003)
2. Bäck, T., Hoffmeister, F., Schwefel, H.: A Survey of Evolution Strategies. In: Belew, R.K., Booker, L.B. (eds.) Proceedings of the Fourth International Conference on Genetic Algorithms, San Mateo, California, pp. 2–9. Morgan Kaufmann Publishers, San Francisco (1991)
3. Bean, J.C., Hadj-Alouane, A.B.: A Dual Genetic Algorithm for Bounded Integer Programs. Technical Report TR 92-53, Technical Report TR 92-53, Department of Industrial and Operations Engineering, The University of Michigan, Ann Arbor, MI, The University of Michigan (1992); to appear in R.A.I.R.O.-R.O. (invited submission to special issue on GAs and OR)
4. Binh, T., Korn, U.: Mobes: A multiobjective evolution strategy for constrained optimization problems. In: Proceedings of the third international Conference on Genetic Algorithms, East Lansing, MI, pp. 176–182 (1997)
5. Cai, Z., Wang, Y.: A multiobjective optimization-based evolutionary algorithm for constrained optimization. IEEE Trans. Evolutionary Computation 10(6), 658–675 (2006)
6. Chafekar, D., Xuan, J., Rasheed, K.: Constrained multi-objective optimization using steady state genetic algorithms. In: Proceedings of Genetic and Evolutionary Computation Conference, pp. 813–824. Springer, Heidelberg (2003)
7. Chankong, V., Haimes, Y.: Multiobjective Decision Making: Theory and Methodology. In: Sage, A.P. (ed.) Systems Science and Engineering, vol. 8, pp. 62–109. North-Holland, Amsterdam (1983)

8. Coello, C., Christiansen, A.: Moses: a multi-objective op-timization tool for engineering design. Engineering Optimization 31(3), 337–368 (1999)
9. Deb, K.: An Efficient Constraint Handling Method for Genetic Algorithms. Computer Methods in Applied Mechanics and Engineering 186(2/4), 311–338 (2000)
10. Deb, K.: Multi-Objective Optimization Using Evolutionary Algorithms. Wiley, Chichester (2001)
11. Deb, K., Pratap, A., Agarwal, S., Meyarivan, T.: A Fast and Elitist Multiobjective Genetic Algorithm: NSGA–II. IEEE Transactions on Evolutionary Computation 6(2)(2), 182–197 (2002)
12. Farmani, R., Wright, J.A.: Self-Adaptive Fitness Formulation for Constrained Optimization. IEEE Transactions on Evolutionary Computation 7(5), 445–455 (2003)
13. Fonseca, C., Fleming, P.: Genetic Algorithms for Multiobjective Optimization: Formulation, Discussion and Generalization. In: Forrest, S. (ed.) Proceedings of the Fifth International Conference on Genetic Algorithms, San Mateo, California, pp. 416–423. Morgan Kauffman Publishers, San Francisco (1993)
14. Fonseca, C., Fleming, P.: Multiobjective optimization and multiple constraint handling with evolutionary algorithms—Part I: A unified formulation. IEEE Transactions on Systems, Man, and Cybernetics, Part A: Systems and Humans 28(1), 26–37 (1998)
15. Goldberg, D.E.: Genetic Algorithms in Search, Optimization and Machine Learning. Addison-Wesley, Reading (1989)
16. Harada, K., Sakuma, J., Ono, I., Kobayashi, S.: Constraint-handling method for multi-objective function optimization: Pareto descent repair operator. In: Obayashi, S., Deb, K., Poloni, C., Hiroyasu, T., Murata, T. (eds.) EMO 2007. LNCS, vol. 4403, pp. 156–170. Springer, Heidelberg (2007)
17. Homaifar, A., Lai, S.H.Y., Qi, X.: Constrained Optimization via Genetic Algorithms. Simulation 62(4), 242–254 (1994)
18. Jiménez, F., Gómez-Skarmeta, A., Sánchez, G., Deb, K.: An Evolutionary Algorithm for Constrained Multi-objective Optimization. In: Proceedings of the Congress on Evolutionary Computation 2002 (CEC 2002), Piscataway, New Jersey, May 2002, vol. 2, pp. 1133–1138. IEEE Service Center (2002)
19. Joines, J., Houck, C.: On the use of non-stationary penalty functions to solve nonlinear constrained optimization problems with GAs. In: Fogel, D. (ed.) Proceedings of the first IEEE Conference on Evolutionary Computation, Orlando, Florida, pp. 579–584. IEEE Press, Los Alamitos (1994)
20. Knowles, J., Thiele, L., Zitzler, E.: A tutorial on performance assess-ment of stochastic multiobjective optimizers. Technical report, TIK-Report No. 214 (revised version), Switzerland Computer Engineering and Networks Laboratory (TIK), Swiss Federal Institute of Technology (ETH), Munich, Switzerland, pp. 1–35 (2006)
21. Knowles, J.D., Corne, D.W.: Approximating the Nondominated Front Using the Pareto Archived Evolution Strategy. Evolutionary Computation 8(2), 149–172 (2000)
22. Mezura, E., Coello-Coello, C.A.: A survey of constraint-handling techniques based on evolutionary multiobjective optimization. Technical report, Technical Report EVOCINV-04-2006, Evolutionary Computation Group at CINVESTAV, Departamento de Computacion, CINVESTAV-IPN, México (2006)
23. Osyczka, A., Kundu, S.: A new method to solve generalized multicriteria optimization problems using the simple genetic algorithm. Structural Optimization 10(2), 94–99 (1995)

24. Oyama, A., Shimoyama, K., Fujii, K.: New constraint-handling method for multi-objective and multi-constraint evolutionary optimization. Transactions of the Japan Society for Aeronautical and Space Sciences 50(167), 56–62 (2007)
25. Raviguururajan, T.S., Bergles, A.E., Reid, D.J.: Genetic algorithm in constrained optimization. Mathematical and Computer Modeling 23(5), 87–111 (1996)
26. Ray, T., Tai, K., Seow, K.: An evolutionary algorithm for multiobjective optimization. Engineering Optimization 33(3), 399–424 (2001)
27. Runarsson, T., Yao, X.: Stochastic Ranking for Constrained Evolutionary Optimization. IEEE Transactions on Evolutionary Computation 4(3), 284–294 (2000)
28. Runarsson, T., Yao, X.: Search biases in constrained evolutionary optimization. IEEE Transactions on Systems, Man, and Cybernetics Part C- Applications and Reviews 35(2), 233–243 (2005)
29. Srinivas, N., Deb, K.: Multiobjective Optimization Using Nondominated Sorting in Genetic Algorithms. Evolutionary Computation 2(3), 221–248 (Fall 1994)
30. Takahama, T., Sakai, S.: Constrained Optimization by Applying the α Constrained Method to the Nonlinear Simplex Method with Mutations. IEEE Transactions on Evolutionary Computation 9(5), 437–451 (2005)
31. Takahama, T., Sakai, S.: Constrained Optimization by the ε Constrained Differential Evolution with Gradient-Based Mutation and Feasible Elites. In: 2006 IEEE Congress on Evolutionary Computation, Vancouver, BC, Canada, July 2006, pp. 308–315. IEEE, Los Alamitos (2006)
32. Tanaka, M., Watanabe, H., Furukawa, Y., Tanino, T.: GA-based decision support system for multicriteria optimization. In: 1995 IEEE International Conference on Systems, Man and Cybernetics. Intelligent Systems for the 21st Century (Cat. No. 9 5CH3576-7), New York, NY, USA, vol. 2, pp. 1556–1561. IEEE, Los Alamitos (1995)
33. Tessema, B., Yen, G.: A Self Adaptative Penalty Function Based Algorithm for Constrained Optimization. In: Proceedings of IEEE Congress on Evolutionary Computation, Vancouver, Canada, July 2006, pp. 950–957. IEEE, Los Alamitos (2006)
34. Venkatraman, S., Yen, G.G.: A Generic Framework for Constrained Optimization Using Genetic Algorithms. IEEE Transactions on Evolutionary Computation 9(4), 424–435 (2005)
35. Young, N.: Blended ranking to cross infeasible regions in constrained multi-objective problems. In: International Conference on Computational Intelligence for Modelling Control and Automation, International Conference on Intelligent Agents, Web Technologies and Internet Commerce, Austria, November 28-30, 2005, pp. 191–196 (2005)
36. Zitzler, E., Thiele, L.: An evolutionary algorithm for multi-objective op-timization: the strength pareto approach. Technical report, Technical Report, Switzerland Computer Engineering and Networks Laboratory (TIK), Swiss Federal In-stitute of Technology (ETH), Munich, Switzerland (1998)
37. Zitzler, E., Thiele, L., Laumanns, M., Foneseca, C.M., Grunert da Fonseca, V.: Performance Assessment of Multiobjective Optimizers: An Analysis and Review. IEEE Transactions on Evolutionary Computation 7(2), 117–132 (2003)

Infeasibility Driven Evolutionary Algorithm for Constrained Optimization

Tapabrata Ray, Hemant Kumar Singh, Amitay Isaacs, and Warren Smith

Abstract. Real life optimization problems often involve one or more constraints and in most cases, the optimal solutions to such problems lie on constraint boundaries. The performance of an optimization algorithm is known to be largely dependent on the underlying mechanism of constraint handling. Most population based stochastic optimization methods prefer a feasible solution over an infeasible solution during their course of search. Such a preference drives the population to feasibility first before improving its objective function value which effectively means that the solutions approach the constraint boundaries from the feasible side of the search space. In this chapter, we introduce an evolutionary algorithm that explicitly maintains a small percentage of infeasible solutions close to the constraint boundaries during its course of evolution. The presence of marginally infeasible solutions in the population allows the algorithm to approach the constraint boundary from the infeasible side of the search space in addition to its approach from the feasible side of the search space via evolution of feasible solutions. Furthermore, "good" infeasible solutions are ranked higher than the feasible solutions, thereby focusing the search for the optimal solutions near the constraint boundaries. The performance of the proposed algorithm is compared with Non-dominated Sorting Genetic Algorithm II (NSGA-II) on a set of single and multi-objective test problems. The results clearly indicate that the rate of convergence of the proposed algorithm is better than NSGA-II on the studied test problems. Additionally, the algorithm provides a set of marginally infeasible solutions which are of great use in trade-off studies.

Keywords: Evolutionary Algorithm, Constrained Handling, Multi-objective Optimization, NSGA-II.

1 Introduction

In real life, one often encounters problems where he/she has to optimize one or more objectives simultaneously, subject to a set of constraints. In single objective

Tapabrata Ray, Hemant Kumar Singh, Amitay Isaacs, and Warren Smith
University of New South Wales, Australian Defence Force Academy, Northcott Drive, Canberra, ACT 2600, Australia
e-mail: {t.ray,h.singh,a.isaacs,w.smith}@adfa.edu.au

E. Mezura-Montes (Ed.): Constraint-Handling in Evolution. Optimi., SCI 198, pp. 145–165.
springerlink.com

optimization problems, the aim is to find one or more solutions which have the best objective function value; whereas in a multi-objective optimization (MO) problem, the aim is to arrive at a set of non-dominated solutions close enough to the Pareto Optimal Set (POS). Population based stochastic optimization algorithms are a preferred choice for MO problems since they are able to generate the POS in a single run. A number of population based stochastic algorithms such as Evolutionary Algorithms (EAs), Differential Evolution (DE), Particle Swarm Optimization (PSO) etc. have been proposed in literature to deal with single and multi-objective optimization problems.

The performance of stochastic optimization methods for constrained optimization problems is known to be largely dependent on the mechanism used for handling constraints. A detailed review of various constraint handling techniques used with evolutionary algorithms is presented in [4, 29]. Penalty function based methods are the most commonly adopted form where the fitness of an infeasible solution is degraded by a weighted sum of the constraint violations. Variants of penalty function based approaches include static penalty models [20, 25], dynamic penalty models [23], annealing penalty models [28, 31], adaptive penalty models [2, 13] or death penalty models [19]. The implementation of any of the above schemes often requires additional parameters and the result of optimization is known to be highly sensitive to the choice of such parameters.

In an attempt to alleviate the problems associated with penalty factors, a number of alternate approaches have been proposed. These include special representation schemes to maintain feasibility [7, 30], use of repair algorithms [32, 34, 43, 44], handling constraints and objective separately [38], incorporation of heuristic rules such as linear ranking [35] and binary tournament [8] to compare individuals in the population. Main drawbacks of the above approaches include the need to develop problem specific repair mechanisms, unavailability of a feasible starting point, and early loss of diversity.

Dominance based approaches have also been proposed to deal with constraints. Ray *et al.* [36] developed an evolutionary algorithm based on non-dominance of solutions in the objective and the constraint space. Ho and Shimizu [18] converted the objective function value and the constraint violation into numerical values with the same order of magnitude. Concepts of dominance have also been used in recent simulated annealing based optimization algorithms:- Hedar and Fukushima [16], and Singh *et al.* [37]. However, simulated annealing based models are known to require a number of associated parameters that need to be identified through trials. A comparison of various Multi-Objective Evolutionary Algorithms (MOEAs) on constrained optimization (single-objective non-linear problems) using concepts of Pareto-dominance can be found in [27].

Most of the evolutionary optimization algorithms focus on generating the best feasible solution of the problem and hence a feasible solution is always considered better than an infeasible solution during the course of the evolution. This fundamental assumption always drives the population towards feasibility. For many constrained optimization problems, the optimal solutions are likely to lie on a constraint boundary. In reality a designer is often interested to look at the solutions that might be marginally infeasible. Hence, a marginally infeasible solution near the constraint boundary is more desirable than a poor feasible solution away from the constraint boundary. The search can be focused around the constraint boundary by treating the constraints as additional objectives and the constraint violation as its fitness. For single objective optimization, Coello Coello proposed to split the population into various sub-populations, each sub-population using either the objective or one of the constraints as the fitness function [3]. For multi-objective problems Vieira *et al.* have used constraints an additional objectives together with a modified Parks & Miller elitist technique [40, 41]. In the above approach a significant amount of time is spent on non-dominated sorting due to a large number of constraints. There is also a risk of generating solutions with excellent objective function values but with poor constraint satisfaction.

Few researchers have proposed maintaining a proportion of infeasible solutions in the population during the evolution. Hamida and Schoenauer [14] developed an adaptive segregational algorithm (ASCHEA), where the proportion of feasible solutions in the population was controlled using an adaptive penalty. The approach used single penalty coefficient for all constraints, and was later extended [15] to incorporate a separate penalty coefficient for each constraint. Hinterding and Michalewicz [17] proposed another approach (CONGA) for constraint handling using effective parent matching, where mating was done between two infeasible solutions satisfying different constraints to create children which would satisfy all constraints. Mezura-Montes and Coello Coello [26] suggested a simple multimembered evolutionary strategy (SMES) where the "best" infeasible solution determined by its objective and function value is allowed to be copied into next generation. Though these algorithms (ASCHEA, CONGA, and SMES) effectively illustrated the benefits of preserving infeasible solutions in the population, their scope was demonstrated on single-objective optimization problems only. An extension to multi-objective domain was not discussed.

In an attempt to simultaneously generate solutions to unconstrained and constrained optimization formulations of a multi-objective problem, Isaacs, Ray and Smith [21] introduced a Constraint Handling Evolutionary Algorithm (CHEA). The evolutionary process used in CHEA is similar to that of NSGA-II [12], but a part of the population is maintained infeasible during the search. The infeasible solutions in the population are ranked using the original objectives along with an additional objective, the number of constraint violations. The incorporation of search through infeasible space improves the efficiency of the algorithm. However, the algorithm does not have any provisions to quantify the amount of constraint violation and the infeasible solutions obtained are not suitable for the trade-off studies.

In this chapter, an evolutionary algorithm is proposed to handle constrained optimization problems along the lines of CHEA. The proposed algorithm is aimed to deliver (a) The set of Optimal Solutions (best function value in case of single-objective and Pareto-Optimal set for multi-objective problems), (b) a few *marginally* infeasible solutions for trade-off studies and (c) an improvement in the rate of convergence. Usually, there are no defining limits on the absolute values of the constraint violations, the term *marginal* is used to denote the solutions that have relatively small constraint violation among the members of the population, and lie close to the constraint boundary. The proposed approach is a modified form of CHEA [21], and is referred as Infeasibility Driven Evolutionary Algorithm (IDEA). The performance of the algorithm is compared with NSGA-II on a number of single objective optimization problems (g-series problems [24, 33]) and multi-objective optimization problems (CTP series problems [9]). The rest of the chapter is organized as follows. The proposed algorithm (IDEA) is described in Section 2. The quantification of constraint violation is discussed in Section 3. In Section 4, the details of the numerical experiments performed are given, followed by the results in Section 5. Finally, the findings of our studies are summarized in Section 6.

2 Infeasibility Driven Evolutionary Algorithm (IDEA)

A multi-objective optimization problem can be formulated as shown in Eq. 1.

$$\begin{aligned} \text{Minimize} \quad & f_1(\mathbf{x}), \ldots, f_k(\mathbf{x}) \\ \text{Subject to} \quad & g_i(\mathbf{x}) \geq 0, \quad i = 1, \ldots, m \end{aligned} \tag{1}$$

where $\mathbf{x} = (x_1, \ldots, x_n)$ is the design variable vector bounded by lower and upper bounds $\mathbf{x} \in \mathscr{S} \subset \mathbf{R}^n$. A single objective optimization problem follows the same formulation with $k = 1$.

The optimal solutions of the constrained optimization problems often lie along the constraint boundary. To effectively search along the constraint boundary, the original k objective constrained optimization problem is reformulated as $k+1$ objective unconstrained optimization problem as given in Equation 2. The first k objectives are the same as in the original constrained problem. The additional objective represents a measure of constraint violation, which is referred to as "violation measure" in our studies.

$$\begin{aligned} \text{Minimize} \quad & f_1'(\mathbf{x}) = f_1(\mathbf{x}), \ldots, f_k'(\mathbf{x}) = f_k(\mathbf{x}) \\ & f_{k+1}'(\mathbf{x}) = \text{Violation measure} \end{aligned} \tag{2}$$

The main steps of IDEA are outlined in Algorithm 1. As in NSGA-II, an offspring population is evolved from parents selected by binary tournament using crossover and mutation. IDEA uses simulated binary crossover (SBX) [10] as given in Eq. 3.

$$\begin{aligned} y_i^1 &= 0.5\left[(1+\beta_{q_i})x_i^1 + (1-\beta_{q_i})x_i^2\right] \\ y_i^2 &= 0.5\left[(1-\beta_{q_i})x_i^1 + (1+\beta_{q_i})x_i^2\right] \end{aligned} \tag{3}$$

where β_{q_i} is calculated as,

$$\beta_{q_i} = \begin{cases} (2u_i)^{1/\eta_c+1}, & \text{if } u_i \leq 0.5, \\ \left(\frac{1}{2(1-u_i)}\right)^{1/\eta_c+1} & \text{if } u_i > 0.5. \end{cases} \tag{4}$$

Algorithm 1. Infeasibility Driven Evolutionary Algorithm (IDEA)

Require: N {Population Size}
Require: $N_G > 1$ {Number of Generations}
Require: $0 < \alpha < 1$ {Proportion of infeasible solutions}
$N_{inf} = \alpha * N$
$N_f = N - N_{inf}$
pop_1 = Initialize()
Evaluate(pop_1)
for $i = 2$ to N_G **do**
  $childpop_{i-1}$ = Evolve(pop_{i-1})
  Evaluate($childpop_{i-1}$)
  (S_f, S_{inf}) = Split($pop_{i-1} + childpop_{i-1}$)
  Rank(S_f)
  Rank(S_{inf})
  $pop_i = S_{inf}(1, N_{inf}) + S_f(1, N_f)$
end for

and where u_i is the uniform random number in the range $[0,1)$ and η_c is the user defined parameter *Distribution for Crossover.* A polynomial mutation operator has been used in this study [11] defined in Eq. 5.

$$y_i = x_i + (\overline{x_i} - \underline{x_i})\,\bar{\delta}_i \tag{5}$$

where $\bar{\delta}_i$ is calculated as,

$$\bar{\delta}_i = \begin{cases} (2r_i)^{1/(\eta_m+1)} - 1, & \text{if } r_i < 0.5, \\ 1 - [2(1-r_i)]^{1/(\eta_m+1)}, & \text{if } r_i \geq 0.5. \end{cases} \tag{6}$$

and where r_i is the uniform random number in the range $[0,1)$ and η_m is the user defined parameter *Distribution Index for mutation.*

For preserving diversity among the population members, crowding distance sorting [12] has been used, which is calculated as shown in Algorithm 2. The main difference between NSGA-II and the IDEA is the mechanism for elite preservation. In IDEA, a few infeasible solutions are retained in the population at every generation. Individual solutions in the population are evaluated as per the original problem definition (Eq. 1) and marked infeasible if any of the constraints are violated. The solutions of the parent and the offspring population are divided into a feasible set (S_f) and an infeasible set (S_{inf}). The solutions in the feasible and the infeasible sets are both ranked using non-dominated sorting and crowding distance sorting of $k+1$ objectives. NSGA-II, on the other hand, uses non-dominated sorting and crowding distance for ranking feasible solutions and ranks infeasible solutions in the increasing value of maximum constraint violation. For the feasible solutions, non-dominated

Algorithm 2. Crowding distance mechanism

Require: F {Non-dominated set}
$N_s = |F|$ {Number of solutions in the non-dominated set}
$M =$ Number of objectives
$F(i).dist = 0 \quad \forall \quad i = 1,2,\ldots N_s$ {Initialize distance}
for $m = 1$ to M **do**
  $F = sort(F,m)$ {Sort based on objective value}
  $F(1).dist = F(N_s).dist = \infty$ {Assign infinity to the corner points}
  for $i = 2$ to $(N_s - 1)$ **do**
    $F(i).dist = F(i).dist + (F(i+1,m) - F(i-1,m))/(f_m^{max} - f_m^{min})$ {calculate $F(i).dist$ based on neighboring points}
  end for
end for
Higher $dist \Rightarrow$ Higher rank

sorting using $k+1$ objectives is equivalent to the non-dominated sorting the original k objectives as the additional objective value (which is based on the constraint violations) for feasible solutions is always 0.

The next step is to choose the solutions that form the population for the next generation. In IDEA, a user-defined parameter α is used to identify the proportion of the infeasible solutions to be retained in the population. The numbers N_f $(=(1-\alpha)\times N)$ and N_{inf} $(=\alpha \times N)$ denote the number of feasible and infeasible solutions in the population respectively, where N is the population size. If the infeasible set S_{inf} has more than N_{inf} solutions, then first N_{inf} solutions are selected based on the rank; otherwise all the solutions from S_{inf} are selected. The rest of the solutions are selected from the feasible set S_f, provided there are at least N_f number of feasible solutions. If S_f has fewer solutions, all the feasible solutions are selected and the rest are filled with infeasible solutions from S_{inf}. The solutions are ranked from 1 to N in the order they are selected. Hence, the infeasible solutions that get selected first (at most N_{inf}), get higher rank than the feasible solutions.

As an example, assuming a population size of 100, during any given generation 100 child solutions will be created. In the pool of 200 (parent + child) solutions, if there are 40 infeasible and 160 feasible solutions, then NSGA-II will select best 100 feasible solutions for the next generation, hence preferring *all* feasible solutions over *all* infeasible solutions. On the other hand, assuming we use $\alpha = 0.2$ for IDEA, it would select 20 best infeasible solutions (based on non-dominated + crowding distance sorting of $k+1$ objectives), and 80 best feasible solutions. Hence, *good* infeasible solutions are preferred over feasible solutions during the course of evolution.

In NSGA-II, the elite preservation mechanism weeds out the infeasible solutions from the population. To retain the infeasible solutions in the population, an alternate mechanism is required. In IDEA, the infeasible solutions are ranked higher than the feasible solutions, thus adding selection pressure to generate *better* infeasible solutions. Presence of infeasible solutions with higher ranks than the feasible solutions translates into an active search through the infeasible space. This feature of IDEA accelerates the movement of solutions towards the constraint boundary. With the modified problem definition and ranking of the infeasible solutions higher than the feasible solutions, IDEA can find the solutions to the original problem more efficiently.

3 Constraint Violation Measure

The additional objective in the modified problem formulation is based on the amount of relative constraint violation among the population members. The constraint violation measure of a solution is based on the constraint violation levels for all constraint values for that solution. Consider one of the constraints (g_i). All the solutions in the population are sorted ascending based on the value of the constraint violation for g_i. The solutions that do not violate the constraint g_i are assigned constraint violation value of 0 (and g_i does not contribute to the Violation measure of those solutions). Rest of the solutions are assigned a constraint violation level for constraint g_i based on the sorted list, starting with rank 1 for the solution with least constraint violation. Solutions with the same value of constraint violation get the same rank. This ranking procedure is repeated for all the constraints. The constraint violation measure for each solution is then calculated as the sum of the ranks (based on constraint violation level) obtained for all the constraints.

The process of determining constraint violation measure is illustrated using following example. Consider an optimization problem with three constraints (C1, C2, C3). A sample population of 10 individuals is shown in Table 1. For each solution, the first three columns list the value of the constraint violation. The constraint violation values are sorted for each constraint and each solution is assigned relative ranks for the constraints. The relative ranks are shown in next three columns. Solutions 3, 7 and 9 do not violate constraint C1 and get a relative rank of 0.

Table 1 Calculation of constraint violation measure

	Violations			Relative ranks			
Solution	C1	C2	C3	C1	C2	C3	Violation Measure
1	3.50	90.60	8.09	3	8	7	18
2	5.76	7.80	6.70	4	6	5	15
3	0.00	3.40	7.10	0	4	6	10
4	1.25	0.00	0.69	1	0	1	2
5	13.75	90.10	5.87	6	7	4	17
6	100.70	2.34	3.20	7	3	2	12
7	0.00	5.09	4.76	0	5	3	8
8	1.90	0.00	0.00	2	0	0	2
9	0.00	0.56	0.00	0	1	0	1
10	8.90	2.30	9.80	5	2	8	15

Solution 4 with the least constraint violation value (1.25) for C1 gets rank 1 and solution 6 with the highest constraint violation value (100.70) gets rank 7. Last column shows the constraint violation measure as the sum of the ranks with respect to each constraint.

It can be seen that the violation measure favors solutions with good ranks for most (or all) of the constraints. As a result, a solution with large violation in only one of the constraints would roughly have a same preference as a solution with marginal violations of multiple constraints. This in a sense incorporates the amount of constraint violation and not just the number of constraints violated as in CHEA. The violation measure is used as the additional objective in IDEA to rank the infeasible solutions using non-dominated sorting. As a result, the final population consists of the solutions with marginal constraint violations.

4 Experimental Setup

4.1 Single Objective Optimization

The g-series problems g01, g02, g04, g06, g07, g08, g09, g10 and g12 (g-series problems without equality constraints) are used as single objective optimization test problems. The detailed formulations of g-series test problems can be found in [24, 33]. All the problems are solved using IDEA and NSGA-II. Multiple runs are performed varying parameters – crossover probability, mutation probability, crossover distribution index and mutation distribution index. The values for the parameters are listed in Table 2, and for each parameter combination, both the algorithms are run with two different random seeds, thus resulting in a total of $2^5 = 32$ runs for each problem.

The population size of 200 is used and both algorithms are run for 1750 generations for all test runs, resulting in 350,000 function evaluations. For IDEA, twenty percent ($\alpha = 0.2$) of the population is maintained infeasible.

Table 2 Parameters used for IDEA and NSGA-II for studies on g-series test functions

Parameter	Values
Crossover Probability	0.8,0.9
Mutation Probability	0.1,0.2
Crossover Distribution Index	15,20
Mutation Distribution Index	20,30

4.2 Multi-objective Optimization

The CTP series of benchmark problems are used as multi-objective optimization test problems. CTP series is a set of seven constrained, bi-objective optimization problems. CTP problems pose various challenges to optimization algorithms – constricted feasible space near the constraint boundaries, discontinuous Pareto optimal fronts, discontinuity in the variable and function space etc. A detailed discussion on CTP test functions can be found in [9].

For CTP problems, 30 independent runs are performed using IDEA and NSGA-II by varying the random seed. The crossover and mutation parameters are fixed as shown in Table 3. A population size of 200 was evolved over 200 generations. For IDEA, twenty percent of the population is maintained infeasible ($\alpha = 0.2$).

Table 3 Parameters used for IDEA and NSGA-II for studies on CTP test functions

Parameter	Value
Crossover Probability	0.9
Mutation Probability	0.1
Crossover Distribution Index	15
Mutation Distribution Index	20

4.2.1 Performance Metrics

To compare the results of multi-objective optimization, various performance metrics have been suggested in the literature. A discussion of some of these metrics can be found in [5]. For our study, two performance metrics are used, *Displacement* and *Hypervolume*.

- **Displacement**
  Displacement [1,6,22] is a measure of how far the non-dominated solutions are from the Pareto optimal front. The *Displacement* metric is defined in Eq. 7 as suggested in [1],

$$\text{Displacement} = \frac{1}{|P|} \times \sum_{i=1}^{|P|} \left(\min_{j=1}^{|Q|} d(i,j) \right) \tag{7}$$

  where P is the known set of Pareto optimal solutions, Q is the non-dominated solution set and $d(i,j)$ is the Euclidean distance between i^{th} solution of set P and j^{th} solution of set Q. A lower value of *Displacement* corresponds to better convergence with respect to the Pareto optimal front.
- **Hypervolume**
  For bi-objective problems, *Hypervolume* for a set S is defined as the area dominated by the set in function space with respect to a reference point. Mathematically, it is defined as given in Eq. 8.

$$H = \left\{ \bigcup_i a_i | v_i \in S \right\} \tag{8}$$

where a_i is the area dominated by the solution v_i with respect to a reference point. A detailed discussion of the *Hypervolume* metric can be found in [5,45, 46]. For test problems CTP6 and CTP8, the reference point used is $(2,20)$ and for the rest of the problems it is $(2,2)$.

5 Results

5.1 Single-Objective Optimization

The results of the runs for g-series test problems have been compared using two performance measures – convergence rate and best solution obtained.

5.1.1 Convergence Rate

An 'average' progress of both the algorithms for g-series test problems is shown in Figure 1. The graphs show the best fitness averaged over all the runs for each generation. The runs with no feasible solutions at any generation are omitted when calculating average. Thus, the graphs indicate average convergence rates for IDEA and NSGA-II.

It is seen from Figure 1 that convergence rate of IDEA is better than NSGA-II for problems g01, g02, g04, g06, g07 and g10. For problems g08, g09 and g12, the average convergence was found to be almost identical and the figures for those problems are not presented.

5.1.2 Converged Values

The average, best and worst objective values across all runs are listed in Table 4. It is seen that for a given number of function evaluations, IDEA consistently obtains a better objective value than NSGA-II for all problems. For the problems g08 and g12, the average objective values obtained by both algorithms are the same.

It is worth mentioning here that although reported values for the g-series functions using IDEA are an improvement over NSGA-II, some recent studies have reported better results for g-series functions [26,39,42]. However, the scope of most of these studies is limited to single objective optimization only.

5.1.3 Trade-Off Solutions

Shown in Figure 2 are the best solutions obtained by IDEA and NSGA-II for problem g06 across all 32 runs. The constraint boundary for g06 is formed by two intersecting circles. The feasible region is the narrow region between the two circles where they intersect. The magnified view of the feasible region near the intersection is shown in Figure 2. The optimum objective value occurs at the intersection $(14.095, 0.84296)$. Along with the best solutions obtained by IDEA and NSGA-II, 5 "best" (based on the violation measure) of the infeasible solutions in the final IDEA

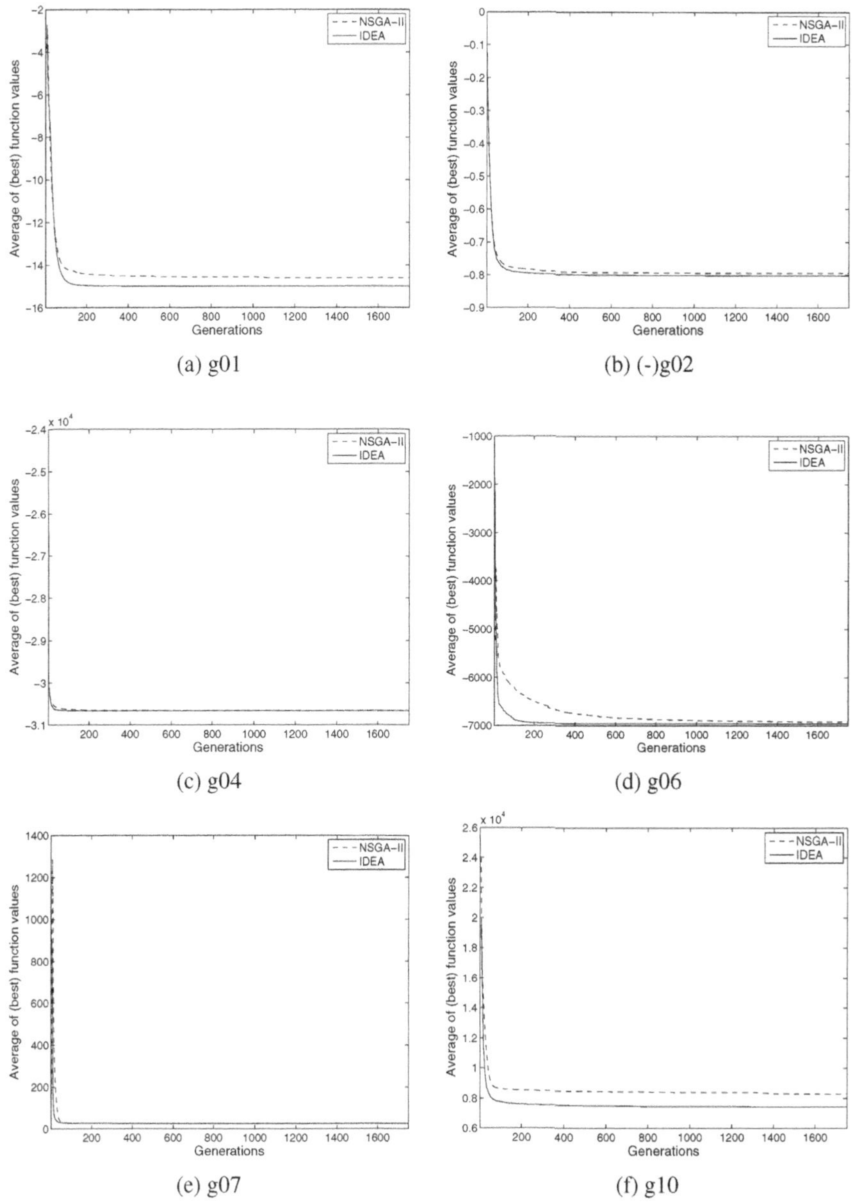

Fig. 1 Average of best feasible function values over 32 runs at each generation, obtained using NSGA-II and IDEA

population are also plotted. All the NSGA-II solutions lie in the feasible region (solutions between the two lines above the intersection point), as expected. Most of the IDEA solutions are concentrated near the intersection point. It is clear that NSGA-II

Table 4 Results for g-series functions

Problem	NSGA-II Results Best	Mean	Worst	IDEA Results Best	Mean	Worst
g01	-15.0000	-14.5979	-12.4457	-15.0000	**-14.9997**	-14.9988
g02	0.8033	0.7943	0.7640	0.8032	**0.8019**	0.7934
g04	-30665.300	-30661.209	-30618.700	-30665.500	**-30665.472**	-30665.300
g06	-6946.5500	-6921.6962	-6892.3900	-6961.7900	**-6961.4731**	-6960.6900
g07	24.4532	25.8522	31.9884	24.3811	**25.0916**	27.1796
g08	-0.09582	-0.09582	-0.09582	-0.09582	-0.09582	-0.09582
g09	680.6450	681.1611	682.2540	680.6670	**680.9331**	681.4170
g10	7355.190	8284.448	10030.100	7113.430	**7434.930**	7778.320
g12	1.000	1.000	1.000	1.000	1.000	1.000

Table 5 Marginally infeasible solutions obtained using IDEA for g06

x1	x2	$f(\mathbf{x})$	Violations C1	C2
14.095100	0.840314	-6964.723300	0	0.023632
14.058700	0.802804	-7007.928303	0.323500	0
14.095100	0.841393	-6963.535085	0	0.014656
14.062200	0.772189	-7041.657047	0.002145	0.063455
14.078800	0.830207	-6976.676602	0.188217	0
14.095200	0.842968	-6961.795875	0	0.003178
14.058900	0.700537	-7121.587900	0	0.621251
14.096500	0.792284	-7017.680063	0	0.448186
14.095100	0.803637	-7005.192343	0	0.330106
14.048200	0.838280	-6969.296140	0.810163	0

has difficulty in searching along the narrow region of the feasible space. The population of IDEA, however, approaches the optimum solution from various directions (as apparent from the distribution of the infeasible solutions around the intersection) and quickly manages to reach the optimum.

Some of the infeasible solutions in the final population of IDEA are listed in Table 5. The optimum objective value for g06 is -6961.81388. The objective can be improved substantially by relaxing one or both the constraints marginally as seen from the table.

5.2 *Multi-objective Optimization*

5.2.1 Evolution

The progress of NSGA-II and IDEA populations up to 200 generations for test problem CTP2 is shown in Figure 3 and Figure 4 respectively. For NSGA-II, the population approaches the POS from the feasible space and has difficulty in searching

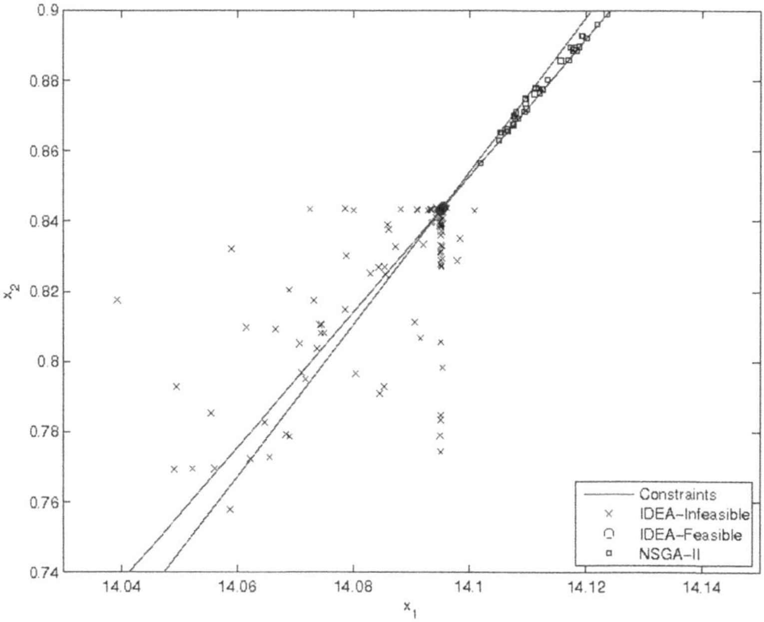

Fig. 2 Final population for g06

close to the constraint boundary. On the other hand, IDEA maintains the population in both the feasible and the infeasible space, thus capturing the entire POS much faster (See Figure 4). IDEA is able to cover most of the disconnected segments of POS for CTP2 by generation 50.

Test problem CTP2 has a disconnected POS. Any population based method that search through the feasible space for disconnected POS, is likely to face difficulty capturing the whole front unless a reasonably large population size is chosen to maintain diversity. Once the entire population converges to fewer regions of the POS, NSGA-II has to rely predominantly on mutation to spread the solutions to other regions of the POS. On the other hand, IDEA population can move through the infeasible regions avoiding 'detours' through feasible space. Hence, even with a small population, IDEA is able to capture the entire POS.

To illustrate this effect, a single run was performed for CTP2 with both NSGA-II and IDEA with a population size of 100 and 200 generations. The crossover and the mutation parameters are kept fixed as in the earlier experiments. The results of the run are shown in Figure 5. It can be seen that IDEA solutions are spread much more evenly across the entire POS as compared to NSGA-II.

5.2.2 Performance Metrics

The performance metrics are calculated using the non-dominated solutions obtained by NSGA-II and IDEA. In case of IDEA, only the feasible solutions in the final

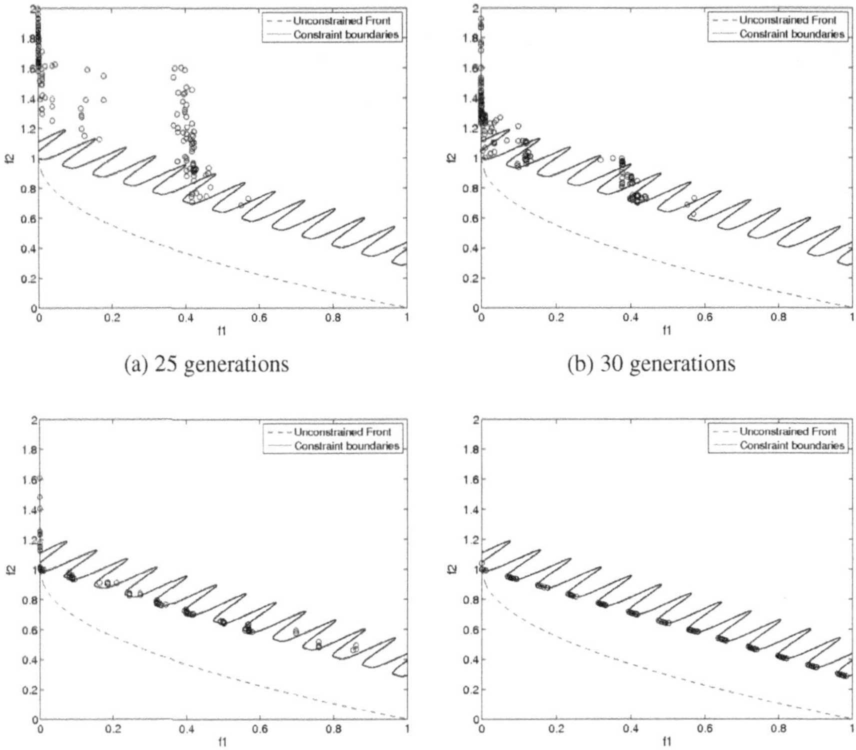

(a) 25 generations (b) 30 generations

(c) 50 generations (d) 200 generations

Fig. 3 Evolution of NSGA-II population over generations for CTP2 test run (Population size is 200)

population are considered. The *Displacement* metric and *Hypervolume* metric results for NSGA-II and IDEA are listed in Tables 6 and 7 respectively for the problems CTP2-CTP8. The tables show the best, mean, and standard deviation (S.D.) of the metric values across 30 runs. From Table 6 it is seen that IDEA has significantly better mean *Displacement* metric values than NSGA-II for all CTP problems. The worse mean values of NSGA-II are due to its tendency to converge to sub-optimal fronts for CTP problems. For the best runs, the values of *Displacement* metrics obtained by both algorithms are fairly close. Also, it is observed that IDEA results have lower S.D., indicating that IDEA obtains more consistent results than NSGA-II over multiple runs. Only for CTP7, however the S.D. value obtained using IDEA is higher than that obtained NSGA-II. It is so because in *one* of the runs, IDEA converged to a sub-optimal front which is far away from the Pareto-optimal front. NSGA-II on the other hand, converged to a sub-optimal that was closer to the Pareto-optimal front, but more number of times than IDEA. Hence, for CTP7, the mean values using IDEA are still better than NSGA-II, where as the S.D. is worse for IDEA.

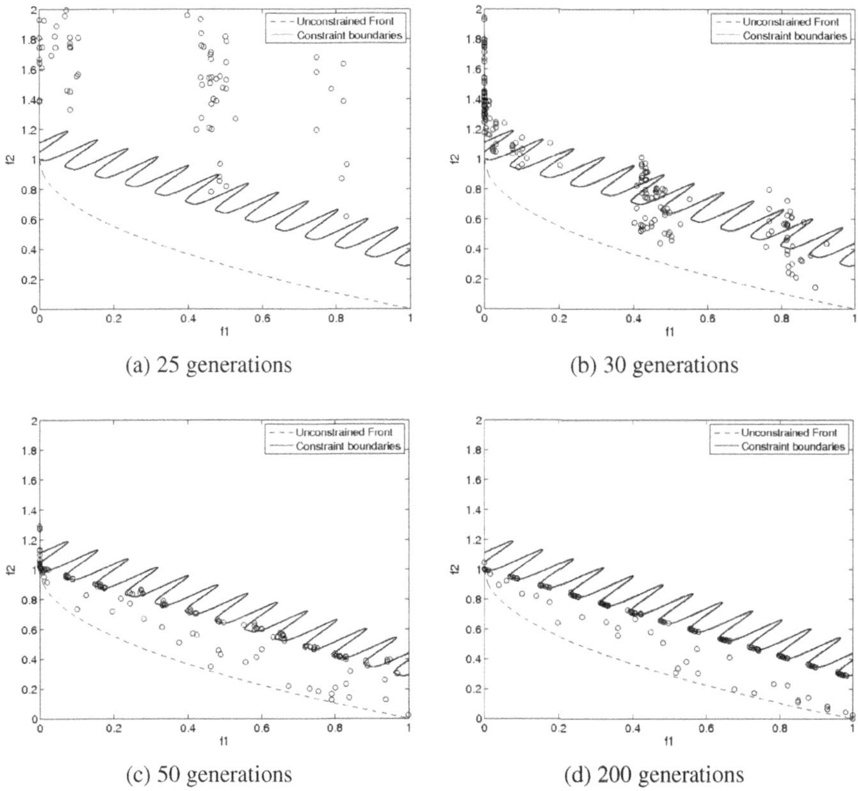

(a) 25 generations (b) 30 generations

(c) 50 generations (d) 200 generations

Fig. 4 Evolution of IDEA population over generations for CTP2 test run (Population size is 200)

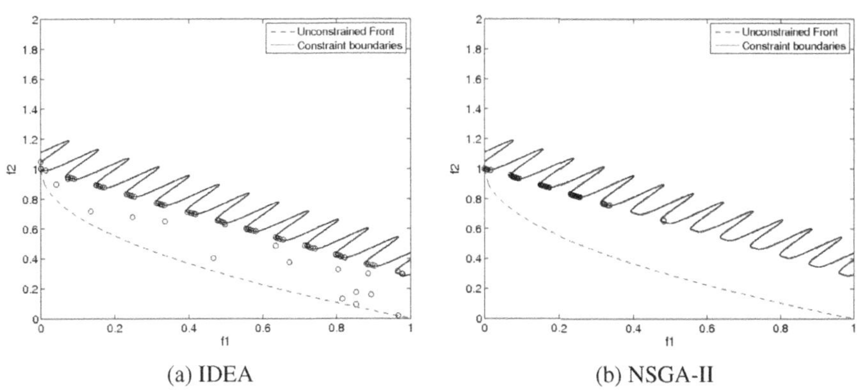

(a) IDEA (b) NSGA-II

Fig. 5 Final front obtained for CTP2 using IDEA and NSGA-II with a population size of 100 evolved over 200 generations

Table 6 *Displacement* metric for CTP problems

	IDEA Results			NSGA-II Results		
	Best	Mean	S.D.	Best	Mean	S.D.
CTP2	0.000101	**0.001254**	0.004286	0.000085	0.006415	0.009226
CTP3	0.001696	**0.006992**	0.017147	0.001820	0.024493	0.035827
CTP4	0.017625	**0.028247**	0.015927	0.019386	0.061003	0.047692
CTP5	0.000266	**0.001369**	0.003965	0.000268	0.003856	0.004762
CTP6	0.000463	**0.000689**	0.000924	0.000395	0.012739	0.050726
CTP7	0.000049	**0.004057**	0.013118	0.000041	0.008308	0.013071
CTP8	0.000251	**0.004592**	0.010069	0.000225	0.104338	0.137467

The values of the best, mean and the standard deviation for the *Hypervolume* metric (across all 30 runs) are shown in Table 7. It is seen that the average metric values obtained by IDEA are higher than those obtained by NSGA-II for all CTP

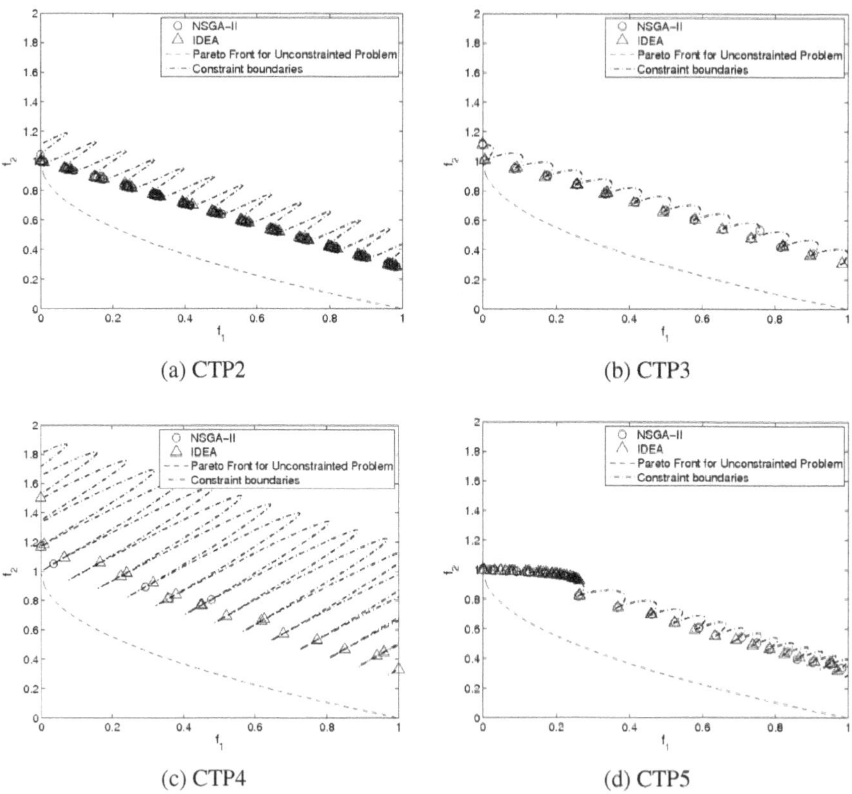

(a) CTP2 (b) CTP3

(c) CTP4 (d) CTP5

Fig. 6 Solutions obtained in a typical run for problems CTP2, CTP3, CTP4 and CTP5 using NSGA-II and IDEA

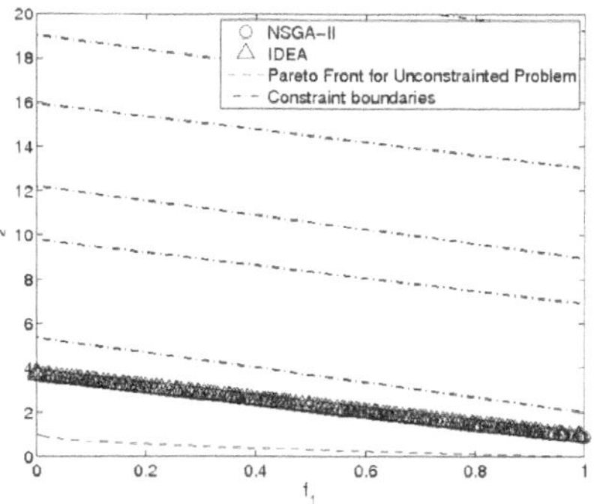

(a) CTP6

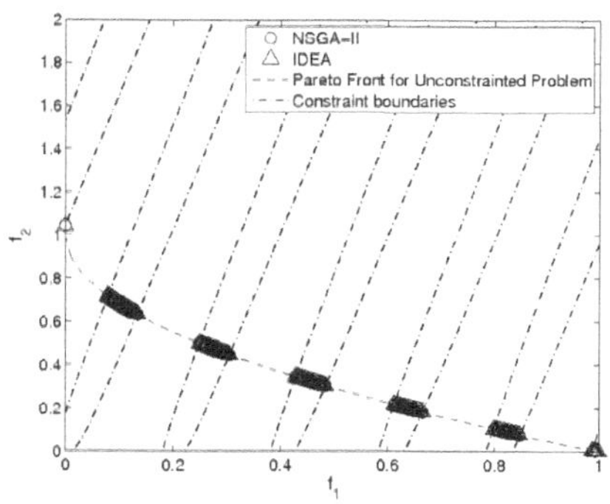

(b) CTP7

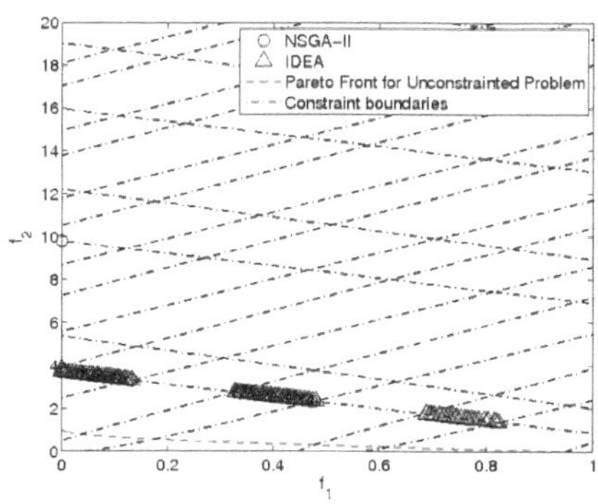

(c) CTP8

Fig. 7 Solutions obtained in a typical run for problems CTP6, CTP7 and CTP8 using NSGA-II and IDEA

problems. The small values of S.D. for IDEA suggest that the performance of the algorithm is consistent. Also, it is seen that the for the case of CTP7 only, the IDEA results have a higher S.D. than NSGA-II results, due the same reason as mentioned in the case of *Displacement* metric.

The non-dominated fronts obtained using NSGA-II and IDEA during a typical run are shown in Figures 6 and 7. It can be visibly seen that IDEA is able to obtain more solutions close to the POS, especially for the problems where the optimum solutions lie in constricted spaces formed by the constrained boundaries, such as CTP3, CTP4 and CTP5. Also, for CTP8, it can be seen that in the given number of evaluations, IDEA was able to obtain all three segments of the POS, whereas NSGA-II was able to get only two of them, for the same reason as explained in 5.2.1. The flexibility of moving through infeasible regions enables IDEA to obtain the front faster.

Table 7 *Hypervolume* metric for CTP problems

	IDEA Results			NSGA-II Results		
	Best	Mean	S.D.	Best	Mean	S.D.
CTP2	3.059180	**3.011390**	0.177100	3.059344	2.870703	0.270056
CTP3	3.015969	**2.960771**	0.163811	3.010435	2.828100	0.254690
CTP4	2.919011	**2.744736**	0.139271	2.848500	2.438106	0.352716
CTP5	3.024724	**2.952929**	0.162089	3.020914	2.723520	0.292627
CTP6	36.819072	**36.787826**	0.075797	36.822693	36.182922	2.187300
CTP7	3.617663	**3.435931**	0.594506	3.617716	3.240162	0.594118
CTP8	36.180362	**35.970564**	0.434540	36.170783	32.085918	5.176305

5.2.3 Trade-Off Solutions

As shown in Figure 4(d), the final population for CTP2 evolved using IDEA contains infeasible points close to the constraint boundary. One can evaluate the benefit in the objective values by relaxing the constraints as illustrated in Section 5.1.3. For multi-objective optimization problems significant benefit may be derived in multiple objectives at the cost of relaxing the constraints marginally.

6 Conclusions

A novel algorithm, Infeasibility Driven Evolutionary Algorithm (IDEA), for constrained optimization problems has been presented in this chapter. The algorithm maintains infeasible solutions during the evolution thereby searching the space through the feasible as well as the infeasible regions. The original constrained optimization problem is reformulated as an unconstrained optimization problem with one additional objective. The additional objective is based on the constraint violation level of the solutions. In addition, the infeasible solutions are ranked higher than the feasible solutions to focus the search near the constraint boundary. The search through the infeasible space enhances the convergence rate of IDEA over NSGA-II, which searches only through the feasible regions.

The proposed algorithm was tested on single objective g-series test problems and multi-objective CTP series problems, and the results were compared with NSGA-II. The results of IDEA show a greater rate of convergence for single objective problems. The results demonstrate the capability of IDEA for handling constraints efficiently for both single and multi-objective problems. Furthermore, IDEA has additional advantage of providing marginally infeasible solutions, which may prove beneficial trade-offs for design considerations.

References

1. Bandyopadhyay, S., Saha, S., Maulik, U., Deb, K.: A simulated annealing-based multi-objective optimization algorithm: AMOSA. IEEE Transactions on Evolutionary Computation 12(3), 269–283 (2008)

2. Bean, J.C., Hadj-Alouane, A.B.: A Dual Genetic Algorithm for Bounded Integer Programs. Technical Report TR 92-53, Department of Industrial and Operations Engineering, The University of Michigan (1992)
3. Coello Coello, C.A.: Constraint-handling using an evolutionary multiobjective optimization technique. Civil engineering and environmental systems 17(4), 319–346 (2000)
4. Coello Coello, C.A.: Theoretical and numerical constraint-handling techniques used with evolutionary algorithms: a survey of the state of the art. Computer Methods in Applied Mechanics and Engineering 191(11-12), 1245–1287 (2002)
5. Collette, Y., Siarry, P.: Multiobjective optimization Principles and Case Studies. Springer, Heidelberg (2003)
6. Czyzak, P., Jaszkiewicz, A.: Pareto simulated annealing - a metaheuristic techinique for multiple-objective combinatorial optimization. Journal of Multi-Criteria Decision Analysis 7(1), 34–47 (1998)
7. Davis, L. (ed.): Handbook of Genetic Algorithms. Van Nostrand Reinhold, New York (1991)
8. Deb, K.: An Efficient Constraint Handling Method for Genetic Algorithms. Computer Methods in Applied Mechanics and Engineering 186(2/4), 311–338 (2000)
9. Deb, K.: Multi-Objective Optimization using Evolutionary Algorithms. John Wiley and Sons Pvt. Ltd., Chichester (2001)
10. Deb, K., Agrawal, S.: Simulated binary crossover for continuous search space. Complex Systems 9, 115–148 (1995)
11. Deb, K., Goyal, M.: A combined genetic adaptive search (GeneAS) for engineering design. Computer Science and Informatics 26, 30–45 (1996)
12. Deb, K., Pratap, A., Agarwal, S., Meyarivan, T.: A fast and elitist multiobjective genetic algorithm: NSGA-II. IEEE Transactions on Evolutionary Computation 6, 182–197 (2002)
13. Hadj-Alouane, A.B., Bean, J.C.: A Genetic Algorithm for the Multiple-Choice Integer Program. Operations Research 45, 92–101 (1997)
14. Hamida, S.B., Schoenauer, M.: An adaptive algorithm for constrained optimization problems. In: Deb, K., Rudolph, G., Lutton, E., Merelo, J.J., Schoenauer, M., Schwefel, H.-P., Yao, X. (eds.) PPSN 2000. LNCS, vol. 1917, pp. 529–538. Springer, Heidelberg (2000)
15. Hamida, S.B., Schoenauer, M.: ASCHEA: new results using adaptive segregational constraint handling. In: Proceedings of the 2002 Congress on Evolutionary Computation, pp. 884–889 (May 2002)
16. Hedar, A., Fukushima, M.: Derivative-free filter simulated annealing method for constrained continuous global optimization. Journal of Global Optimization 35(4), 291–308 (2006)
17. Hinterding, R., Michalewicz, Z.: Your brains and my beauty: parent matching for constrained optimisation. In: Proceedings of 1998 IEEE Conference on Evolutionary Computaion, May 1998, pp. 810–815 (1998)
18. Ho, P.Y., Shimizu, K.: Evolutionary constrained optimization using an addition of ranking method and a percentage-based tolerance value adjustment scheme. Information Sciences 177, 2985–3004 (2007)
19. Hoffmeister, F., Sprave, J.: Problem-independent handling of constraints by use of metric penalty functions. In: Fogel, L.J., Angeline, P.J., Bäck, T. (eds.) Proceedings of the Fifth Annual Conference on Evolutionary Programming (EP 1996), San Diego, California, February 1996, pp. 289–294. MIT Press, Cambridge (1996)

20. Homaifar, A., Lai, S.H.Y., Qi, X.: Constrained Optimization via Genetic Algorithms. Simulation 62(4), 242–254 (1994)
21. Isaacs, A., Ray, T., Smith, W.: Blessings of maintaining infeasible solutions for constrained multi-objective optimization problems. In: Proceedings of 2008 IEEE Congress on Evolutionary Computation, Hong Kong, June 2008, pp. 2785–2792 (2008)
22. Ishibuchi, H., Yoshida, T., Murata, T.: Balance between genetic search and local search in memetic algorithms for multiobjective permutation flowshop scheduling. IEEE Transactions on Evolutionary Computing 7(2), 204–223 (2003)
23. Joines, J., Houck, C.: On the use of non-stationary penalty functions to solve nonlinear constrained optimization problems with GAs. In: Fogel, D. (ed.) Proceedings of the first IEEE Conference on Evolutionary Computation, Orlando, Florida, pp. 579–584. IEEE Press, Los Alamitos (1994)
24. Koziel, S., Michalewicz, Z.: Evolutionary Algorithms, Homomorphous Mappings, and Constrained Parameter Optimization. Evolutionary Computation 7(1), 19–44 (1999)
25. Kuri-Morales, A., Quezada, C.V.: A Universal Eclectic Genetic Algorithm for Constrained Optimization. In: Proceedings 6th European Congress on Intelligent Techniques & Soft Computing, EUFIT 1998, Aachen, Germany, September 1998, pp. 518–522. Verlag Mainz (1998)
26. Mezura-Montes, E., Coello Coello, C.: A simple multimembered evolution strategy to solve constrained optimization problems. IEEE Transactions on Evolutionary Computation 9(1), 1–17 (2005)
27. Mezura-Montes, E., Coello Coello, C.A.: Constrained Optimization via Multiobjective Evolutionary Algorithms. In: Knowles, J., Corne, D., Deb, K. (eds.) Multiobjective Problems Solving from Nature: From Concepts to Applications. Natural Computing Series. Springer, Heidelberg (2008)
28. Michalewicz, Z.: Genetic Algorithms, Numerical Optimization, and Constraints. In: Eshelman, L.J. (ed.) Proceedings of the Sixth International Conference on Genetic Algorithms (ICGA 1995), July 1995, pp. 151–158. University of Pittsburgh, Morgan Kaufmann Publishers (1995)
29. Michalewicz, Z.: A Survey of Constraint Handling Techniques in Evolutionary Computation Methods. In: McDonnell, J.R., Reynolds, R.G., Fogel, D.B. (eds.) Proceedings of the 4th Annual Conference on Evolutionary Programming, pp. 135–155. The MIT Press, Cambridge (1995)
30. Michalewicz, Z.: Genetic Algorithms + Data Structures = Evolution Programs, 3rd edn. Springer, Heidelberg (1996)
31. Michalewicz, Z., Attia, N.F.: Evolutionary Optimization of Constrained Problems. In: Proceedings of the 3rd Annual Conference on Evolutionary Programming, pp. 98–108. World Scientific, Singapore (1994)
32. Michalewicz, Z., Nazhiyath, G.: Genocop III: A co-evolutionary algorithm for numerical optimization with nonlinear constraints. In: Fogel, D.B. (ed.) Proceedings of the Second IEEE International Conference on Evolutionary Computation, pp. 647–651. IEEE Press, Los Alamitos (1995)
33. Michalewicz, Z., Schoenauer, M.: Evolutionary Algorithms for Constrained Parameter Optimization Problems. Evolutionary Computation 4(1), 1–32 (1996)
34. Michalewicz, Z., Xiao, J.: Evaluation of Paths in Evolutionary Planner/Navigator. In: Proceedings of the 1995 International Workshop on Biologically Inspired Evolutionary Systems, Tokyo, Japan, May 1995, pp. 45–52 (1995)

35. Powell, D., Skolnick, M.M.: Using genetic algorithms in engineering design optimization with non-linear constraints. In: Forrest, S. (ed.) Proceedings of the Fifth International Conference on Genetic Algorithms (ICGA 1993), San Mateo, California, July 1993, pp. 424–431. University of Illinois at Urbana-Champaign, Morgan Kaufmann Publishers (1993)
36. Ray, T., Tai, K., Seow, K.: Multiobjective design optimization by an evolutionary algorithm. Engineering Optimization 33(4), 399–424 (2001)
37. Singh, H., Isaacs, A., Ray, T., Smith, W.: A simulated annealing algorithm for constrained multi-objective optimization. In: Proceedings of 2008 IEEE Congress on Evolutionary Computation, Hong Kong, June 2008, pp. 1655–1662 (2008)
38. Surry, P.D., Radcliffe, N.J.: The COMOGA method: Constrained optimisation by multi-objective genetic algorithms. Control and Cybernetics 26(3) (1997)
39. Takahama, T., Sakai, S.: Constrained optimization by applying the /spl alpha/ constrained method to the nonlinear simplex method with mutations. IEEE Transactions on Evolutionary Computation 9(5), 437–451 (2005)
40. Vieira, D.A.G., Adriano, R.L.S., Krahenbuhl, L., Vasconcelos, J.A.: Handing constraints as objectives in a multiobjective genetic based algorithm. Journal of Microwaves and Optoelectronics 2(6), 50–58 (2002)
41. Vieira, D.A.G., Adriano, R.L.S., Vasconcelos, J.A., Krahenbuhl, L.: Treating constraints as objectives in multiobjective optimization problems using niched pareto genetic algorithm. IEEE Transactions on Magnetics 40(2) (March 2004)
42. Wang, Y., Cai, Z., Guo, G., Zhou, Y.: Multiobjective optimization and hybrid evolutionary algorithm to solve constrained optimization problems. IEEE Transactions on Systems, Man and Cybernetics - Part B: Cybernetics 37(3), 560–575 (2007)
43. Xiao, J., Michalewicz, Z., Trojanowski, K.: Adaptive Evolutionary Planner/Navigator for Mobile Robots. IEEE Transactions on Evolutionary Computation 1(1), 18–28 (1997)
44. Xiao, J., Michalewicz, Z., Zhang, L.: Evolutionary Planner/Navigator: Operator Performance and Self-Tuning. In: Proceedings of the 3rd IEEE International Conference on Evolutionary Computation, Nagoya, Japan, May 1996. IEEE Press, Los Alamitos (1996)
45. Zitzler, E.: Evolutionary Algorithms for Multiobjective Optimization: Methods and Applications. Shaker Verlag, Germany (1999)
46. Zitzler, E., Thiele, L.: Multiobjective optimization using evolutionary algorithms - A comparative case study. In: Eiben, A.E., Bäck, T., Schoenauer, M., Schwefel, H.-P. (eds.) PPSN 1998. LNCS, vol. 1498, pp. 292–301. Springer, Heidelberg (1998)

On GA-AIS Hybrids for Constrained Optimization Problems in Engineering

Heder S. Bernardino, Helio J.C. Barbosa, Afonso C.C. Lemonge, and Leonardo G. Fonseca

Abstract. A genetic algorithm (GA) is hybridized with an artificial immune system (AIS) as an alternative to tackle constrained optimization problems in engineering. The AIS is inspired by the clonal selection principle and is embedded into a standard GA search engine in order to help move the population into the feasible region. The resulting GA-AIS hybrid is tested in a suite of constrained optimization problems with continuous variables, as well as structural and mixed integer reliability engineering optimization problems. In order to improve the diversity of the population, a variant of the algorithm is developed with the inclusion of a clearing procedure. The performance of the GA-AIS hybrids is compared with that of alternative techniques, such as the Adaptive Penalty Method, and the Stochastic Ranking technique, which represent two different types of constraint handling techniques that have been shown to provide good results in the literature.

Keywords: genetic algorithm, artificial immune system, hydrid algorithm.

1 Introduction

Nature-inspired population-based algorithms (NPAs) , which can be readily applied to unconstrained optimization problems, must be usually equipped with an additional constraint handling procedure whenever the constraints cannot be automatically satisfied by all candidate solutions in the population. The many available techniques to handle constrained optimization problems within NPAs can be classified as *direct* (feasible or interior), when only feasible elements are considered, or

H.S. Bernardino and A.C.C. Lemonge.
Universidade Federal de Juiz de Fora, Brazil e-mail: hedersb@gmail.com, e-mail: afonso.lemonge@ufjf.edu.br

H.J.C. Barbosa and L.G. Fonseca.
Laboratório Nacional de Computação Científica, Av. Getúlio Vargas 333, 25651-075 Petrópolis, RJ, Brazil e-mail: hcbm@lncc.br, e-mail: goliatt@lncc.br

E. Mezura-Montes (Ed.): Constraint-Handling in Evolution. Optimi., SCI 198, pp. 167–192.
springerlink.com

indirect (exterior), when both feasible and infeasible elements are used during the search process. Direct techniques comprise: a) special *closed* genetic operators [40], b) special decoders [29], c) repair techniques [36], and d) "death penalty". Direct techniques are problem dependent (with the exception of the "death penalty") and actually of extremely reduced practical applicability. Indirect techniques include: a) the use of Lagrange multipliers [1], which may lead to the introduction of a population of multipliers and to the use of the concept of coevolution [2], b) combining fitness and constraint violation in a multi-objective optimization setting [12,44], c) the use of special selection techniques [39], d) assigning to any infeasible offspring a very low fitness value [26], and e) penalty techniques [3,4,25,31,32,38,42]. Other strategies proposed in the NPA literature can be found in [23,27,28,34,40], and in the repository [10].

However, of particular interest here is the application of ideas from artificial immune systems (AIS) in constrained optimization problems. A hybrid GA-AIS is proposed to solve constrained optimization problems involving continuous and/or discrete variables. This chapter is organized as follows. The formulation of the constrained optimization problem and some techniques are described in Section 2, the AIS heuristic and previous work are summarized in Section 3. Section 4 presents the GA-AIS hybrid technique, and numerical experiments are discussed in Section 5. The numerical experiments comprise a well known suite of continuous constrained optimization problems, a mixed-integer non-linear programming (MINLP) problem, two non-linear integer programming (NLIP) problems, a discrete structural engineering optimization problem, and, finally, a mixed-integer reliability problem (considering three objective functions). Finally, in Section 6, some conclusions are drawn.

2 Constrained Optimization Problems

A standard constrained optimization problem in R^n can be thought of as the minimization of a given objective function $f(x)$, where $x \in R^n$ is the vector of design/decision variables, subject to inequality constraints $g_p(x) \geq 0$, $p = 1,2,\ldots,\bar{p}$ as well as equality constraints $h_q(x) = 0$, $q = 1,2,\ldots,\bar{q}$. Additionally, the variables are usually subject to bounds $x_i^L \leq x_i \leq x_i^U$. Due to the standard encoding techniques used by GAs and AISs, bound constraints do not require special treatement since they will be trivially enforced here.

Very often the design variables are further constrained to belong to a given finite set of pre-defined values, as in design optimization problems when parts must be selected from commercially available types. A mixed discrete-continuous constrained optimization problem arises. Often, in optimization problems arising from multidisciplinary design tasks, the constraints are in fact a complex *implicit* function of the design variables, and the check for feasibility may be computationally expensive. Constraint handling techniques which do not require the explicit form of the constraints and do not require additional objective function evaluations are thus most valuable.

In this paper two techniques to handle constrained optimization problems are selected in order to provide comparisons with the performance of the proposed GA-AIS. The first one is the Adaptive Penalty Method (APM) [3,32] and the second one is the Stochastic Ranking technique (SR) presented in [39]. Both techniques have been shown to be capable of producing very good solutions in the literature.

3 Artificial Immune Systems

Computer science has, for some time now, searched for inspiration in nature in order to improve the way of solving computational problems. The Artificial Immune Systems (AISs) are composed by intelligent methodologies, inspired by biological immune system, to solve real world problems [15].

In nature, when an animal is exposed to antigens, relatively specific antibodies are produced to combat them. The immune response has to be efficient in order to defend the organism from foreign agents (antigens). To do this, the best antibodies are cloned, hypermutated, and selected. Sometimes, random antibodies are generated to improve diversity to the population (produced by the bone marrow). Also, if the organism is again attacked by that antigen a quicker immune response is developed. This skill of adaptation is known as clonal selection and affinity maturation by hypermutation, or more simply, clonal selection [18] .

The natural immunity comprises innate and adaptive immunities [47]. The innate immune system is composed by cells and mechanisms that defend the host from attacks by other organisms, in a non-specific manner. The adaptive immune system comprises highly specialized cells and processes that defend the organism from antigens. The main adaptive immunity feature is to distinguish between proteins produced by cells of the body (self) and the ones produced by intruders or by cells under virus control (non-self). The clonal selection belongs to the adaptive immune system.

AISs are used in various applications [8]: pattern recognition, scheduling, control, machine-learning, information systems security, and optimization. For optimization problems, different techniques based in AISs can be found in the literature such as the immune network theory [45] or the clonal selection principle [7,9].

The clonal selection algorithm, used by our hybrid method, is similar to other stochastic search methods. The individuals of the population (or candidate solutions) are the antibodies. The clonal selection evolution is based on the principle that each individual is cloned, hypermutated, and those with higher affinity are selected. For unconstrained problems, affinity is usually associated with a fitness value, but here affinity will be given by the genotypic distance between antibodies and antigens. The mutation rate is, normally, inversely proportional to the affinity of the antibody with respect to the antigens. AISs usually do not use recombination operators (such as crossover in GAs). The resulting AIS technique has a good balance between global and local exploration and is simple to implement [46].

3.1 Previous Work Using AIS

In 1995, Hajela and co-workers [20–22, 51], aiming at increasing the similarity (or reducing the distance) between infeasible elements (playing the role of antibodies) and feasible ones (antigens), proposed the idea of using another GA, embedded into the original one. The inner GA uses as fitness function a genotypical distance in order to evolve better (hopefully feasible) antibodies. In this way there is no need for additional expensive evaluations of the original fitness function of the problem, which only happen during the search performed by the external GA. The internal GA uses a relatively inexpensive fitness based on Hamming distance calculations.

Coello and Cruz-Cortés [11] proposed an extension of Hajela's algorithm, together with a parallel version, and tested them in a larger problem set. In this paper some test functions of the well-known benchmark problems were used including optimization problems in the mechanical engineering field. Comparisons were made using several penalty techiniques in order to check the performance of the proposed aproach.

An adaptive clonal selection (ACS) was proposed by Garrett [18]. His work suggests some new ways for defining the value of user defined paramenters, such as mutation rate and number of clones, along the run.

A different approach was followed by Cruz-Cortés et al. [13] where CLONALG (see [7, 9]), an existing AIS, is modified in order to deal with constrained optimization problems. Binary as well as real representations were considered although results for the real-coded version of CLONALG were disappointing. So, an alternative mutation operator was proposed by the authors in order to improve the performance of the AIS with real representation. In that approach, the mutation operator depends on the affinity of the antibodies, range of each decision variable, and antibody population size.

Wu [47] combined two techniques in his approach: clonal selection and idiotypic network theory, the later being used to control the number of good solutions. The clonal selection operator explores the search space looking for good solutions and maintaining the diversity of the antibodies population. The performance of that algorithm is evaluated in constrained optimization problems with continuous variables.

Based on the work by Hajela and Yoo [22], Rajasekaran and Lavanya [43] proposed an immune network for constraint handling in GAs. The technique was applied to obtain optimal sectional areas for minimum weight of structures such as space trusses in civil engineering subject to static loading and earthquake ground motion. Also, test problems were conducted for the design of the optimal mix of high-performance concrete (HPC), which is still based on trial mix and for which no rigorous mathematical approach is available.

A survey discussing the major works in the immunity-based techniques is presented in [24], in particular, reviewing the existing works, methods, and new initiatives from 1999 to 2003 years. Similarly, Garrett [19] presented a survey of different AIS techniques and their application to different types of computational problems.

4 The Hybrid GA-AIS

Hybrid algorithms are designed by combining existing methods in such way that the resulting performance is superior to that of each method acting alone. Yen et al. [49] describe four forms of hybrid NPAs : (i) pipelining hybrids, (ii) asynchronous hybrids, (iii) hierarchical hybrids, and (iv) the use of additional operators. In pipelining hybrids, a NPA and some other technique are applied sequentially: one generating data to be used by the other. The NPA can be applied before (more frequently) or after the local search procedure, and a third possibility is an interleaved or staged form of application of such search procedures.

An asynchronous hybrid NPA maintains a population which is shared by the NPA and the other search technique. Both procedures work cooperatively and asynchronously, posting and retrieving improved solutions from the shared population.

A hierarchical hybrid uses an NPA and another search procedure at different levels of the search problem. For example, an NPA can specify the topology of a neural network, while back-propagation computes the weights.

Finally, a hybrid algorithm can also be constructed by introducing a given search procedure as an additional move operator which improves one or more individuals by operating on them.

Here, we integrate an AIS with a binary-coded GA in a staged pipeline fashion.

In previous work [5, 6], following the idea of Hajela and co-workers, a hybrid GA was proposed where an AIS is called to help the GA in increasing the number of feasible individuals in the population. However, instead of embedding another GA into the main search cycle, a simple technique, inspired in the clonal selection principle, is used inside the GA cycle.

The hybrid consists in an outer (GA) search loop where the current population is checked for constraint violation and then divided into feasible (antigens) and infeasible individuals (antibodies). If there are no feasible individuals, the two (or other user defined quantity) “least infeasible” ones (those with the lowest constraint violation) are moved to the antigen population.

The AIS is introduced as an inner loop with the objective of bringing infeasible individuals into the feasible search space. In an AIS iteration, antibodies are first cloned and then hypermutated. Next, the distances (affinities) between antibodies and antigens are computed. Those with higher affinity (smaller sum of distances) are selected thus defining the new antibodies (closer to the feasible region). More precisely, each selected antibody is the best one (the one with highest affinity) among a given parent and its offspring (generated by cloning and hypermutating the parent). Also, the affinity is given by the sum of genotypical (Hamming) distances between a given antibody and the antigens. The AIS iteration is repeated a number of times.

It is important to notice that each new antibody in the population inherits its parent fitness, so that no fitness function evaluation is required at this point. As a result, (new) infeasible individuals have their genotype and fitness temporarily incompatible.

At the end of the AIS loop, the two populations (antibodies and antigens) are combined to generate the GA population. The selection operation is then performed in order to apply recombination and mutation operators to the selected parents producing a new population and finishing the external (GA) loop.

The selection procedure in the GA consists in binary tournaments where each individual is selected once and its opponent is randomly draw, with replacement, from the population, ensuring that all individuals will participate in at least one tournament. The rules of the tournament are: (i) any feasible individual is preferred to any infeasible one, (ii) between two feasible individuals, the one with the higher fitness value is chosen, and (iii) between two infeasible individuals, the one with the smaller constraint violation is chosen. After the selection procedure, an intermediary population is ready to be subject to the crossover operator.

A pseudo-code for the proposed hybrid (GA-AIS) is given in Algorithms 1 and 2.

Algorithm 1. The Hybrid GA-AIS Algorithm

```
procedure HybridGA(nGenerationsGA,nIterationsAIS)
    ComputeFitnessViolation(population)
    for i = 1 : nGenerationsGA do
        Divide(population, antibodies, antigens)
        ComputeDistance(antigens, antibodies)
        for j = 1 : nIterationsAIS do
            Clone(antibodies, temp)
            Mutation(temp)
            ComputeDistance(antigens, temp)
            SelectBetter(temp, antibodies)
        end for
        Union(antibodies, antigens, population)
        TournamentSelection(population, temp)
        Crossover(temp)
        Mutation(temp)
        ComputeFitnessViolation(temp)
        ChangePopulation(temp, population)
    end for
end procedure
```

However, it has been observed that the inner loop tends to decrease the diversity of the population, negatively affecting the solution quality. This motivated the introduction of another technique in order to improve the diversity of the population (improving the exploration of the search space) leading to a GA-AIS hybrid variant. In the GA-AIS^C variant, a modified version of Petrowski's clearing procedure [37] is introduced. The clearing procedure, originally used for multimodal problems, is a niching method inspired by the principle of sharing limited resources within subpopulations of individuals characterized by some similarities [41]. The clearing procedure leaves those resources to the better individuals of each subpopulation. According to [41], that procedure is normally applied after evaluating the fitness of individuals and before applying the selection operator. The individuals are

Algorithm 2. Auxiliary Functions

```
function COMPUTEFITNESSVIOLATION(population)
    Evaluate fitness and constraint violation for each individual in population.
end function

function DIVIDE(population, antibodies, antigens)
    Segregate population into antibodies (infeasible) and antigens (feasible) populations.
end function

function COMPUTEDISTANCE(antigens, antibodies)
    For each antibody in antibodies population, compute the sum of its genotypical distance
    to each antigen in antigens population.
end function

function CLONE(antibodies, temp)
    Clone each antibody in antibodies population.
    Put parents and clones in temp population.
end function

function MUTATION(temp)
    Mutate individuals in temp population by randomly changing the bits of each genotype.
end function

function SELECTBETTER(temp, antibodies)
    For each parent and its offspring in temp population, select the one with highest affinity
    and store in antibodies population.
end function

function UNION(antibodies, antigens, population)
    Combine feasible individuals from antigens population and infeasible ones from
    antibodies population into population.
end function

function INSERT(g, pop)
    Insert individual g in population pop.
end function

function TOURNAMENTSELECTION(population, temp)
    for i = 1 : population.size do
        GETBEST(population[i], population[random()], best)
        INSERT(best, temp)
    end for
end function

function CROSSOVER(temp)
    Perform uniform crossover on the individuals in temp.
end function

function CHANGEPOPULATION(temp, population)
    Keep elite individuals in population and complete it with the best individuals from temp.
end function
```

Algorithm 3. Function changePopulation

```
function CHANGEPOPULATION^C(temp, population)
    UNION(population, temp, tmp)
    SORT(tmp)
    for i = 1 : tmp.size do
        for j = i+1 : tmp.size do
            if not ISCLEARING(tmp[j]) then
                CALCDISTANCE(tmp[i], tmp[j], d)
                if d < criticalDistance then
                    SETCLEAR(tmp[j])
                end if
            end if
        end for
    end for
    CLEAR(population)
    for i = 1 : tmp.size do
        if not ISCLEARING(tmp[i]) then
            if temp.size! = population.size then
                INSERT(tmp[i], population)
            end if
        end if
    end for
    SORT(temp)
    i ← 1
    while temp.size! = population.size do
        if ISCLEARING(tmp[i]) then
            INSERT(tmp[i], population)
            i ← i+1
        end if
    end while
end function
```

sorted from best to worst and all solutions having a distance from each pivot solution in the population smaller than a given threshold (clearing radius) have their fitness values set to zero. The pivot is the best individual not cleared in the sequence. This procedure is continued until all solutions are considered, that is, either to be a pivot or to be cleared.

Differently from [41], the clearing procedure is applied here when a new population substitutes the previous one. A new set of individuals is composed from the union of both populations (previous and offspring). The procedure of clearing is then executed on that union. However, the fitness values are not set to zero as in [41]. Instead, the individuals that would have been cleared are actually tagged. The new population is made up of non-cleared individuals and, if necessary, completed with the best tagged individuals found in the offspring population (see Algorithm 3).

In [41], the clearing procedure when applied alone to unconstrained multimodal optimization did not produce good results. Here, in order to favor the maintenance

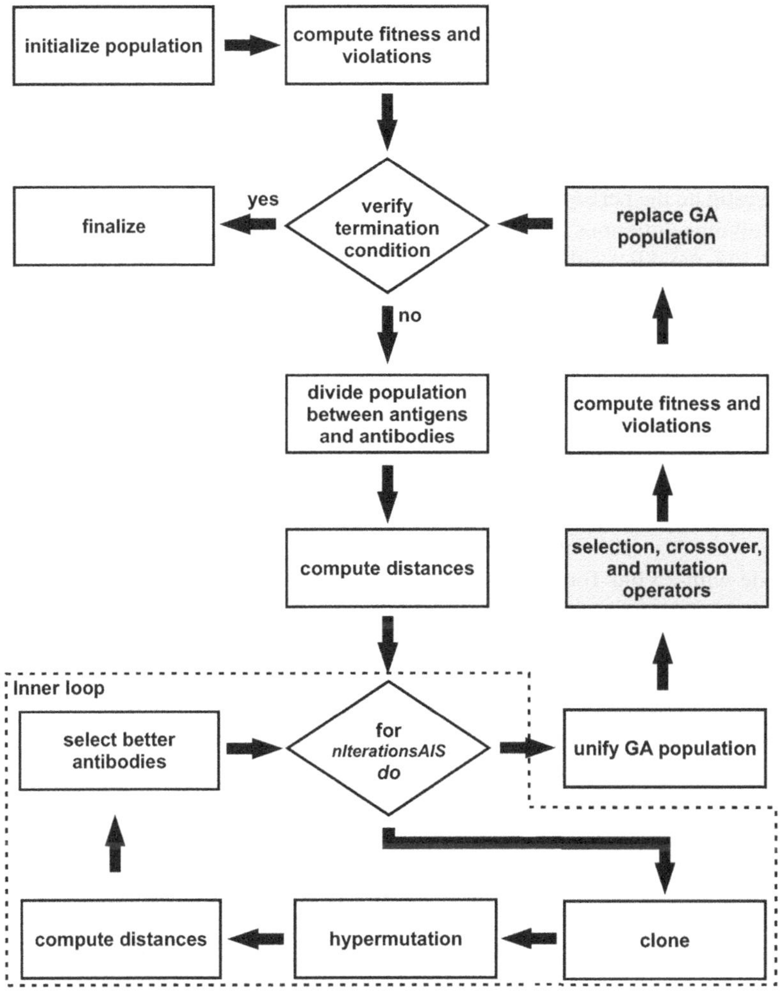

Fig. 1 Fluxogram of the GA-AIS hybrid algorithm

of the niches, the crossover operator is applied to similar individuals. The mate of a given individual from the intermediary population is the closest one, according to the genotypical distance, in that population. This mate-restriction process improved the performance of the clearing procedure, although increasing the computational complexity. The remaining steps of the GA-AIS^C hybrid are the same as those in GA-AIS.

The Figure 1 shows a fluxogram of the GA-AIS hybrid algorithm. For the GA-AIS with clearing procedure the fluxogram is the same. However, as described above, the steps "selection, crossover, and mutation operators" and "replace GA population" are not the same. On the first step, the crossover procedure is applied to similar individuals. On the second one, the elitism is replaced by the modified clearing method detailed above in the Algorithm 3.

5 Numerical Experiments

In order to investigate the performance of the proposed hybrid, several optimization problems, involving continuous and/or discrete variables, are considered.

The APM and the SR method used a population size of 100 individuals, a crossover probability equal to 0.9, and a mutation rate equal to 0.004. Also, elitism is implemented: the copies of the two best individuals remain in the next generation. For the GA-AIS and the GA-AIS^C (GA-AIS hybrid with clearing procedure), the population size was set equal to 20, the mutation rate was set equal to 0.04, uniform crossover was applied with probability 1, the bits being swapped with probability 0.5. Elitism was implemented exactly as in the APM and SR procedures. Also, each antibody generates three clones, and the number of inner (AIS) generations is set equal to 20. For the GA-AIS^C hybrid the minimum distance (clearing radius) to avoid clearing was set to 10% of the chromosome length. All techniques use a binary Gray code with 25 bits for each continuous variable. A total of 25 runs were performed for each test-case in the 24-function suite, and 50 runs for the remaining optimization problems.

In [5], the GA-AIS hybrid was tested in 13 functions from the G-Suite [29]. Also, other experiments, not included here, can be found in [6], where six constrained optimization problems for engineering, with continuous, discrete, and mixed variables, were used to evaluate the performance of GA-AIS and GA-AIS^C. The presented algorithms also show good results for the problems in that reference.

5.1 Test 1 - The G-Suite

A suite of test-problems with continuous variables introduced in [29] and recently enlarged in [33] has been often used as a test-bed for constraint handling techniques in the evolutionary computation literature. Three levels for the maximum number of function evaluations allowed are considered here. For each problem, the Tables 11 to 31, presented in Appendix, show the results obtained for the three techniques using 5000, 50000, and 500000 evaluations of the objective function. In each table the first three lines correspond to the results obtained using 5000 function evaluations. The second and the third blocks of three lines correspond to 50000 and 500000 function evaluations, respectively. The values of the objective functions for the best, median, and worst run are presented together with the average, standard deviation (St.Dev), the number of runs where feasible results were found (FR), and ne is the maximum number of function evaluations. The best results in each case are displayed in **boldface**.

Results for the functions g_{20}, g_{21} and g_{22} are not presented since all techniques were unable to produce feasible solutions.

Table 1 Best performing technique in each problem with 5000 objective function evaluations

	Best	Median	Average	Worst	FR
g_1	GA-AIS^C	GA-AIS^C	GA-AIS^C	GA-AIS^C	GA-AIS/GA-AIS^C
g_2	GA-AIS	SR	SR	SR	–
g_3	APM	APM	APM	APM	APM
g_4	GA-AIS	GA-AIS	GA-AIS^C	GA-AIS^C	–
g_5	SR	SR	SR	SR	SR
g_6	SR	GA-AIS^C	SR	SR	APM
g_7	GA-AIS^C	GA-AIS^C	SR	SR	–
g_8	–	GA-AIS/GA-AIS^C SR	GA-AIS^C	GA-AIS^C	GA-AIS/GA-AIS^C APM
g_9	GA-AIS^C	GA-AIS^C	GA-AIS^C	GA-AIS^C	–
g_{10}	GA-AIS^C	GA-AIS^C	GA-AIS^C	GA-AIS^C	GA-AIS
g_{11}	SR	SR	SR	SR	APM
g_{12}	GA-AIS/GA-AIS^C SR	GA-AIS/GA-AIS^C SR	GA-AIS	GA-AIS	GA-AIS/GA-AIS^C SR
g_{13}	–	–	–	–	–
g_{14}	–	–	–	–	–
g_{15}	APM	APM	APM	APM	APM
g_{16}	GA-AIS^C	GA-AIS^C	GA-AIS^C	GA-AIS^C	GA-AIS/GA-AIS^C
g_{17}	–	–	–	–	–
g_{18}	GA-AIS^C	GA-AIS^C	GA-AIS^C	GA-AIS^C	GA-AIS^C
g_{19}	GA-AIS^C	GA-AIS^C	GA-AIS^C	GA-AIS^C	–
g_{23}	–	–	–	–	–
g_{24}	GA-AIS^C	GA-AIS	GA-AIS	GA-AIS^C	–
APM	33.33%	28.57%	28.57%	28.57%	**71.43%**
SR	42.86%	42.86%	42.86%	42.86%	57.14%
GA-AIS	33.34%	38.10%	28.57%	23.81%	**71.43%**
GA-AIS^C	**66.67%**	**66.67%**	**57.14%**	**61.90%**	**71.43%**

The Tables 1, 2, and 3 present the best performing technique in each problem considering 5000, 50000, and 500000 objective function evaluations, respectively. A "–" indicates that all techniques attained the same performance indicator in the corresponding problem.

Using 5000 function evaluations one can observe from the Table 1 that the GA-AIS technique with clearing produced the best overall performance. The APM and the GA-AIS produced the same result with respect to FR.

Considering 50000 function evaluations, the APM produced the best performance when the best value found, average and FR are considered (Table 2). The Median is equal for the APM, SR , and GA-AIS^C techniques. When the worst value found is considered the APM method has only a slightly better performance against the value found by GA-AIS technique with clearing procedure.

Considering 500000 function evaluations (Table 3), the APM produced better results for all indicators. The GA-AIS^C is the second one here.

Table 2 Best performing technique in each problem with 50000 objective function evaluations

	Best	Median	Average	Worst	FR
g_1	SR	SR	SR	SR	GA-AIS/GA-AIS^C APM
g_2	APM	SR	APM	SR	–
g_3	APM	APM	APM	APM	GA-AIS/GA-AIS^C APM
g_4	APM	GA-AIS^C	GA-AIS^C	GA-AIS^C	–
g_5	SR	SR	SR	SR	SR
g_6	GA-AIS	GA-AIS^C	GA-AIS	GA-AIS	GA-AIS/GA-AIS^C
g_7	APM	GA-AIS^C	GA-AIS^C	GA-AIS^C	GA-AIS/GA-AIS^C APM
g_8	–	–	GA-AIS/GA-AIS^C APM	GA-AIS/GA-AIS^C APM	–
g_9	APM	GA-AIS^C	GA-AIS^C	GA-AIS^C	–
g_{10}	APM	APM	APM	APM	GA-AIS
g_{11}	GA-AIS	SR	SR	SR	GA-AIS/GA-AIS^C APM
g_{12}	–	–	–	–	–
g_{13}	APM	APM	APM	APM	APM
g_{14}	APM	APM	APM	GA-AIS	APM
g_{15}	APM	SR	SR	SR	APM
g_{16}	GA-AIS^C	GA-AIS^C	GA-AIS^C	GA-AIS^C	GA-AIS/GA-AIS^C APM
g_{17}	APM	APM	APM	APM	APM
g_{18}	SR	SR	APM	APM	GA-AIS
g_{19}	APM	GA-AIS^C	GA-AIS^C	GA-AIS^C	GA-AIS/GA-AIS^C APM
g_{23}	APM	APM	APM	APM	APM
g_{24}	GA-AIS/GA-AIS^C APM	GA-AIS	GA-AIS	GA-AIS	GA-AIS/GA-AIS^C APM
APM	**71.43%**	**38.10%**	**47.62%**	**38.10%**	**80.95%**
SR	23.81%	**38.10%**	23.81%	28.57%	28.57%
GA-AIS	23.81%	14.29%	19.05%	23.81%	71.43%
GA-AIS^C	19.05%	**38.10%**	33.33%	33.33%	61.90%

In general, the GA-AIS^C hybrid produced better results when a reduced (5000) number of function evaluations are used. This is due to the exploration effect generated by the clearing method.

It is also observed that in problems with equality constraints, such as g_3, g_5, g_{11}, and g_{15}, the GA-AIS hybrids did not produce good results.

5.2 *Test 2 - A MINLP Problem*

This problem, from [3, 14], corresponds to the maximization of a nonlinear function with three continuous, two integer variables, and three nonlinear inequalities. This

Table 3 Best performing technique in each problem with 500000 objective function evaluations

	Best	Median	Average	Worst	FR
g_1	APM	APM	SR	SR	–
g_2	APM	APM	APM	APM	–
g_3	APM	APM	APM	APM	GA-AIS/GA-AIS^C APM
g_4	APM	APM	APM	APM	GA-AIS/GA-AIS^C APM
g_5	SR	SR	SR	GA-AIS	APM
g_6	GA-AIS^C/SR APM	SR	SR	SR	GA-AIS/GA-AIS^C
g_7	SR	GA-AIS^C	GA-AIS^C	GA-AIS^C	GA-AIS/GA-AIS^C APM
g_8	GA-AIS GA-AIS^C/APM	GA-AIS APM/GA-AIS^C	GA-AIS APM/GA-AIS^C	GA-AIS APM/GA-AIS^C	–
g_9	APM	APM	APM	GA-AIS^C	GA-AIS/GA-AIS^C APM
g_{10}	GA-AIS^C	GA-AIS^C	GA-AIS^C	SR	GA-AIS/GA-AIS^C
g_{11}	GA-AIS	GA-AIS	GA-AIS	GA-AIS	GA-AIS/GA-AIS^C APM
g_{12}	–	–	–	–	–
g_{13}	APM	SR	SR	APM	APM
g_{14}	APM	APM	APM	APM	APM
g_{15}	GA-AIS	SR	SR	SR	APM
g_{16}	APM	APM	GA-AIS^C	GA-AIS^C	GA-AIS/GA-AIS^C APM
g_{17}	APM	APM	APM	APM	APM
g_{18}	GA-AIS^C	APM	APM	APM	GA-AIS
g_{19}	GA-AIS^C	GA-AIS^C	GA-AIS^C	GA-AIS^C	–
g_{23}	GA-AIS^C	APM	APM	APM	GA-AIS^C
g_{24}	–	–	–	–	–
APM	**61.90%**	**61.90%**	**52.38%**	**52.38%**	**80.95%**
SR	23.81%	28.57%	33.33%	28.57%	28.57%
GA-AIS	23.81%	19.05%	19.05%	23.81%	71.43%
GA-AIS^C	38.01%	28.57%	33.33%	33.33%	71.43%

is the Problem 6 in [14] and Test-problem 14 in [3]. Fifty independent runs were performed against ten in [14].

A comparison was done with the APM [3] and Stochastic Ranking. The statistical results are presented in Table 4. The SR and GA-AIS techniques used 5000 function evaluations against 30000 in the APM [3] case. The results are similar to those obtained by the APM in [3]. However, the number of function evaluations used here is one sixth of that used by the APM in [3].

Table 4 Results for the Test 2. Optimal value is −32217.42778

Method	Best	Median	Average	St.Dev	Worst	FR	ne
APM [3]	**32217.43**	–	**32217.43**	–	**32217.43**	–	30000
SR	**32217.43**	32217.35	32217.05	1.20	32209.27	**50**	**5000**
GA-AIS	**32217.43**	**32217.43**	**32217.43**	**5.89E − 3**	32217.39	**50**	**5000**
GA-AIS^C	**32217.43**	**32217.43**	32217.42	$7.80E-3$	32217.37	**50**	**5000**

5.3 *Test 3 - A NLIP Problem*

This is a nonlinear programming problem (Problem 8 in [35]) with five integer variables and eight inequality constraints. Fifty independent runs were performed against thirty in [35]. A comparison between the GA-AIS hybrids and the SR technique is given in Table 5, when 20000 function evaluations are allowed. The GA-AIS technique produced the best overall performance, except for the worst value found, where the GA-AIS with clearing procedure found the best value.

Table 5 Results for the Test 3. Optimal value is 807

Method	Best	Median	Average	St.Dev	Worst	FR	ne
SR	**807**	860	872.76	61.65	1036	**50**	**20000**
GA-AIS	**807**	**807**	**808.96**	**5.80**	842	**50**	**20000**
GA-AIS^C	**807**	**807**	812.12	7.84	**833**	**50**	**20000**

5.4 *Test 4 - A NLIP Problem*

This problem (Problem 16 in [35]) has thirteen integer variables and nine inequality constraints. Fifty independent runs were performed against thirty in [35]. In Table 6 a comparison of the results found by the GA-AIS hybrids and the SR technique is presented when 1500 evaluations are available. The GA-AIS techniques produced the best overall results.

5.5 *Test 5 - The 25-Bar Truss Design*

This is a well known structural weight minimization problem for a truss structure. The statistical results for GA-AIS and APM [32] can be seen in Table 7. The results from [30] were obtained using a binary-coded GA. Both GA-AIS hybrids and the APM [32] found a final weight of 484.854 lb. Using only 800 function evaluations the GA-AIS hybrids from [6] found very good results when compared to those presented by the references [30] and [52].

It should be remarked that: (i) all solutions are feasible, (ii) correspond to distinct designs, and (iii) the GA-AIS provides the smallest weight.

Table 6 Results for the Test 4. Optimal value is −15

Method	Best	Median	Average	St.Dev	Worst	FR	ne
SR	**−15**	**−15**	−14.86	$4.05E-1$	−13	**50**	**1500**
GA-AIS	**−15**	**−15**	**−15**	**0.00**	**−15**	**50**	**1500**
GA-AIS^C	**−15**	**−15**	**−15**	**0.00**	**−15**	**50**	**1500**

Table 7 Values found for the final weight of the 25-bar Truss design using 20000 function evaluations

	Best	Median	Average	St.Dev	Worst	FR	nfe
Ref. [30]	546.01	–	–	–	–	–	**800**
Ref. [52]	562.93	–	–	–	–	–	**800**
GA-AIS	487.329	511.554	512.506	15.73	562.222	**50**	**800**
GA-AIS^C	487.718	514.850	515.520	11.31	548.501	**50**	**800**
Ref. [48]	486.29	–	–	–	–	–	40000
Ref. [17]	493.80	–	–	–	–	–	30000
APM [32]	**484.854**	–	**485.967**	–	**490.742**	–	20000
GA-AIS	**484.854**	**486.023**	486.196	**1.49**	492.554	**50**	20000
GA-AIS^C	**484.854**	**486.023**	486.485	2.10	496.783	**50**	20000

5.6 *Test 6 - Reliability Problems*

In this section a reliability problem is solved and compared with results from the literature. Various objectives can be considered, as one would like to optimize system reliability (R_s), total system cost (C), and total system weight (W). The formulation can be found in [16,50].

Dhingra [16] used a nonlinear mathematical programming (NMP) method and Li et al [50] used a GA. We run 50 independent tests against 10 from [50].

We present in Tables 8, 9, and 10 the results found with 15000 function evaluations for all three optimization problems. When the objective is to maximize the reliability, the GA-AIS^C produced the best overall results. When the minimization

Table 8 Results found by the GA-AIS hybrids for the reliability problem with 15000 function evaluations (300 generations). The objective is to maximize the reliability R_s

	Best	Median	Average	St.Dev	Worst	FR
NMP [16]	0.99961	-	-	-	-	-
GA ($\gamma = 1$) [50]	**0.999955**	-	-	-	-	-
GA ($\gamma = 2$) [50]	0.999954	-	-	-	-	-
GA ($\gamma = 4$) [50]	0.999954	-	-	-	-	-
GA-AIS	0.999954	0.999945	0.999939	$1.98E-5$	0.999880	**50**
GA-AIS^C	**0.999955**	**0.999946**	**0.999946**	**6.60E − 6**	**0.999917**	**50**

of cost is considered, the GA-AIS found best results. All algorithms produced the same result when the objective was to minimize the weight.

The proposed GA-AIS found the best results for all optimization cases.

Table 9 Results found by the GA-AIS for the reliability problem with 15000 function evaluations (300 generations). The objective is to minimize the cost C

	Best	Median	Average	St.Dev	Worst	FR
NMP [16]	20.7252	-	-	-	-	-
GA-AIS	**20.14028**	**20.37048**	**20.35425**	**1.82E − 1**	**20.80367**	**50**
GA-AIS^C	20.14380	20.40202	20.38474	$1.87E-1$	20.84020	**50**

Table 10 Results found by the GA-AIS for the reliability problem with 15000 function evaluations (300 generations). The objective is to minimize the weight W

	Best	Median	Average	St.Dev	Worst	FR
NMP [16]	34.6687	-	-	-	-	-
GA-AIS	**34.668686**	**34.668686**	**34.668686**	**0.00**	**34.668686**	**50**
GA-AIS^C	**34.668686**	**34.668686**	**34.668686**	**0.00**	**34.668686**	**50**

6 Conclusions

A genetic algorithm hybridized with an artificial immune system was tested in a set of continuous, integer, and mixed integer-continuous constrained optimization problems. The experiments included the G-Suite with 24 functions, a MINLP problem, two NLIP problems, a discrete structural engineering optimization problem, and, three mixed-integer reliability problems. Two techniques to handle constraints are used in order to provide comparisons for the G-Suite, i.e., the Adaptive Penalty Method and the Stochastic Ranking technique. For the other problems, the results were compared with the literature. The hybrid GA-AIS performed very well in various problems presenting continuous, discrete, and mixed design variables. Its efficacy and generality indicate its applicability to other constrained optimization problems.

Acknowledgements. The authors thank the support from CNPq (311651/2006-2 and 154674/2006-0), CAPES, FAPERJ (E-26/100.101/2008 and E-26/102.825/2008), and FAPEMIG, as well as the corrections and suggestions from the reviewers.

Appendix

This appendix presents Tables 11 to 31 with the results found by the discussed algorithms when applied to the G-Suite of test-problems, commented in Section 5.1.

Table 11 Results for the g_1 test-problem. Optimal value is -15

Method	Best	Median	Average	St.Dev	Worst	FR	ne
APM	−12.54560	−9.66313	−9.63801	1.83	−3.54334	23	5000
SR	−12.53710	−8.89068	−8.50374	2.15	−5.51379	9	5000
GA-AIS	−12.42924	−9.97209	−9.70580	1.69	−3.11102	**25**	5000
GA-AIS^C	**−13.58816**	**−12.07793**	**−11.80723**	**1.31**	**−8.12925**	**25**	5000
APM	−14.99420	−14.98342	−14.97430	$3.26E-2$	−14.83784	**25**	50000
SR	**−14.99857**	**−14.99300**	**−14.99313**	**3.81E − 3**	**−14.98008**	24	50000
GA-AIS	−14.78591	−14.65819	−14.60320	$2.16E-1$	−13.25278	**25**	50000
GA-AIS^C	−14.86625	−14.76883	−14.76332	$4.81E-2$	−14.64413	**25**	50000
APM	**−14.99996**	**−14.99954**	−14.96378	**9.79E − 2**	−14.57225	**25**	500000
SR	−14.99977	−14.99951	**−14.99931**	$5.42E-4$	**−14.99757**	**25**	500000
GA-AIS	−14.98813	−14.98056	−14.97966	$5.66E-3$	−14.96613	**25**	500000
GA-AIS^C	−14.98887	−14.97595	−14.97514	$8.06E-3$	−14.93905	**25**	500000

Table 12 Results for the g_2 test-problem. Optimal value is -0.8036191

Method	Best	Median	Average	St.Dev	Worst	FR	ne
APM	−0.5944454	−0.4985342	−0.4854191	$5.91E-2$	−0.3512636	**25**	5000
SR	−0.5864694	**−0.5283935**	**−0.5312732**	$3.37E-2$	**−0.4541653**	**25**	5000
GA-AIS	**−0.6110063**	−0.5275892	−0.5284940	$3.61E-2$	−0.4487732	**25**	5000
GA-AIS^C	−0.5732360	−0.5236345	−0.5213102	**2.90E − 2**	−0.4498942	**25**	5000
APM	**−0.7745439**	−0.6517182	**−0.6663961**	$6.24E-2$	−0.5706031	**25**	50000
SR	−0.7486565	**−0.6644478**	−0.6653669	$7.01E-2$	**−0.5154426**	**25**	50000
GA-AIS	−0.7332407	−0.6444320	−0.6526506	$4.90E-2$	−0.5351180	**25**	50000
GA-AIS^C	−0.7482104	−0.6469565	−0.6455348	**4.70E − 2**	−0.5435568	**25**	50000
APM	**−0.8015561**	**−0.7787246**	**−0.7724801**	**2.26E − 2**	**−0.7261164**	**25**	500000
SR	−0.7854180	−0.7405743	−0.7256381	$5.57E-2$	−0.6079553	**25**	500000
GA-AIS	−0.7934579	−0.7428760	−0.7383075	$3.73E-2$	−0.6111616	**25**	500000
GA-AIS^C	−0.7834507	−0.7437401	−0.7351923	$3.51E-2$	−0.6567412	**25**	500000

Table 13 Results for the g_3 test-problem. Optimal value is -1.0005001. It is important to notice that the SR technique was unable to produce feasible solutions for this test-problem

Method	Best	Median	Average	St.Dev	Worst	FR	ne
APM	**−0.4770853**	**−0.1529288**	**−0.2313726**	$9.65E-2$	**−0.1104597**	**24**	5000
GA-AIS	−0.2367439	−0.0035223	−0.0253767	$5.03E-2$	−0.0000028	20	5000
GA-AIS^C	−0.1749013	−0.0029394	−0.0166668	$3.39E-1$	−0.0000003	22	5000
APM	**−0.9173612**	**−0.5289950**	**−0.5508323**	**1.44E − 1**	**−0.3099690**	**25**	50000
GA-AIS	−0.2853399	−0.0199313	−0.0437244	$6.24E-2$	−0.0001201	**25**	50000
GA-AIS^C	−0.4007668	−0.0172259	−0.0517864	$8.53E-2$	−0.0000029	**25**	50000
APM	**−1.0004896**	**−1.0004466**	**−1.0004036**	**1.11E − 4**	**−0.9999571**	**25**	500000
GA-AIS	−0.7615700	−0.2659286	−0.2976153	$1.93E-1$	−0.0267321	**25**	500000
GA-AIS^C	−0.8879961	−0.4592956	−0.4908986	$2.12E-2$	−0.0406379	**25**	500000

Table 14 Results for the g_4 test-problem. Optimal value is −30665.5386717

Method	Best	Median	Average	St.Dev	Worst	FR	ne
APM	−30566.3455	−30383.8927	−30375.9930	$1.24E+2$	−30103.4912	**25**	5000
SR	−30574.5977	−30396.2109	−30374.8434	$1.22E+3$	−30170.9648	**25**	5000
GA-AIS	**−30642.9014**	**−30520.1760**	−30491.8207	$9.42E+1$	−30225.0856	**25**	5000
GA-AIS^C	−30641.3563	−30515.4392	**−30495.9788**	**8.06E+1**	**−30281.1385**	**25**	5000
APM	**−30664.6630**	−30564.6626	−30541.6923	$8.76E+1$	−30367.2095	**25**	50000
SR	−30660.2910	−30508.0762	−30491.9935	$1.47E+3$	−29985.3633	**25**	50000
GA-AIS	−30661.1218	−30630.4060	−30622.2918	$3.87E+1$	−30416.7505	**25**	50000
GA-AIS^C	−30664.4368	**−30636.4277**	**−30632.8975**	**1.99E+1**	**−30581.5809**	**25**	50000
APM	**−30665.5238**	**−30665.4709**	**−30665.0947**	**8.55E−1**	**−30661.7610**	**25**	500000
SR	−30664.7051	−30651.1426	−30633.9923	$4.79E+1$	−30453.1055	23	500000
GA-AIS	−30665.4397	−30663.4553	−30662.6091	3.21	−30651.3820	**25**	500000
GA-AIS^C	−30665.3889	−30663.7472	−30662.9301	2.28	−30655.5270	**25**	500000

Table 15 Results for the g_5 test-problem. Optimal value is 5126.4967140. Only the techniques presenting at least one feasible solution in at least one run are shown

Method	Best	Median	Average	St.Dev	Worst	FR	ne
SR	**5141.8721**	**5193.0991**	**5221.2872**	**8.76E+1**	**5421.4668**	**9**	5000
APM	5137.3998	5262.4268	5420.4905	$3.54E+2$	5993.0124	9	50000
SR	**5126.5215**	**5149.3882**	**5157.0944**	**3.507E+1**	**5258.4805**	**16**	50000
APM	5127.3606	5244.5322	5312.6175	$2.45E+2$	5993.0113	**24**	500000
SR	**5126.5195**	**5137.3833**	**5155.8345**	$3.87E+1$	5258.4736	13	500000
GA-AIS	5166.0885	5189.4730	5204.1544	**3.85E+1**	**5256.9017**	3	500000

Table 16 Results for the g_6 test-problem. Optimal value is −6961.8138755

Method	Best	Median	Average	St.Dev	Worst	FR	ne
APM	−6834.5344	−3836.6674	−5120.8467	$1.58E+3$	−1276.4607	**24**	5000
SR	**−6922.1831**	−5787.4535	**−5605.5055**	**1.09E+3**	**−2215.5444**	18	5000
GA-AIS	−6650.0106	−5626.4849	−5318.8276	$1.22E+3$	−1372.5897	23	5000
GA-AIS^C	−6584.6677	**−5847.3395**	−5553.2601	$1.22E+3$	−1394.2647	20	5000
APM	−6939.3001	−3984.0864	−6413.3729	$1.11E+3$	−1319.0707	24	50000
SR	−6956.6471	−6682.5586	−6612.7806	$2.96E+2$	−5772.6938	15	50000
GA-AIS	**−6960.8910**	−6904.2614	**−6901.5555**	**2.97E+1**	**−6786.0714**	**25**	50000
GA-AIS^C	−6955.0718	**−6907.1951**	−6896.5274	$6.85E+1$	−6464.7069	**25**	50000
APM	**−6961.7961**	−6961.7710	−6961.7742	$1.42E−2$	−6961.7592	24	500000
SR	**−6961.7961**	**−6961.7827**	**−6961.7863**	**5.19E−3**	**−6961.7827**	13	500000
GA-AIS	−6961.7894	−6961.7659	−6961.7682	$7.87E−3$	−6961.7491	**25**	500000
GA-AIS^C	**−6961.7961**	−6961.7558	−6961.7574	$1.23E−2$	−6961.7424	**25**	500000

Table 17 Results for the g_7 test-problem. Optimal value is 24.3062090

Method	Best	Median	Average	St.Dev	Worst	FR	ne
APM	43.2072211	86.2003283	93.1306621	$3.28E+1$	152.181957	**25**	5000
SR	f31.9050388	50.1892052	**51.5728580**	**1.33E+1**	**74.3306732**	**25**	5000
GA-AIS	34.7181969	72.7931753	110.996260	$1.25E+2$	776.563144	**25**	5000
GA-AIS^C	**28.6809702**	**45.0236259**	55.4336500	$2.90E+1$	211.927963	**25**	5000
APM	**24.5673615**	28.3454008	29.2488815	3.67	38.5389701	**25**	50000
SR	25.0301437	28.3579140	30.0073784	5.61	46.1200676	24	50000
GA-AIS	25.4217483	31.6841466	33.9099435	$1.02E+1$	95.1135317	**25**	50000
GA-AIS^C	24.9646997	**26.0109090**	**26.1673499**	**9.77E−1**	**28.9968472**	**25**	50000
APM	24.4860035	26.8199501	27.0714049	2.27	34.4138410	**25**	500000
SR	**24.4150887**	28.1761951	28.4540093	4.00	41.6882782	21	500000
GA-AIS	24.5600669	28.4441414	29.8455249	4.35	41.0650770	**25**	500000
GA-AIS^C	24.4482393	**24.7079128**	**24.8274137**	**3.20E−1**	**25.9281175**	**25**	500000

Table 18 Results for the g_8 test-problem. Optimal value is −0.0958250

Method	Best	Median	Average	St.Dev	Worst	FR	ne
APM	**−0.0958250**	−0.0958248	−0.9582019	$1.17E-5$	−0.0957811	**25**	5000
SR	**−0.0958250**	**−0.0958250**	−0.0874181	$2.25E-2$	−0.0291438	24	5000
GA-AIS	**−0.0958250**	**−0.0958250**	−0.0944914	$9.34E-3$	−0.0291438	**25**	5000
GA-AIS^C	**−0.0958250**	**−0.0958250**	**−0.0958250**	**0.00**	**−0.0958250**	**25**	5000
APM	**−0.0958250**	**−0.0958250**	**−0.0958250**	**0.00**	**−0.0958250**	**25**	50000
SR	**−0.0958250**	**−0.0958250**	−0.0851560	$2.49E-2$	−0.0291438	**25**	50000
GA-AIS	**−0.0958250**	**−0.0958250**	**−0.0958250**	**0.00**	**−0.0958250**	**25**	50000
GA-AIS^C	**−0.0958250**	**−0.0958250**	**−0.0958250**	**0.00**	**−0.0958250**	**25**	50000
APM	**−0.0958250**	**−0.0958250**	**−0.0958250**	**0.00**	**−0.0958250**	**25**	500000
SR	−0.0958251	−0.0958251	−0.0878233	$2.21E-2$	−0.0291438	**25**	500000
GA-AIS	**−0.0958250**	**−0.0958250**	**−0.0958250**	**0.00**	**−0.0958250**	**25**	500000
GA-AIS^C	**−0.0958250**	**−0.0958250**	**−0.0958250**	**0.00**	**−0.0958250**	**25**	500000

Table 19 Results for the g_9 test-problem. Optimal value is 680.6300573

Method	Best	Median	Average	St.Dev	Worst	FR	ne
APM	684.6761547	703.3695986	714.3836304	$2.96E+1$	825.0185225	**25**	5000
SR	684.8146362	696.2620239	702.4067383	$1.38E+1$	734.6710205	**25**	5000
GA-AIS	682.9336332	690.1082142	693.8352805	9.39	726.2259899	**25**	5000
GA-AIS^C	**681.8268793**	**686.3363980**	**686.9471871**	**3.20**	**694.0967521**	**25**	5000
APM	**680.7658016**	681.6898093	682.1195215	1.36	687.0250951	**25**	50000
SR	680.8317261	683.3109131	685.3920581	6.91	715.9301758	**25**	50000
GA-AIS	680.8432734	683.2076576	683.8952635	2.42	690.3016947	**25**	50000
GA-AIS^C	680.7949849	**681.4916867**	**681.5876354**	**4.48E−1**	**682.6530750**	**25**	50000
APM	**680.6474288**	**680.7799403**	**680.7979077**	**1.72E−2**	681.4545510	**25**	500000
SR	680.7388916	681.9906006	682.1157532	1.31	687.0321655	22	500000
GA-AIS	680.6545927	681.3924261	681.4811963	$5.75E-1$	683.0550587	**25**	500000
GA-AIS^C	680.6801414	680.8877626	680.9005951	$1.30E-1$	**681.2122556**	**25**	500000

Table 20 Results for the g_{10} test-problem. Optimal value is 7049.2480205. Only the techniques presenting at least one feasible solution in at least one run are shown

Method	Best	Median	Average	St.Dev	Worst	FR	ne
GA-AIS	8821.3618	11666.3139	12543.9797	**3.70E+3**	24155.3658	**18**	5000
GA-AIS^C	**7769.5966**	**10143.4935**	**11684.9899**	$3.95E+3$	**24097.0972**	15	5000
APM	**7080.4052**	**7581.6931**	**7683.2236**	**6.59E+2**	**8387.5723**	3	50000
GA-AIS	7157.2501	8297.4039	8630.7599	$1.42E+3$	14944.8744	**25**	50000
GA-AIS^C	7142.5267	7675.6942	7873.9504	$7.17E+2$	10682.2014	23	50000
APM	7068.6338	8181.6974	8154.6199	$7.75E+2$	9769.1018	9	500000
SR	7538.5068	7538.5068	7538.5068	–	**7538.5068**	1	500000
GA-AIS	7054.8350	8123.7119	8417.9070	$1.27E+3$	13205.017	**25**	500000
GA-AIS^C	**7053.5055**	**7314.0347**	**7361.5796**	**2.69E+2**	8701.1528	**25**	500000

Table 21 Results for the g_{11} test-problem. Optimal value is 0.7499

Method	Best	Median	Average	St.Dev	Worst	FR	ne
APM	0.7499568	0.8204312	0.8380892	$7.94E-2$	0.9972218	**25**	5000
SR	**0.7499117**	**0.7596881**	**0.7707811**	**3.51E−2**	**0.8982297**	18	5000
GA-AIS	0.7501970	0.8125605	0.8489786	$8.62E-2$	0.9968189	24	5000
GA-AIS^C	0.7500200	0.8214822	0.8463420	$8.83E-2$	0.9995183	22	5000
APM	0.7499568	0.8204122	0.8377019	$7.92E-2$	0.9970093	**25**	50000
SR	0.7499093	**0.7531331**	**0.7697072**	**3.07E−2**	**0.8891737**	24	50000
GA-AIS	**0.7499000**	0.7549079	0.7824960	$5.07E-2$	0.9782222	**25**	50000
GA-AIS^C	0.7499013	0.7659117	0.7999625	$6.49E-2$	0.9904011	**25**	50000
APM	0.7499540	0.8118332	0.8334051	$7.65E-2$	0.9936831	**25**	500000
SR	0.7499090	0.7583824	0.7708550	$2.60E-2$	0.8309575	23	500000
GA-AIS	**0.7499000**	**0.7499317**	**0.7499695**	**1.62E−4**	**0.7510275**	**25**	500000
GA-AIS^C	0.7499001	0.7499380	0.7508844	$3.90E-3$	0.7709872	**25**	500000

Table 22 Results for the g_{12} test-problem. Optimal value is -1. Only the techniques presenting at least one feasible solution in at least one run are shown

Method	Best	Median	Average	St.Dev	Worst	FR	ne
SR	−1	−1	−0.9985502	$2.66E-3$	−0.9924536	**25**	5000
GA-AIS	−1	−1	−1	**0.00**	−1	**25**	5000
GA-AIS^C	−1	−1	−0.9999999	$2.04E-7$	−0.999999	**25**	5000
APM	−1	−1	−1	**0.00**	−1	**25**	50000
SR	−1	−1	−1	**0.00**	−1	**25**	50000
GA-AIS	−1	−1	−1	**0.00**	−1	**25**	50000
GA-AIS^C	−1	−1	−1	**0.00**	−1	**25**	50000
APM	−1	−1	−1	**0.00**	−1	**25**	500000
SR	−1	−1	−1	**0.00**	−1	**25**	500000
GA-AIS	−1	−1	−1	**0.00**	−1	**25**	500000
GA-AIS^C	−1	−1	−1	**0.00**	−1	**25**	500000

Table 23 Results for the g_{13} test-problem. Optimal value is 0.0539415. The results with 5000 function evaluations are not presented since all techniques were unable to produce feasible solutions

Method	Best	Median	Average	St.Dev	Worst	FR	ne
APM	**0.0556113**	**0.0966101**	**0.8095851**	**2.93E − 1**	**1.3202477**	**25**	50000
SR	0.0670723	0.3804254	0.5788253	$6.24E-1$	2.4329371	13	50000
GA-AIS	0.2970864	0.9747390	1.0930967	$8.92E-1$	4.6175239	9	50000
GA-AIS^C	0.5150024	0.9535153	1.3761583	1.25	4.8781921	5	50000
APM	**0.0556111**	0.9288099	0.8829213	**3.83E − 1**	**1.8784189**	**25**	500000
SR	0.1003231	**0.5086704**	**0.7698459**	$7.59E-1$	2.4842145	12	500000
GA-AIS	0.2970188	0.9754438	1.1859940	1.11	7.2200664	24	500000
GA-AIS^C	0.5148793	0.9138921	1.2220168	1.14	4.8769912	6	500000

Table 24 Results for the g_{14} test-problem. Optimal value is −47.7648885. It is important to notice that the SR technique was unable to produce feasible solutions for this test-problem. All techniques were unable to produce feasible solutions for 5000 function evaluations

Method	Best	Median	Average	St.Dev	Worst	FR	ne
APM	**−46.890438**	**−43.194835**	**−43.027409**	2.41	−38.807925	**25**	50000
GA-AIS	−42.961579	−42.961579	−42.961579	–	**−42.961579**	1	50000
GA-AIS^C	−44.439144	−42.838209	−42.838209	**1.60**	−41.237273	2	50000
APM	**−46.890438**	**−43.203376**	**−43.044209**	2.40	**−38.807998**	**25**	500000
GA-AIS	−45.689376	−42.445971	−42.315373	**1.77**	−37.332921	22	500000
GA-AIS^C	−46.373549	−42.848299	−42.496643	1.83	−38.419681	22	500000

Table 25 Results for the g_{15} test-problem. Optimal value is 961.7150222. Only the techniques presenting at least one feasible solution in at least one run are shown

Method	Best	Median	Average	St.Dev	Worst	FR	ne
APM	**968.965583**	**969.924408**	**969.924408**	**1.35**	**970.883266**	**2**	5000
APM	**962.640839**	965.298033	966.220198	3.06	971.286895	**25**	50000
SR	962.973328	**963.785889**	**963.965942**	**1.02**	**965.377930**	5	50000
GA-AIS	963.051821	966.700903	967.167665	3.39	972.217032	3	50000
APM	962.640483	965.297403	966.218809	3.06	971.285135	**25**	500000
SR	962.482971	**962.698608**	**963.219477**	**8.15E − 1**	**964.853943**	21	500000
GA-AIS	**961.767880**	964.695332	965.472313	3.16	972.216474	12	500000

Table 26 Results for the g_{16} test-problem. Optimal value is −1.9051553

Method	Best	Median	Average	St.Dev	Worst	FR	ne
APM	−1.8598855	−1.4981970	−1.4960260	$2.60E-1$	−1.0315077	17	5000
SR	−1.7976446	−1.5929062	−1.5427779	$1.81E-1$	−1.1406803	22	5000
GA-AIS	−1.8691629	−1.6957917	−1.6466281	$1.66E-1$	−1.1723105	**25**	5000
GA-AIS^C	**−1.8847265**	**−1.7729120**	**−1.7525989**	**8.77E − 2**	**−1.5163932**	**25**	5000
APM	−1.8968595	−1.8343355	−1.8293013	$4.45E-2$	−1.7492531	**25**	50000
SR	−1.8618839	−1.7203965	−1.6604525	$1.89E-1$	−0.9743471	24	50000
GA-AIS	−1.9020257	−1.8343978	−1.8201434	$6.43E-2$	−1.5358576	**25**	50000
GA-AIS^C	**−1.9046090**	**−1.8963019**	**−1.8938212**	**8.09E − 3**	**−1.8700456**	**25**	50000
APM	**−1.9051516**	**−1.9050654**	−1.9032724	$5.45E-3$	−1.8818044	**25**	500000
SR	−1.8696414	−1.7357190	−1.7042652	$1.78E-1$	−1.0286503	24	500000
GA-AIS	−1.9049337	−1.8935411	−1.8863406	$1.93E-2$	−1.8168770	**25**	500000
GA-AIS^C	−1.9051032	−1.9044968	**−1.9042373**	**6.93E − 4**	**−1.9025437**	**25**	500000

Table 27 Results for the g_{17} test-problem. Optimal value is 8853.5396748. Only the techniques presenting at least one feasible solution in at least one run are shown. All techniques were unable to produce feasible solutions for 5000 function evaluations

Method	Best	Median	Average	St.Dev	Worst	FR	ne
APM	**8940.168230**	**8940.168230**	**8940.168230**	–	**8940.168230**	**1**	50000
APM	**8875.515487**	**8948.681152**	**8980.116781**	**1.10E + 2**	**9276.896157**	**24**	500000

Table 28 Results for the g_{18} test-problem. Optimal value is −0.8660254. Only the techniques presenting at least one feasible solution in at least one run are shown

Method	Best	Median	Average	St.Dev	Worst	FR	ne
GA-AIS^C	**−0.5348792**	**−0.2637329**	**−0.3002225**	**1.22E − 1**	**−0.1420471**	**4**	5000
APM	−0.7869344	−0.7869344	**−0.7869344**	–	**−0.7869344**	1	50000
SR	**−0.8621180**	**−0.8423864**	−0.7611128	$1.35E-1$	−0.4947852	15	50000
GA-AIS	−0.8546679	0.5379352	−0.5779785	$1.26E-1$	−0.3574985	**18**	50000
GA-AIS^C	−0.8586420	−0.6608961	−0.7214536	**1.09E − 1**	−0.5130703	9	50000
APM	−0.8641388	**−0.8640182**	**−0.8619589**	**2.34E − 3**	**−0.8575919**	8	500000
SR	−0.8646700	−0.8491147	−0.7718814	$1.46E-1$	−0.4999466	19	500000
GA-AIS	−0.8641044	−0.6547800	−0.6715419	$1.29E-1$	−0.4882556	**23**	500000
GA-AIS^C	**−0.8652776**	−0.6698892	−0.7442860	$1.04E-1$	−0.5615551	9	500000

Table 29 Results for the g_{19} test-problem. Optimal value is 32.6555929

Method	Best	Median	Average	St.Dev	Worst	FR	ne
APM	255.724401	437.289256	490.432396	$2.04E+2$	1115.49838	**25**	5000
SR	184.100540	307.581970	341.076149	$1.38E+2$	780.532165	**25**	5000
GA-AIS	121.955285	212.209075	215.377929	$5.96E+1$	437.843290	**25**	5000
GA-AIS^C	**108.631591**	**198.263474**	**200.606560**	**4.41E+1**	**301.788178**	**25**	5000
APM	**55.5094357**	90.2558150	100.584946	$2.83E+1$	156.224591	**25**	50000
SR	64.2224197	97.0017242	110.557444	$3.61E+1$	210.558655	24	50000
GA-AIS	65.9034885	117.350395	118.912289	$3.32E+1$	208.347350	**25**	50000
GA-AIS^C	63.5711752	**83.4115213**	**89.0866357**	**2.11E+1**	**155.190918**	**25**	50000
APM	47.5302335	67.7416098	67.5369704	$1.34E+1$	101.205100	**25**	500000
SR	49.4397850	78.6502228	79.2372183	$2.10E+1$	126.887253	**25**	500000
GA-AIS	45.2934737	84.4632468	90.4197446	$2.50E+1$	152.756103	**25**	500000
GA-AIS^C	**40.6118492**	**65.5883385**	**66.0103369**	**1.19E+1**	**94.8934192**	**25**	500000

Table 30 Results for the g_{23} test-problem. Optimal value is −400.0551000. It is important to notice that the SR technique was unable to produce feasible solutions for this test-problem. All techniques were unable to produce feasible solutions for 5000 function evaluations

Method	Best	Median	Average	St.Dev	Worst	FR	ne
APM	**16.670943**	**41.018660**	**41.018660**	**3.44E+1**	**65.366380**	**2**	50000
APM	9.7424748	**37.553526**	**37.553526**	**3.93E+1**	**65.364580**	2	500000
GA-AIS	286.83437	286.83437	286.83437	–	286.83437	1	500000
GA-AIS^C	**4.2437108**	89.368312	118.06984	$1.11E+2$	289.29904	**4**	500000

Table 31 Results for the g_{24} test-problem. Optimal value is −5.5080133

Method	Best	Median	Average	St.Dev	Worst	FR	ne
APM	−5.5076836	−5.4848828	−5.4474105	$7.47E-2$	−5.2005319	**25**	5000
SR	−5.5050306	−5.3985375	−5.3958694	$9.90E-2$	−5.0622700	**25**	5000
GA-AIS	−5.5079887	**−5.5068253**	**−5.5055916**	**3.11E−3**	−5.4901054	**25**	5000
GA-AIS^C	**−5.5079953**	−5.5064843	−5.5052758	$3.20E-3$	**−5.4943009**	**25**	5000
APM	**−5.5080131**	−5.5079922	−5.5051182	$8.38E-3$	−5.4703893	**25**	50000
SR	−5.5080125	−5.4986755	−5.4598721	$7.11E-2$	−5.2227953	23	50000
GA-AIS	**−5.5080131**	**−5.5080119**	**−5.5080083**	**8.89E−6**	**−5.5079661**	**25**	50000
GA-AIS^C	**−5.5080131**	−5.5080106	−5.5080055	$1.33E-5$	−5.5079409	**25**	50000
APM	**−5.5080131**	**−5.5080131**	**−5.5080131**	**0.00**	**−5.5080131**	**25**	500000
SR	**−5.5080131**	**−5.5080131**	**−5.5080131**	**0.00**	**−5.5080131**	**25**	500000
GA-AIS	**−5.5080131**	**−5.5080131**	**−5.5080131**	**0.00**	**−5.5080131**	**25**	500000
GA-AIS^C	**−5.5080131**	**−5.5080131**	**−5.5080131**	**0.00**	**−5.5080131**	**25**	500000

References

1. Adeli, H., Cheng, N.T.: Augmented Lagrangian Genetic Algorithm for Structural Optimization. Journal of Aerospace Engineering 7(1), 104–118 (1994)
2. Barbosa, H.J.C.: A coevolutionary genetic algorithm for constrained optimization problems. In: Proc. of the 1999 Congress on Evolutionary Computation, pp. 1605–1611. IEEE Service Center, Washington (1999)
3. Barbosa, H.J.C., Lemonge, A.C.C.: A new adaptive penalty scheme for genetic algorithms. Information Sciences 156, 215–251 (2003)
4. Bean, J., Alouane, A.: A dual genetic algorithm for bounded integer programs. Tech. Rep. TR 92-53, Departament of Industrial and Operations Engineering, University of Michigan (1992)
5. Bernardino, H.S., Barbosa, H.J.C., Lemonge, A.C.C.: Constraint handling in genetic algorithms via artificial immune systems. In: J. Grahl (ed.) Late breaking paper at Genetic and Evolutionary Computation Conference (GECCO 2006). Seattle, WA, USA (2006), `http://www.cs.bham.ac.uk/~wbl/biblio/gecco2006etc/papers/lbp134.pdf`
6. Bernardino, H.S., Barbosa, H.J.C., Lemonge, A.C.C.: A hybrid genetic algorithm for constrained optimization problems in mechanical engineering. In: Proceedings of the 2007 IEEE Congress on Evolutionary Computation. IEEE Press, Singapore (2007)
7. de Castro, L.N., Timmis, J.: An artificial immune network for multimodal function optimization. In: Proc. of the 2002 IEEE World Congress on Computational Intelligence, Honolulu, Hawaii, USA, vol. I, pp. 669–674 (2002)
8. de Castro, L.N., Timmis, J.: Artificial Immune Systems: A New Computational Intelligence Approach. Springer, Heidelberg (2002)
9. de Castro, L.N., Zuben, F.J.V.: Learning and optimization using the clonal selection principle. IEEE Transactions on Evolutionary Computation 6(3), 239–251 (2002)
10. Coello, C.A.C.: List of references on constraint-handling techniques used with evolutionary algorithms, `http://www.cs.cinvestav.mx/~constraint/`
11. Coello, C.A.C., Cortés, N.C.: Hybridizing a genetic algorithm with an artificial immune system for global optimization. Engineering Optimization 36(5), 607–634 (2004)
12. Coello, C.A.C., Montes, E.M.: Constraint-handling in genetic algorithms through the use of dominance-based tournament selection. In: Advanced Engineering Informatics, pp. 193–203 (2002)
13. Cortés, N.C., Trejo-Pérez, D., Coello, C.A.C.: Handling constraints in global optimization using an artificial immune system. In: Jacob, C., Pilat, M.L., Bentley, P.J., Timmis, J.I. (eds.) ICARIS 2005. LNCS, vol. 3627, pp. 234–247. Springer, Heidelberg (2005)
14. Costa, L., Oliveira, P.: Evolutionary algorithm approach to the solution of mixed integer non-linear programming problems. Computers and Chemical Engineering 25, 257–266 (2001)
15. Dasgupta, D.: Artificial Immune Systems and Their Applications, 1st edn. Springer, Heidelberg (1998)
16. Dhingra, A.: Optimal apportionment of reliability and redundancy in series systems under multiple objectives. IEEE Transactions on Reliability 41(4), 576–582 (1992)
17. Erbatur, F., Hasançebi, O., Tütüncü, I., Kilç, H.: Optimal design of planar and space structures with genetic algorithms. Computers & Structures 75, 209–224 (2000)
18. Garrett, S.M.: Parameter-free, adaptive clonal selection. In: Congress on Evolutionary Computation, CEC 2004, vol. 1, pp. 1052–1058 (2004)

19. Garrett, S.M.: How do we evaluate artificial immune systems? Evolutionary Computation 13(2), 145–177 (2005)
20. Hajela, P., Lee, J.: Constrained genetic search via schema adaptation. An immune network solution. In: 1st World Congress of Stuctural and Multidisciplinary Optimization, pp. 915–920. Pergamon Press, Goslar (1995)
21. Hajela, P., Lee, J.: Constrained genetic search via schema adaptation. An immune network solution. Structural Optimization 12, 11–15 (1996)
22. Hajela, P., Yoo, J.S.: Immune network modelling in design optimization. In: Corne, D., Dorigo, M., Glover, F. (eds.) New Ideas in Optimization, pp. 167–183. McGraw-Hill, New York (1999)
23. Hinterding, R., Michalewicz, Z.: Your brains and my beauty: Parent matching for constrained optimization. In: Proc. of the Fifth Int. Conf. on Evolutionary Computation, Alaska, pp. 810–815 (1998)
24. Ji, D.D.Z., González, F.: Artificial immune system (ais) research in the last five years. In: McKay, B., et al. (eds.) Congress on Evolutionary Computation, pp. 123–130. IEEE Press, Canberra (2003)
25. Joines, J., Houck, C.R.: On the use of non-stationary penalty methods to solve nonlinear constrained optimization problems with GAs. In: Fogel, D., Michalewicz, Z. (eds.) Proc. of 1994 IEEE Conf. on Evolutionary Computation, pp. 579–585 (1994)
26. van Kampen, A., Strom, C., Buydens, L.: Lethalization, penalty and repair functions for constraint handling in the genetic algorithm methodology. Chemometrics and Intelligent Laboratory Systems 34, 55–68 (1996)
27. Kim, J.H., Myung, H.: Evolutionary programming techniques for constrained optimization problems. IEEE Transactions on Evolutionary Computation 2(1), 129–140 (1997)
28. Koziel, S., Michalewicz, Z.: A decoder-based evolutionary algorithm for constrained parameter optimization problems. In: Eiben, A.E., Bäck, T., Schoenauer, M., Schwefel, H.-P. (eds.) PPSN 1998. LNCS, vol. 1498, p. 231. Springer, Heidelberg (1998)
29. Koziel, S., Michalewicz, Z.: Evolutionary algorithms, homomorphous mappings, and constrained parameter optimization. Evolutionary Computation 7(1), 19–44 (1999)
30. Krishnamoorty, C., Rajeev, S.: Discrete optimization of structures using genetic algorithms. Journal of Structural Engineering 118(5) (1992)
31. Lai, H.H.S.Y., Qi, X.: Constrained optimization via genetic algorithms. Simulation 62(4), 242–254 (1994)
32. Lemonge, A.C.C., Barbosa, H.J.C.: An adaptive penalty scheme for genetic algorithms in structural optimization. Int. J. for Numerical Methods in Engineering 59(5), 703–736 (2004)
33. Liang, J.J., Runarsson, T.P., Montes, E.M., Clerc, M., Suganthan, P.N., Coello, C.A.C., Deb, K.: Problem definitions and evaluation criteria for the CEC 2006 special session on constrained real-parameter optimization. Tech. rep., School of EEE, Nanyang Technological University, Singapure, 639798 (2006)
34. Michalewicz, Z., Shoenauer, M.: Evolutionary algorithms for constrained parameter optimization problems. Evolutionary Computation 4(1), 1–32 (1996)
35. Mohan, C., Nguyen, H.T.: A controlled random search technique incorporating the simulated annealing concept for solving integer and mixed integer global optimization problems. Comput. Optim. Appl. 14(1), 103–132 (1999)
36. Orvosh, D., Davis, L.: Using a Genetic Algorithm to Optimize Problems with Feasibility Constraints. In: Proc. of the First IEEE Conf. on Evolutionary Computation, pp. 548–553 (1994)

37. Petrowski, A.: A clearing procedure as a niching method for genetic algorithms. In: Proc. Third IEEE International Conf. on Evolutionary Computation, pp. 798–803 (1996)
38. Riche, R.L., Knopf-Lenoir, C., Haftka, R.: A segregated genetic algorithm for constrained structural optimization. In: Eshelman, L. (ed.) Proc. of the Sixth Int. Conf. on Genetic Algorithms, Pittsburgh, PA, pp. 558–565 (1995)
39. Runarsson, T.P., Yao, X.: Stochastic ranking for constrained evolutionary optimization. IEEE Transactions on Evolutionary Computation 4(3), 284–294 (2000)
40. Shoenauer, M., Michalewicz, Z.: Evolutionary computation at the edge of feasibility. In: Ebeling, W., Rechenberg, I., Voigt, H.-M., Schwefel, H.-P. (eds.) PPSN 1996. LNCS, vol. 1141, pp. 245–254. Springer, Heidelberg (1996)
41. Singh, G., Deb, K.: Comparison of multi-modal optimization algorithms based on evolutionary algorithms. In: Genetic And Evolutionary Computation Conference, GECCO 2006, Seattle, WA, USA (2006)
42. Smith, D.C.A., Tate, D.: Adaptive penalty methods for genetic optimization of constrained combinatorial problems. INFORMS Journal on Computing 6(2), 173–182 (1996)
43. Rajasekaran, S., Lavanya, S.: Hybridization of genetic algorithm with immune system for optimization problems in structural engineering. Structural and Multidisciplinary Optimizationn 34(5), 415–429 (2007)
44. Surry, P., Radcliffe, N.: The COMOGA Method: Constrained Optimisation by Multiobjective Genetic Algorithms. Control and Cybernetics 26(3), 391–412 (1997)
45. Toma, N., Endo, S., Yamada, K., Miyagi, H.: Evolutionary optimization algorithm using mhc and immune network. In: 26th Annual Conference of the IEEE Industrial Electronics Society, 2000. IECON 2000, vol. 4, pp. 2849–2854 (2000)
46. Watanabe, K., Campelo, F., Igarashi, H.: Topology optimization based on immune algorithm and multigrid method. IEEE Trans. on Magnetics 43(4), 1637–1640 (2007)
47. Wu, J.Y.: Artificial immune system for solving constrained global optimization problems. In: Artificial Life 2007, ALIFE 2007, Honolulu, HI, pp. 92–99 (2007)
48. Wu, S., Chow, P.: Steady-state genetic algorithms for discrete optimization of trusses. Computers & Structures 56(6), 979–991 (1995)
49. Yen, J., Liao, J., Lee, B., Randolph, D.: A hybrid approach to modeling metabolic systems using a genetic algorithm and simplex method. IEEE Transactions on Systems, Man and Cybernetics – Part B: Cybernetics 28(2), 173–191 (1998)
50. Yin-Xiu, L., Gen, M.: Nonlinear mixed integer programming problems using genetic algorithm and penalty function. In: IEEE International Conference on Systems, Man, and Cybernetics, 1996, October 14-17, 1996, vol. 4, pp. 2677–2682 (1996)
51. Yoo, J.S., Hajela, P.: Immune network simulations in multicriterion design. Structural Optimization 18, 85–94 (1999)
52. Zhu, D.: An improved Templeman's algorithm for optimum design of trusses with discrete member sizes. Engineering Optimization 9, 303–312 (1986)

Constrained Optimization Based on Quadratic Approximations in Genetic Algorithms

Marcella C. Araujo, Elizabeth F. Wanner, Frederico G. Guimarães, and Ricardo H.C. Takahashi

Abstract. An aspect that often causes difficulties when using Genetic Algorithms for optimization is that these algorithms operate as unconstrained search procedures and most of the real-world problems have constraints of different types. There is a lack of efficient constraint handling technique to bias the search in constrained search spaces toward the feasible regions. We propose a novel methodology to be coupled with a Genetic Algorithm to solve optimization problems with inequality constraints. This methodology can be seen as a local search operator that uses quadratic and linear approximations for both objective function and constraints. In the local search phase, these approximations define an associated problem with a quadratic objective function and quadratic and/or linear constraints that is solved using an LMI (linear matrix inequality) formulation. The solution of this associated problems is then re-introduced in the GA population. We test the proposed methodology with a set of analytical function and the results show that the hybrid algorithm has a better performance when compared to the same Genetic Algorithm without the proposed local search operator. The tests also suggest that the proposed methodology is at least equivalent, and sometimes better than other methods that have been reported recently in literature.

Keywords: Evolutionary algorithms, genetic algorithms, quadratic approximation, linear matrix inequality, local search operators.

Marcella C. Araujo
Departamento de Física, Universidade Federal de Ouro Preto, Ouro Preto, MG, Brazil
e-mail: `marcellaop@gmail.com`

Elizabeth F. Wanner
Departamento de Matemática, Universidade Federal de Ouro Preto, Ouro Preto, MG, Brazil
e-mail: `efwanner@iceb.ufop.br`

Frederico G. Guimarães
Departamento de Engenharia Elétrica, Universidade Federal de Minas Gerais, Belo Horizonte, MG, Brazil
e-mail: `frederico.guimaraes@yahoo.com.br`

Ricardo H. C. Takahashi
Departamento de Matemática, Universidade Federal de Minas Gerais, Belo Horizonte, MG, Brazil
e-mail: `taka@mat.ufmg.br`

E. Mezura-Montes (Ed.): Constraint-Handling in Evolution. Optimi., SCI 198, pp. 193–217.
springerlink.com

1 Introduction

The field of search and optimization has changed over the last decades by the introduction of a number of non-classical and stochastic search and optimization algorithms: the *Evolutionary Algorithms*. Inspired by the natural evolution principles, the Evolutionary Algorithms (EAs) utilize a collective learning process of a population of individuals. The EAs provide a framework of tools that includes Genetic Algorithms, Evolutionary Strategies, Genetic Programming and Evolutionary Programming, among others. Each evolutionary method is designed along a different approach but despite their differences, all methods are heuristic population-based search procedures that incorporate random variation and selection.

The three most significant differences between the EAs and the classical search methods are:

- EAs search a population of solutions in each iteration, instead of a single point;
- EAs do not require extra knowledge or gradient information;
- EAs use probabilistic transition rules, not deterministic ones;

These characteristics permit the EAs to be applied to rather different kinds of functions. Besides, the EAs, as a population-based search strategies, have a ability of escaping from local optima, an ability very unlike to be observed in the classical search methods.

EAs refine a population of initial solutions, using a combination of operators to produce better approximations to a solution. This process leads to the evolution of populations of individuals that are better suited to their environment than the individuals that they were created from. If the problem has a single optimum, the EA population members can be expected to converge to that optimal point. In cases where a problem does not have one optimal point, as is the case in multiobjective optimization and scheduling problems, then the EA can be used to capture multiple optimal solutions.

An aspect that often causes difficulties when using them for optimization in the case of continuous-variable problems is that these algorithms operate as unconstrained search procedures and most of the real-world problems have constraints of different types. There is a lack of efficient constraint handling technique to bias the search toward the feasible regions, in constrained search spaces. The most common approach of incorporating constraints into EAs has been the use of penalty functions. Penalty functions have, in general, several drawbacks, for instance the lack of well-defined rules for defining the penalty parameters. Due to the limitations associated with the penalty functions, several researches have developed alternative approaches to handle constraints such as fitness approximations, incorporation of knowledge with cultural techniques in constrained problems, and so on. For a comprehensive survey of existing constraint handling methods, see [5].

The usage of quadratic approximations for dealing with constraints was initially exposed in [29], where the authors developed a direct approach to tackle problems with one non-linear equality constraint. In [30] the authors also use quadratic approximations, and develop a *linear matrix inequality* (LMI) methodology [4] for dealing with such approximations, in the local search phase of multiobjective optimization problems.

The methodology proposed in this paper is a further development of an idea that has been suggested in [30] but that has not been exploited there: the usage of an LMI formulation for dealing with an arbitrary number of inequality constraints. The arbitrary non-linear constraint functions are approximated by quadratic or linear functions, depending on their convexity. This methodology can be seen as a local search operator that uses approximations for both objective function and constraints. In the local search phase, these approximations define an associated problem with a quadratic objective function and quadratic and/or linear constraints. This associated problem is solved using a formulation based in the linear matrix inequality (LMI) formulation [4]. Such operator guides the EA toward the feasible region of the search space and, in this way, helps the EA to find the optimal point with more accuracy and convergence velocity. For testing our technique, the novel operator was coupled with a real-coded Genetic Algorithm (GA). The main focus of this work is not to compare the algorithms but to show how the new operator, coupled with simple EA procedures, can produce better results once applied in non-linear constrained problems.

This paper is organized as follow. In Section II, we define the general non-linear optimization problem that we aim to solve. After that, in Section III we present a detailed description of our approximation technique. Section IV shows the method based on the LMI formulation to solve the associated problem. Section V hybridizes a Genetic Algorithm with the local search operator. Then, in Section VI the performance for the hybrid algorithm is tested in a set of problems. Finally, in Section VII some conclusions are established.

2 Statement of the Problem

In this section the non-linear constrained problem is stated. The non-linear programming problem with k inequality constraints is formulated as:

$$\begin{aligned} &\text{Minimize} f(\mathbf{x}) \\ &\text{subject to: } g_i(\mathbf{x}) \leq 0, \;\; i = 1, \cdots, k \end{aligned} \tag{1}$$

where $\mathbf{x}$ is the vector of decision variables $\mathbf{x} = [x_1, x_2, x_3, \cdots, x_n]^T$ such that $x \in \mathscr{S} \subseteq \mathbb{R}^n$ and k is the number of inequality constraints. The constraints can be linear or non-linear and the objective function is non-linear.

The search region $\mathscr{S}$ is an n-dimensional rectangle formed by the upper and lower bounds for the variables, $x_j^l \leq x_j \leq x_j^u$, where $j = 1, 2, \ldots, n$. If we denote the feasible region (the region for which the constraints are satisfied) with $\mathscr{F}$, then it is clear that $\mathscr{F} \subseteq \mathscr{S}$. The optimal solution is denoted by $\mathbf{x}^*$ and a constraint is said to be active at the point $\mathbf{x}^*$ if $g_i(\mathbf{x}^*) = 0$.

Since the methodology that will be discussed in this work uses a local approximation, then all the involved functions must be smooth almost everywhere.

3 Approximation Methodology

Motivated by the fact that Evolutionary Algorithms work with a population of solutions and generate a number of samples of the objective and constraints evaluations along the process, we attempt to explore all information available of the EAs to create approximation models for all functions of the problem. Each approximation model (or meta-model) is a quadratic or a linear function that will used inside the optimization process.

With the goal of avoiding extra computational cost, we utilize the current and past samples generated by the population of the algorithm to fit suitable models for each function. If the function is linear, we can use the analytical expression of the function, or if this expression is not known, we fit a linear approximation for it. If the function is non-linear, but locally convex, we construct a quadratic approximation for the function. In the case of approximating locally non-convex regions of a function, the linear function approximation is employed (the best convex approximation in such cases). Any approximation models are determined in the least square sense. In the next sub-section, we discuss the quadratic approximation model in a detailed way.

3.1 *Quadratic Approximations*

Let h be a real-valued function. Given distinct points $z_1, z_2, \cdots, z_n$, we consider the problem of finding a convex quadratic real-valued function

$$f(z) = \mathbf{z}^T.H.\mathbf{z} + \mathbf{r}^T\mathbf{z} + \gamma \tag{2}$$

for some suitable symmetric $n \times n$ matrix H, $n \times 1$ vector $\mathbf{r}$ and some scalar γ, such that

$$h(z_i) \cong f(z_i) \tag{3}$$

for $i = 1, 2, \cdots, N$ where N is the number of available points.

Hence, the problem of finding f such that (3) holds can be restated as to find H, $\mathbf{r}$ and γ such that

$$E_i = {z_i}^T.H.z_i + \mathbf{r}^T z_i + \gamma - h(z_i) \tag{4}$$

for $i = 1, 2, \cdots, N$.

This is a linear system of N equations in the unknown entries of H, $\mathbf{r}$ and γ. The number of unknowns in H is equal to

$$n + \tfrac{n^2-n}{2},$$

hence the total number of unknowns is given by

$$n + \frac{n^2-n}{2} + n + 1 = \frac{(n+1)(n+2)}{2} \tag{5}$$

If

$$ {z_i}^{T} \cdot H.z_i + \mathbf{r}^T z_i + \gamma = 0 \Rightarrow H = 0, \mathbf{r} = 0, \gamma = 0 \tag{6}$$

for $i = 1, 2, \cdots, N$, and $N = \frac{(n+1)(n+2)}{2}$ then there exists a unique quadratic function f such that (3) holds. This is an *interpolation case*. When $N > \frac{(n+1)(n+2)}{2}$, the linear system (4) is over-determined and we can find a least norm solution:

$$\min_{H,\mathbf{r},\gamma} || \sum_i E_i || \tag{7}$$

If the norm is the Euclidean norm, then the function f is the quadratic least squares approximation.

Convexity of f is equivalent to the matrix H in (2) being positive semidefinite , yielding the conditions $Q \succeq 0$. It is impossible to guarantee convexity if we want f to coincide with h [13]. We can minimize the Euclidean norm of $f - h$, and f can be found by solving the objective

$$\min \left(t : \sqrt{\sum_i E_i^2} \leq t, Q \succeq 0 \right). \tag{8}$$

Since the constraint of positive-definiteness of Q is a Lorentz (or second order) cone ,

$$L^m = \left\{ \mathbf{x} \in \mathbb{R}^m \ : \ x_m \geq \sqrt{\sum_{i=1}^{m-1} x_i^2} \right\} \tag{9}$$

the problem (8) is a semidefinite programming, a special case of optimization over symmetric cones, and can be efficiently solved using SeDuMi [26].

SeDuMi is an add-on for MATLAB, which lets you solve optimization problems with linear, quadratic and semi-definiteness constraints. SeDuMi stands for *Self-Dual-Minimization* and it implements the self-dual embedding technique for optimization over self-dual homogeneous cone, or more concisely, optimization over symmetric cones. SeDuMi takes full advantage of sparsity, leading to significant speed benefits, and has a theoretically proved $\mathscr{O}\left(\sqrt{n} \log \frac{1}{\varepsilon}\right)$ worst-case iteration bound.

Once we have obtained a quadratic approximation

$$f(z) = \mathbf{z}^T.H.\mathbf{z} + \mathbf{r}^T\mathbf{z} + \gamma \tag{10}$$

we can find the point of minimum of$f$

$$\mathbf{z_f} = -\frac{1}{2} H^{-1} \mathbf{r} \tag{11}$$

and then, we can rewrite the analytical expression of the function

$$f(\mathbf{z}) = (\mathbf{z} - \mathbf{z_f})^T.H.(\mathbf{z} - \mathbf{z_f}) - C \tag{12}$$

where $C = 0.25\mathbf{r}^T H^{-1}\mathbf{r} - \gamma$.

4 The LMI Formulation for Solving the Associated Problem

With the local approximation for each function, we can write an associated problem for the problem (1):

$$\min(\mathbf{x}-\mathbf{x_f})^T.H.(\mathbf{x}-\mathbf{x_f})$$

$$\text{subject to:}\begin{cases} g_i(\mathbf{x}) = (\mathbf{x}-\mathbf{x_{gi}})^T.Q_i.(\mathbf{x}-\mathbf{x_{gi}})- \\ \qquad -C_i \leq 0 \\ \qquad i = 1,\cdots,p \\ g^l_j(\mathbf{x}) = \mathbf{ax}+\mathbf{c} \leq 0 \\ \qquad j = 1,\cdots,q \end{cases} \tag{13}$$

where $\mathbf{x_f}$ and H are, respectively, the unconstrained optimal point and the Hessian matrix of the approximated objective function, $\mathbf{x_{gi}}$ and Q_i, for $i = 1,\cdots,m$ are, respectively, the center point of the level curves and the Hessian matrix of each non-linear locally convex constraint function, and g^l_j is the linear approximation for each non-linear locally non-convex constraint in the original problem, and $k = p+q$.

Using the associated problem (13), we can estimate a solution for (1) through LMI formulation. In this section we present the mathematical formulation that allows the usage of the LMI to solve (13).

Linear Matrix Inequalities (LMIs) and LMI techniques have emerged as a powerful tool in areas such as control engineering, systems identification and structural design. A wide variety of problems can be formulated using LMI and, once stated in terms of LMIs, a problem can be solved *exactly* by efficient convex optimization algorithms (LMIs solvers). These solvers are significantly faster than classical convex optimization algorithms.

A linear matrix inequality is any constraint of the form:

$$A(\mathbf{x}) = x_1A_1 + x_2A_2 + \cdots + x_nA_n < A_0 \tag{14}$$

where $\mathbf{x} = (x_1,\cdots,x_n)$ is a vector of unknown scalars, $A_i \in \mathscr{S}^n = \{X \in \mathbb{R}^{n\times n} | X = X^T\}, i = 1,2,\cdots,n$, are symmetric matrices and < 0 stands for *definite negative*, i.e., the largest eigenvalue of $A(\mathbf{x})$ is negative. Observe that $A(\mathbf{x}) > 0$ and $A(\mathbf{x}) < B(\mathbf{x})$ can be rewritten, respectively, as $-A(\mathbf{x}) < 0$ and $A(\mathbf{x}) - B(\mathbf{x}) < 0$.

In other words, the LMI (14) is a affine functional mapping a vectorial space onto a cone of negative definite matrices. An important property of LMIs is that its solution space, denoted by $\{x \in \mathbb{R}^m : A(x) < A_0\}$, is a convex set. In fact, defining an affine function $f : \mathbb{R}^n \to \mathscr{S}^n$, where $f(x) = A_0 - A(x)$, the set

$$\{x|f(x) \in \mathscr{C} \subseteq \mathscr{S}^n{}_{\geq}\} = \{x \in \mathbb{R}^n | A_0 - A(x) \in \mathscr{S}^n{}_{\geq}\}$$

represents the inverse image of the cone of semidefinite positive matrices and it is a convex set. In this way, the task of finding a solution for (14) is a convex programming problem.

We will employ the following lemma and some results to derive our methodology:

Theorem 1. (Schur's complement) The statements (15) and (16) bellow are equivalent:

$$\begin{bmatrix} Q & S \\ S' & R \end{bmatrix} > 0 \tag{15}$$

$$\begin{cases} R > 0 \\ Q - SR^{-1}S' > 0 \end{cases} \tag{16}$$

in which R and Q are symmetric matrices, S is a matrix with compatible dimensions, and $(\cdot) > 0$ denotes that the matrix argument is positive definite.
The proof of Lemma (4) can be found in [4].

The first result, based on the Schur's Lemma, is stated as:

Theorem 2. Consider the following optimization problem with a quadratic objective function and quadratic constraints:

$$\begin{aligned} \mathbf{x}^* = &\ \arg\ \min\ (\mathbf{x}-\mathbf{x}_0)'Q_0(\mathbf{x}-\mathbf{x}_0) \\ \text{s.t.} &\begin{cases} (\mathbf{x}-\mathbf{x}_1)'Q_1(\mathbf{x}-\mathbf{x}_1) - C_1 \le 0 \\ (\mathbf{x}-\mathbf{x}_2)'Q_2(\mathbf{x}-\mathbf{x}_2) - C_2 \le 0 \\ \vdots \\ (\mathbf{x}-\mathbf{x}_m)'Q_m(\mathbf{x}-\mathbf{x}_m) - C_m \le 0 \end{cases} \end{aligned} \tag{17}$$

The optimization problem (17) can be re-stated as:

$$\begin{aligned} \mathbf{x}^* = &\ \arg_{\mathbf{x}}\ \min_{\mathbf{x},\varepsilon}\ \varepsilon \\ \text{s.t.} &\begin{cases} \begin{bmatrix} \varepsilon & (\mathbf{x}-\mathbf{x}_0)' \\ \mathbf{x}-\mathbf{x}_0 & Q_0^{-}1 \end{bmatrix} > 0 \\ \begin{bmatrix} C_1 & (\mathbf{x}-\mathbf{x}_1)' \\ \mathbf{x}-\mathbf{x}_1 & Q_1^{-}1 \end{bmatrix} > 0 \\ \begin{bmatrix} C_2 & (\mathbf{x}-\mathbf{x}_2)' \\ \mathbf{x}-\mathbf{x}_2 & Q_2^{-}1 \end{bmatrix} > 0 \\ \vdots \\ \begin{bmatrix} C_m & (\mathbf{x}-\mathbf{x}_m)' \\ \mathbf{x}-\mathbf{x}_m & Q_m^{-}1 \end{bmatrix} > 0 \end{cases} \end{aligned} \tag{18}$$

Proof. Replace:

$$\min\ (\mathbf{x}-\mathbf{x}_0)'Q_0(\mathbf{x}-\mathbf{x}_0)$$

by:

$$\begin{aligned} &\min\ \varepsilon \\ &\text{s.t. } (\mathbf{x}-\mathbf{x}_0)'Q_0(\mathbf{x}-\mathbf{x}_0) < \varepsilon \end{aligned}$$

The remainder operations are direct applications of Schur's complement to the quadratic inequalities. □

In the same way, another result can be derived for problems with quadratic and linear constraints:

Theorem 3. Consider the following optimization problem with a quadratic objective function and constraints that are quadratic and linear:

$$\mathbf{x}^* = \arg\ \min\ (\mathbf{x}-\mathbf{x}_0)'Q_0(\mathbf{x}-\mathbf{x}_0)$$

$$\text{s.t.} \begin{cases} (\mathbf{x}-\mathbf{x}_1)'Q_1(\mathbf{x}-\mathbf{x}_1) - C_1 \leq 0 \\ (\mathbf{x}-\mathbf{x}_2)'Q_2(\mathbf{x}-\mathbf{x}_2) - C_2 \leq 0 \\ \vdots \\ (\mathbf{x}-\mathbf{x}_p)'Q_p(\mathbf{x}-\mathbf{x}_p) - C_p \leq 0 \\ a_1\mathbf{x} - b_1 \leq 0 \\ a_2\mathbf{x} - b_2 \leq 0 \\ \vdots \\ a_q\mathbf{x} - b_q \leq 0 \end{cases} \tag{19}$$

The optimization problem (19) can be re-stated as:

$$\mathbf{x}^* = \arg_{\mathbf{x}}\ \min_{\mathbf{x},\varepsilon}\ \varepsilon$$

$$\text{s.t.} \begin{cases} \begin{bmatrix} \varepsilon & (\mathbf{x}-\mathbf{x}_0)' \\ \mathbf{x}-\mathbf{x}_0 & Q_0^-1 \end{bmatrix} > 0 \\ \begin{bmatrix} C_1 & (\mathbf{x}-\mathbf{x}_1)' \\ \mathbf{x}-\mathbf{x}_1 & Q_1^-1 \end{bmatrix} > 0 \\ \begin{bmatrix} C_2 & (\mathbf{x}-\mathbf{x}_2)' \\ \mathbf{x}-\mathbf{x}_2 & Q_2^-1 \end{bmatrix} > 0 \\ \vdots \\ \begin{bmatrix} C_p & (\mathbf{x}-\mathbf{x}_p)' \\ \mathbf{x}-\mathbf{x}_p & Q_p^-1 \end{bmatrix} > 0 \\ 2a_1 - c_1\mathbf{x} - \mathbf{x}'c_1' > 0 \\ 2a_2 - c_2\mathbf{x} - \mathbf{x}'c_2' > 0 \\ \vdots \\ 2a_q - c_q\mathbf{x} - \mathbf{x}'c_q' > 0 \end{cases} \tag{20}$$

The first LMI accounts for the quadratic objective function, the next p LMIs account for the locally convex inequality constraints of the original problem and the last q LMIs for the locally non-convex inequality constraints of the original problem. Notice that *Theorem (1)* and *Theorem (2)* produce a point that is the exact solution for the quadratic problem (13) which is associated to the original problem (1). The optimization problem (20) can be efficiently solved with any LMI solver based on interior point methods. In this work, we used, as in the construction of the quadratic approximation, SeDuMi to solve this problem.

The solution of the associated problem (20), which was constructed using the approximations, provides a locally improved individual. How we introduce this new improved solution in the current population of the Genetic Algorithm is a particular issue and will be addressed latter.

5 Hybridizing the Genetic Algorithm with the Local Approximations Operator

Genetic Algorithms, like any other evolutionary technique, are especially well tuned for solving a wide class of problems due to their ability to explore vast solutions spaces and search from a family of candidate solutions rather than from just a single point. However, these algorithms are less suited to fine-tuning structures that are already close to optimal solutions. As stated by Davis [6] and re-illustrated by Knowles [16], for improving optimization results achieved by Genetic Algorithms one should: "Hybridize where possible".

Hybrid Genetic Algorithms (HGAs) or *Memetic Algorithms* (MAs) denote the association of local and global search operators inside GAs. The term Memetic Algorithm was first used by Moscato [22], denoting algorithms that use some kind of structured information, that is obtained and refined as the algorithm evolves, and is transmitted from one generation to the other, for enhancing the search. In this work, the author associated a simulated-annealing-based operator as a local search into a genetic algorithm. Since then, this idea has gained wide acceptance and has been applied in a large class of problems [15, 17, 18, 23].

It is argued that the success of MAs is due to the trade-off between exploration abilities of the underlying GA and the exploitation abilities of the local searchers used. The price to be paid is a greater number of extra fitness evaluations and often a swift loss of diversity within the population. The required cost by local search is an important issue in hybrid algorithms. This point often becomes more important in real-valued optimization problems which involve expensive-to-evaluate black-box function.

The local search operator presented in the previous section is coupled with a simple real-coded Genetic Algorithm, inside a normal cycle of the algorithm. We used a version of a simple real-coded Genetic Algorithm with the basic following characteristics:

- Gaussian Mutation;
- Selection by roulette-wheel with ranking-based linear fitness;

- Elitism mechanism;

The hybridization interface between the newly implemented local search and the GA is located prior to the selection for recombination and the recombination steps. This choice of location is designed to make any beneficial effects of the local search operator available to the selection and recombination process. Then the local search fine-tunes the parents of the population instead of the offspring with the goal of producing fitter offspring. This concept has its analogy with the heredity mechanism in human biology: healthy parents are more probable to produce healthy offspring.

The new local search operator improves the solution because it allows all the constraints to be reached with more precision. Considering that the local search phase does not require any extra function evaluation, this operator alleviates the computational burden associated to local search in MAs. We show below the basic sketch of the Hybrid GA:

Initialize parameters
Initialize population
WHILE no stop criterion

- each σ generations: local search
- selection
- crossover
- mutation

END-WHILE

An implementation of the proposed methodology could be performed in several different ways. For coupling the local search operator with the GA, we have established the following arbitrary definitions:

- The GA is executed for the optimization of a modified objective function with a penalty term that takes into account the inequality constraint:

$$F(\mathbf{x}) = f(\mathbf{x}) + 100.\sum_i |g_i(\mathbf{x})| \tag{21}$$

- The local search operator will be run every 5 generations. We have performed experiments on executing the operator every 1, 2, 5 and 10 generations – these tests have shown that the interval of 5 generations represents a good trade-off between the effort spent in the local search and the effort spent in the usual GA operations.
- As this operator is a local search one, only points in a neighborhood of the current best point will be used to build the approximation model. This neighbourhood is arbitrarily defined here as:

$$\mathcal{N}(\mathbf{x}_0) = \{\mathbf{x} : (\mathbf{x} - \mathbf{x}_0)^T R(\mathbf{x} - \mathbf{x}_0) \leq 1\} \tag{22}$$

and

$$R_{ij} = \begin{cases} [0.1(u_i - l_i)]^{-1}, & i = j \\ 0, & i \neq j \end{cases} \tag{23}$$

where l_i and u_i are respectively the minimum and maximum value for the i-th variable. Thus the neighbourhood $\mathscr{N}(\mathbf{x}_0)$ is an ellipsoidal region in which the dimension of each axis is a function of the parameter range. As a mathematical condition, the number of points inside this neighborhood must be greater than or equal to

$$\frac{(n+1)(n+2)}{2} \tag{24}$$

where n is the problem dimension. The higher the number of points inside this neighborhood, the more accurate the quadratic approximation becomes.

- Finally, the output point of the new operator will deterministically replace the worst point of the current population.

Notice that the output point also can be used as an additional stop criterion for the algorithm: the stabilization of such point can be interpreted as the algorithm finding the solution.

6 Experiments and Results

6.1 Analytical Functions

For evaluating the performance of the proposed algorithm, we used a set of test functions. Some functions of this set present characteristics that can be considered difficult, while other ones define easy problems: the idea is to compare the different algorithms in diverse situations. Their analytical expressions are provided below.

Test problem (T1): This is a simple 2-dimensional quadratic problem stated as:

$$\begin{aligned} &\min f(\mathbf{x}) = (x_1-4)^2 + (x_2-6)^2 \\ &\text{subject to: } \begin{cases} g_1(\mathbf{x}) = 2(x_1-2)^2 + (x_2-2)^2 - 1 \leq 0 \\ g_2(\mathbf{x}) = 2(x_1-1)^2 + 4(x_2-2)^2 - 1 \leq 0 \\ -4 \leq x_i \leq 4, \quad i = 1,2 \end{cases} \end{aligned} \tag{25}$$

The optimal solution to this problem is $\mathbf{x}^* = [\,1.4932\ 2.3583\,]^T$, which gives $f^* = 19.5440$. At the solution the constraint g_2 is active.

Test problem (T2): The problem stated below is a bi-dimensional quasi-quadratic problem:

$$\begin{aligned} &\min f(\mathbf{x}) = (x_1-2)^4 + (x_1-2x_2)^2 \\ &\text{subject to: } \begin{cases} g_1(\mathbf{x}) = (x_1)^2 + x_2 \leq 0 \\ g_2(\mathbf{x}) = (x_1-2)^2 + (x_2-1)^2 - 1 \leq 0 \\ -4 \leq x_i \leq 4 \quad i = 1,2 \end{cases} \end{aligned} \tag{26}$$

The optimal solution to this problem is $\mathbf{x}^* = [\,1\ 1\,]^T$ which gives $f^* = 2$. At the solution the constraint g_2 is active.

Test problem (T3): Despite of the fact of being a quadratic problem, this problem has 15 variables and 12 constraints. The high dimension of the variable space is a hard issue for optimization algorithms. The problem is stated below:

$$\begin{aligned} &\min f(\mathbf{x}) = (\mathbf{x}-\mathbf{v})^T(\mathbf{x}-\mathbf{v}) \\ &\text{subject to: } \begin{cases} g_i(\mathbf{x}) = (\mathbf{x}-\mathbf{c}_i)^T.Q_i.(\mathbf{x}-\mathbf{c}_i) - 1 \le 0 \\ -4 \le x_i \le 4 \quad i = 1,\cdots,12 \end{cases} \end{aligned} \tag{27}$$

where

$$\mathbf{a} = [a_j], \quad a_j = 1 \quad j = 1,2,\cdots,15 \;\; \mathbf{v} = 5.\mathbf{a}, \;\; \mathbf{c}_1 = \left(1+\tfrac{\sqrt{15}}{15}\right).\mathbf{a}, \;\; \mathbf{c}_2 = \left(1+\tfrac{\sqrt{30}}{30}\right).\mathbf{a},$$
$$\mathbf{c}_3 = \left(1+\tfrac{\sqrt{15}}{30}\right).\mathbf{a}, \;\; \mathbf{c}_4 = \left(1+\tfrac{\sqrt{15}}{15}\right).\mathbf{a}, \;\; \mathbf{c}_5 = \left(1+\tfrac{\sqrt{29}}{29}\right).\mathbf{a}, \;\; \mathbf{c}_6 = \tfrac{6}{5}.\mathbf{a},$$
$$\mathbf{c}_7 = \left(1+\tfrac{\sqrt{6}}{12}\right).\mathbf{a}, \;\; \mathbf{c}_8 = \left(1+\tfrac{\sqrt{23}}{23}\right).\mathbf{a}, \;\; \mathbf{c}_9 = \left(1+\tfrac{\sqrt{10}}{20}\right).\mathbf{a}, \;\; \mathbf{c}_{10} = \left(1+\tfrac{\sqrt{35}}{35}\right).\mathbf{a},$$
$$\mathbf{c}_{11} = \left(1+\tfrac{\sqrt{10}}{20}\right).\mathbf{a}, \;\; \mathbf{c}_{12} = \left(1+\tfrac{\sqrt{3}}{15}\right).\mathbf{a}$$

$$\begin{aligned}
&\text{Considering } i = j = 1,2,\cdots,15 \\
&Q_1 = (a_{ij}) = \begin{cases} a_{ij} = 0, \text{ if } i \neq j \\ a_{ij} = 1, \text{ otherwise} \end{cases} \\
&Q_2 = 2.Q_1,\; Q_3 = 4.Q_1,\; Q_4 = 3.Q_1 \\
&Q_5 = (a_{ij}) = \begin{cases} a_{ij} = 0, \text{ if } i \neq j \\ a_{ij} = 1, \text{ if } i = j \text{ and } i = 1 \\ a_{ij} = 2, \text{ otherwise} \end{cases} \\
&Q_6 = (a_{ij}) = \begin{cases} a_{ij} = 0, \text{ if } i \neq j \\ a_{ij} = 1, \text{ if } i = j \text{ and } i = 1,\cdots,5 \\ a_{ij} = 2, \text{ otherwise} \end{cases} \\
&Q_7 = (a_{ij}) = \begin{cases} a_{ij} = 0, \text{ if } i \neq j \\ a_{ij} = 1, \text{ if } i = j \text{ and } i = 1,\cdots,6 \\ a_{ij} = 2, \text{ otherwise} \end{cases} \\
&Q_8 = (a_{ij}) = \begin{cases} a_{ij} = 0, \text{ if } i \neq j \\ a_{ij} = 1, \text{ if } i = j \text{ and } i = 1,\cdots,8 \\ a_{ij} = 2, \text{ otherwise} \end{cases} \\
&Q_9 = (a_{ij}) = \begin{cases} a_{ij} = 0, \text{ if } i \neq j \\ a_{ij} = 4, \text{ if } i = j \text{ and } i = 1,\cdots,5 \\ a_{ij} = 2, \text{ otherwise} \end{cases} \\
&Q_{10} = (a_{ij}) = \begin{cases} a_{ij} = 0, \text{ if } i \neq j \\ a_{ij} = 3, \text{ if } i = j \text{ and } i = 1,\cdots,5 \\ a_{ij} = 2, \text{ otherwise} \end{cases} \\
&Q_{11} = (a_{ij}) = \begin{cases} a_{ij} = 0, \text{ if } i \neq j \\ a_{ij} = 3, \text{ if } i = j \text{ and } i = 1,\cdots,10 \\ a_{ij} = 2, \text{ otherwise} \end{cases} \\
&Q_{12} = 5.Q_1
\end{aligned} \tag{28}$$

The optimal solution, $\mathbf{x}^*$, to this problem is $[1\;1\;1\;1\;1\;1\;1\;1\;1\;1\;1\;1\;1\;1\;1]^T$, which gives $f^* = 213.0872$. At the solution the constraint g_{12} is active.

Test problem (T4): The fourth problem is stated as:

$$\min f(\mathbf{x}) = (x_1 - 10)^3 + (x_2 - 20)^3$$
$$\text{subject to:}$$
$$\begin{cases} g_1(\mathbf{x}) = -(x_1-5)^2 - (x_2-5)^2 + 100 \leq 0 \\ g_2(\mathbf{x}) = (x_1-6)^2 + (x_2-5)^2 - 82.81 \leq 0 \\ 13 \leq x_1 \leq 100, \\ 0 \leq x_2 \leq 100 \end{cases} \tag{29}$$

The optimal solution to this problem is $\mathbf{x}^* = [14.095\ 0.84296]^T$, which gives $f^* = -6961.81388$. At the solution all constraints are active.

Test problem (T5): This problem is stated as:

$$\begin{aligned} \min f(\mathbf{x}) = {} & (x_1-10)^2 + 5(x_2-12)^2 + x_3^4 + \\ & +3(x_4-11)^2 + 10x_5^6 + 7x_6^2 + x_7^4 - \\ & -4x_6x_7 - 10x_6 - 8x_7 \end{aligned}$$
$$\text{subject to:}$$
$$\begin{cases} g_1(\mathbf{x}) = -127 + 2x_1^2 + 3x_2^4 + x_3 + 4x_4^2 + 5x_5 \leq 0 \\ g_2(\mathbf{x}) = -282 + 7x_1 + 3x_2 + 10x_3^2 + x_4 - x_5 \leq 0 \\ g_3(\mathbf{x}) = -196 + 23x_1 + x_2^2 + 6x_6^2 - 8x_7 \leq 0 \\ g_4(\mathbf{x}) = 4x_1^2 + x_2^2 - 3x_1x_2 + 2x_3^2 + 5x_6 - 11x_7 \leq 0 \\ -10 \leq x_i \leq 10, \quad i = 1,2,\cdots,7 \end{cases} \tag{30}$$

This problem is found in [8, 20]. The best value reported so far [8] is $[2.330499\ 1.951372\ -0.4775414\ 4.365723\ -0.6244870\ 1.038131\ 1.594227]^T$ which gives $f^* = 680.6300$. At the solution the constraints g_1 and g_4 are active.

Test problem (T6): This problem was first presented in [14] and it has been used in [5] and in [9] to evaluate the performance of various GAs for constrained optimization.

$$\begin{aligned} \min f(\mathbf{x}) = {} & 5.3578547x_3^2 + 0.8356891x_1x_5 + \\ & +37.293239x_1 - 40792.141 \end{aligned}$$

$$\text{subject to:} \begin{cases} g_1(\mathbf{x}) = -85.334407 - 0.0056858x_2x_5 - \\ \qquad -0.0006262x_1x_4 + 0.0022053x_3x_5 \leq 0 \\ g_2(\mathbf{x}) = 85.334407 + 0.0056858x_2x_5 + \\ \qquad +0.0006262x_1x_4 - 0.0022053x_3x_5 - 91 \leq 0 \\ g_3(\mathbf{x}) = -80.51249 - 0.0071317x_2x_5 - \\ \qquad -0.0029955x_1x_2 - 0.0021813x_3^2 \leq 0 \\ g_4(\mathbf{x}) = 80.51249 + 0.0071317x_2x_5 + \\ \qquad +0.0029955x_1x_2 + 0.0021813x_3^2 - 110 \leq 0 \\ g_5(\mathbf{x}) = -9.300961 - 0.0047026x_3x_5 - \\ \qquad -0.0012547x_1x_3 - 0.0019085x_3x_4 \leq 0 \\ g_6(\mathbf{x}) = 9.300961 + 0.0047026x_3x_5 + \\ \qquad +0.0012547x_1x_3 + 0.0019085x_3x_4 - 25 \leq 0 \\ 78 \leq x_1 \leq 102, \\ 33 \leq x_2 \leq 45, \\ 27 \leq x_i \leq 45, \quad i = 3,4,5 \end{cases} \tag{31}$$

The best known solution to this problem is $\mathbf{x}^* = [78\ 33\ 29.995\ 45\ 36.776]^T$, which gives $f^* = -30665.5$. At the solution the constraints g_2 and g_5 are active.

Test problem (T7): This problem is stated as:

$$\begin{aligned} \min f(\mathbf{x}) &= -\frac{sin^3(2\pi x_1)sin(2\pi x_2)}{x_1^3(x_1+x_2)} \\ \text{subject to: } &\begin{cases} g_1(\mathbf{x}) = x_1^2 - x_2 + 1 \leq 0 \\ g_2(\mathbf{x}) = 1 - x_1 + (x_2-4)^2 \leq 0 \\ 0 \leq x_i \leq 10 \quad i = 1,2 \end{cases} \end{aligned} \tag{32}$$

The best optimal solution to this problem is $\mathbf{x}^* = [1.2279713\ 4.2453733]^T$ which gives $f^* = -0.095821$.

Test problem (T8): This problem is stated as:

$$\begin{aligned} \min f(\mathbf{x}) &= \frac{\sum_{i=1}^{5} 0.01((x_i+0.5)^4 - 30x_i^2 - 20x_i)}{5} \\ \text{subject to: } &\begin{cases} g_1(\mathbf{x}) = (\mathbf{x}-\mathbf{v})'.(\mathbf{x}-\mathbf{v}) \leq 0 \\ g_2(\mathbf{x}) = (\mathbf{x}-\mathbf{c})'.H.(\mathbf{x}-\mathbf{c}) \leq 0 \\ -6 \leq x_i \leq 6 \\ i = 1,\cdots,5 \end{cases} \end{aligned} \tag{33}$$

where

$$\mathbf{v} = [6\,6\,6\,6\,6]^T$$

$$\mathbf{c} = [5\,6\,6\,6\,6]^T$$

$$H = \begin{pmatrix} 4\,0\,0\,0\,0 \\ 0\,1\,0\,0\,0 \\ 0\,0\,1\,0\,1 \\ 0\,0\,0\,1\,0 \\ 0\,0\,0\,0\,1 \end{pmatrix}$$

The objective function is a multimodal function with a global minimum at $\mathbf{x}^* = [-4.4538 \cdots -4.4538]$ and other local minima located at the corner of the hyper-square $X = [\pm 4.4538 \cdots \pm 4.4538]$. The feasible region is located near one of the local minimum of the objective function.

Test problem (T9): This problem is stated as:

$$\begin{aligned} \min f(\mathbf{x}) &= \mathbf{x}^T.A^T.A.\mathbf{x} - 10[1\ \ 1]cos(2\pi A\mathbf{x}) \\ \text{subject to: } &\begin{cases} g_1(\mathbf{x}) = (x_1-2)^2 + (x_2-2)^2 - 1 \leq 0 \\ -4 \leq x_i \leq 4 \\ i = 1,2 \end{cases} \end{aligned} \tag{34}$$

where $A = \begin{pmatrix} 1\,0 \\ 0\,4 \end{pmatrix}$.

It is known that the optimal point lies in the boundary of the feasible region however, we cannot find any related solution for this problem in the literature.

The best obtained result to this problem is $\mathbf{x}^* = [\,1.989999\ 1.0001\,]^T$ which gives $f^* = -0.169$.

It is possible to estimate the size of the feasible region of a general problem through the metric ρ , as suggested in [21]. The ρ-metric, given by,

$$\rho = \frac{|F|}{|T|}, \tag{35}$$

where $|T|$ is the number of random solutions generated and $|F|$, represents the number of feasible solutions found out of the total. This metric gives us an estimate of the size of the feasible region and, consequently, indicates how difficult is to generate a feasible solution through a random process.

The table 1 presents a summary of the chosen problems, where n is the number of variables, LI is the number of linear inequalities, NLI is the number of non-linear inequalities and ρ is the size of the feasible region. The maximum number of generation used in each problem test and the respective population size are also listed in the same table. For all experiments, we used the following parameters:

- Crossover Rate = 0.8
- Mutation Rate = 0.05
- Mutation Size = 0.01
- Dispersion Factor in Fitness Function = 1.8
- number of generations = 800
- population size = 300

Due to stochastic nature of the GA, a well-based judgment concerning the performance of a specific algorithm cannot be stated unless the whole optimization process is repeated a number of times. In the case of this work, we performed 30 independent runs for both algorithm using each test function.

Table 1 Characteristics of each test problem

Problem	Type of Function	n	LI	NLI	ρ
T1	quadratic	2	0	2	0.00045
T2	non-linear	2	0	2	0.00010
T3	quadratic	15	0	12	0
T4	non-linear	2	0	2	0.00001
T5	non-linear	7	0	4	0.00058
T6	quadratic	5	0	6	0.02765
T7	non-linear	10	3	5	0.00086
T8	non-linear	5	0	2	0
T9	non-linear	2	0	1	0.00510

The tables 2 and 3 show the performance of the simple and hybrid GAs in the test problems. The symbol "–" means that the algorithm was not able to find any feasible solution. All the other solutions used to evaluate the algorithm were feasible

Table 2 Statistical results obtained by the GA Hybrid for the test functions

Problem	Opt. Value	Best Value	Mean	Worst
T1	19.5440	19.5440	19.5440	19.5440
T2	2	2	2.0073	2.0313
T3	213.0872	213.0872	213.0872	213.0872
T4	-6961.8139	-6811.8	-6533.99	-5987.64
T5	680.6300	681.4534	681.6135	682.0012
T6	-30665.5	-30665.530	-30654.998	-30654.297
T7	-0.095821	-0.0958	-0.0951	-0.0899
T8	3.0615	3.0615	3.0615	3.0615
T9	-0.0169	-0.0169	-0.0147	0.0223

Table 3 Statistical results obtained by the pure GA for the test functions

Problem	Opt. Value	Best Value	Mean	Worst
T1	19.5440	19.6010	19.7101	20.2102
T2	2	2.0053	2.2025	2.7126
T3	213.0872	243.9816	245.1588	251.6680
T4	-6961.8139	-6298.6	-6398.6	-5503.6
T5	680.632	691.4714	697.8759	699.2311
T6	-30665.5	–	–	–
T7	-0.095821	-0.09057	-0.0872	-0.0469
T8	3.0615	3.0862	4.0590	9.7844
T9	-0.0169	-0.0162	0.0088	0.1324

ones. We can see that the GA Hybrid was able to find the exact global minimum in six problems, (T1), (T2), (T3), (T7), (T8) and (T9). In the other problems, the GA Hybrid found solutions closer to the optimal value when compared with pure GA.

We compare these results with the approach described in [19]. In table 4, we summarize the best results obtained in this paper and compare that with the best reported results found in [19]. The symbol (#) indicates that there are not results available for comparison, for the respective problem, in the literature. As we can see, the methodology proposed in this work has shown a very competitive performance with respect to the pure GA and to the results in [19].

At the end of 30 executions of the simple and hybrid algorithms, we obtained the mean convergence line which corresponds to the mean value of the best individual throughout the first 100 generations. Figures 1, 2, 3, 4, 5, 6, 7, 8 and 9 show the convergence line for each problem. In the graphs, the x-axes represent the generation and the y-axes represent the base 10 logarithm of the objective function value of the best individual. The base 10 logarithm was used, except in the problem (T4), (T6), (T7) and (T9), only to enforce the difference between the lines.

These figures reveal a pattern in which the GA Hybrid converges faster (in terms of number of function evaluations) than the pure GA to the problem optimum: the mean convergence line of GA Hybrid becomes systematically below the one of pure

Table 4 Summary results of this work

Prob.	x^*	Best-Known			Results of this work		
		Best	Mean	Worst	Best	Mean	Worst
T1	19.5440	#	#	#	19.5440	19.5440	19.5440
T2	2	#	#	#	2	2	2
T3	213.0872	#	#	#	213.0872	213.0872	213.0872
T4	-6961.8139	-6961.814	-6961.284	-6952.482	-6811.8	-6533.99	-5987.64
T5	680.632	680.632	680.634	680.719	681.4534	681.6135	682.0012
T6	-30665.5	-30655.539	-30655.539	-30655.539	-30665.530	-30654.998	-30654.297
T7	-0.095821	0.095825	0.095825	0.095825	-0.0958	-0.0951	-0.0899
T8	3.0615	#	#	#	3.0615	3.0615	3.0615
T9	-0.0169	#	#	#	-0.0169	-0.0147	0.0223

GA in all tests, except (T5), in which an alternation occurs. Adding the data about the final solution that is found by each algorithm, that has been analysed above, the conclusion is that the GA Hybrid usually converges to better feasible solutions, for the same number of function evaluations, when compared with the pure GA.

6.2 *Case Study: TEAM Benchmark Problem 22*

Since the number of function evaluations needed for finding the optimum of the original problem is significantly reduced with the proposed procedure, we can say that the methodology is suitable for dealing with costly black-box optimization problems. A case study is presented: the well-known TEAM 22 benchmark problem, an expensive problem of electromagnetic design.

The TEAM benchmark problem 22 with three variables deals with the optimization of the geometric parameters of a superconducting magnetic energy storage (SMES) configuration. The objectives are to maintain a prescribed level for the stored energy on the device and to minimize the strayed field evaluated along lines a and b while not violating the quench condition that assures the superconductivity state. For a description of this benchmark problem, see reference [1,28].

For the purpose of defining the benchmark problem, these objectives are expressed as a single-objective problem:

$$\min f(\mathbf{x}) = \sqrt{\frac{1}{21}\sum_{i=1}^{21} B_{stray,i}(\mathbf{x})}$$

$$\text{subject to: } \begin{cases} g_1(\mathbf{x}) = B_{\max} - 4.92 \le 0 \\ g_2(\mathbf{x}) = \dfrac{|E - 180\text{MJ}|}{180\text{MJ}} - 0.05 \le 0 \end{cases} \tag{36}$$

Table 5 Parameters for the Problem 22

Var	R_2	h_2	d_2
Unit	(m)	(m)	(m)
min	2.6	0.408	0.1
max	3.4	2.2	0.4

Table 6 Fixed variables for the Problem 22

Var	R_1	h_1	d_1	J_1	J_2
Unit	(m)	(m)	(m)	(MA/m^2)	(MA/m^2)
Value	2.0	0.8	0.27	22.5	-22.5

where $B_{stray,j}$ is the magnetic flux density evaluated in each of the 21 evaluation points for the strayed field along lines a and b, the maximum value of the magnitude of the flux density in the domain, and E is the stored energy in the device. Table 5 shows the upper and lower limits of the variables (R_2, h_2, d_2) and table 6 shows the values of the fixed ones.

In electromagnetic design problems, the evaluations of f and g involve the implicit solution of a field problem, which is described by one (or a system of) partial differential equation(s). The analysis step in the optimization process can be computationally intensive (this situation can become even worse in the case of 3D problems). Therefore, the total number of function calls required by an optimization algorithm is of primary concern.

We performed 10 independent runs for each algorithm (pure GA and hybrid GA), using the parameters described in the tests of subsection 6.1. The maximum number of generation (50 generations) was the only termination criteria and the population size was set to 40 individuals. The hybrid GA has found feasible solutions in all tests, while the pure GA has been able to find feasible solutions only in 80% of the tests.

Table 7 shows the mean value and the standard deviation of the objective function for each algorithm, considering only the runs in which feasible solutions were found. We can observe that the hybrid GA was able to find a better solution, with a better value of the standard deviation, when compared with the pure GA. The best solution achieved by the hybrid algorithm is shown in Table 9. For this solution we have $g_1 < 0$ and $g_2 = 0$. The value of the stored energy achieved is somewhat below the target (180MJ). This is due to the formulation used in this work, which considered the stored energy as a constraint and not as the main objective of problem 22.

With the goal of assessing the time required for the hybrid algorithm, we measured the time (in seconds) spent by each algorithm in each test. Table 8 shows the mean time and the standard deviation for all tests using the pure and the hybrid algorithm. It should be noticed that the computation time spent by the hybrid algorithm is about 1.0% greater than the computation time spent by the basic algorithm.

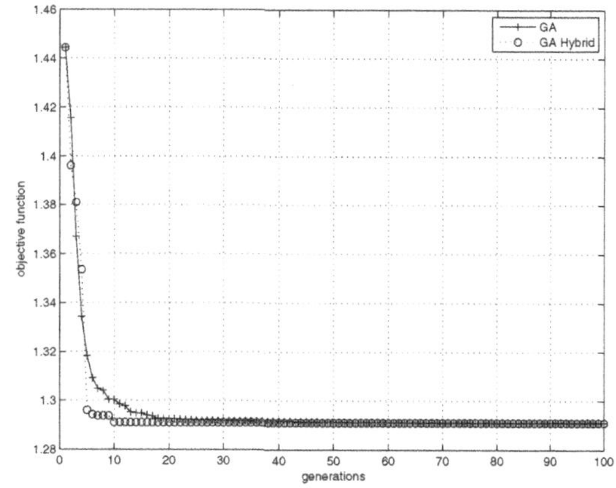

Fig. 1 Convergence Line for the Problem Test (T1) using the GAs

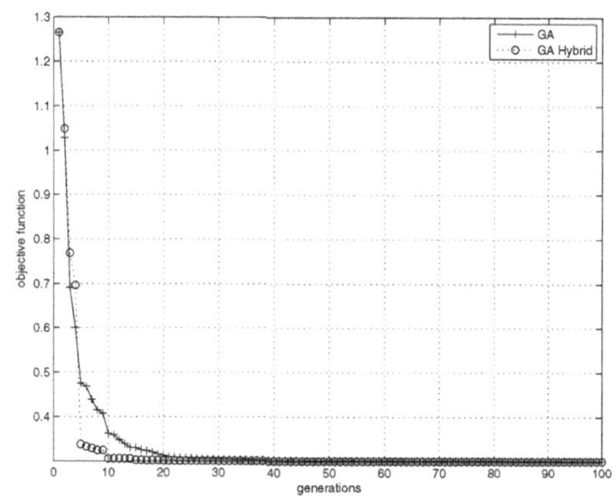

Fig. 2 Convergence Line for the Problem Test (T2) using the GAs

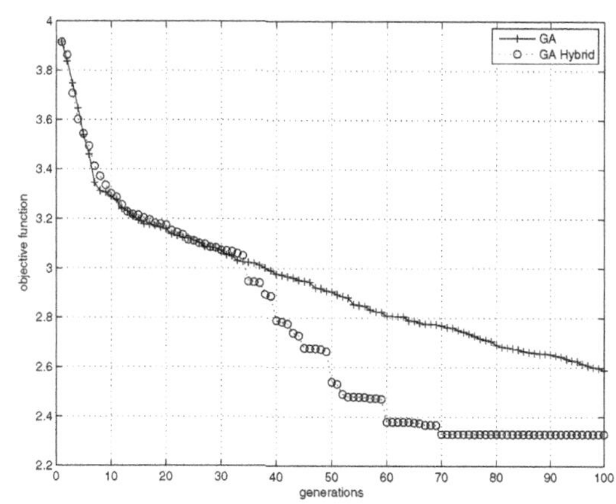

Fig. 3 Convergence Line for the Problem Test (T3) using the GAs

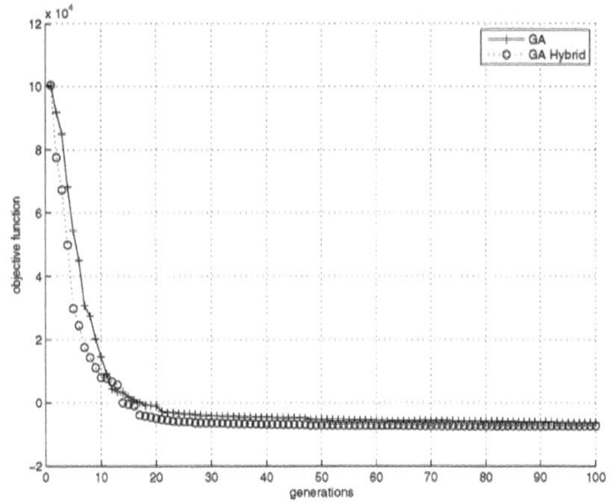

Fig. 4 Convergence Line for the Problem Test (T4) using the GAs

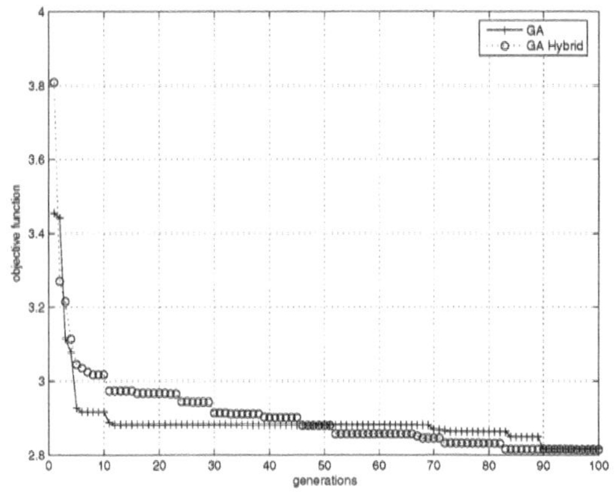

Fig. 5 Convergence Line for the Problem Test (T5) using the GAs

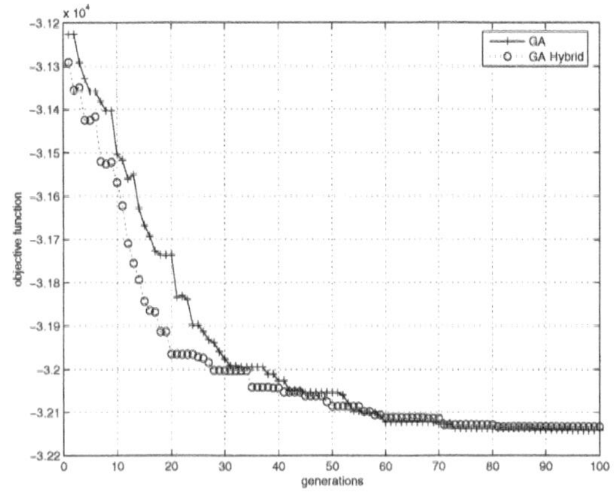

Fig. 6 Convergence Line for the Problem Test (T6) using the GAs

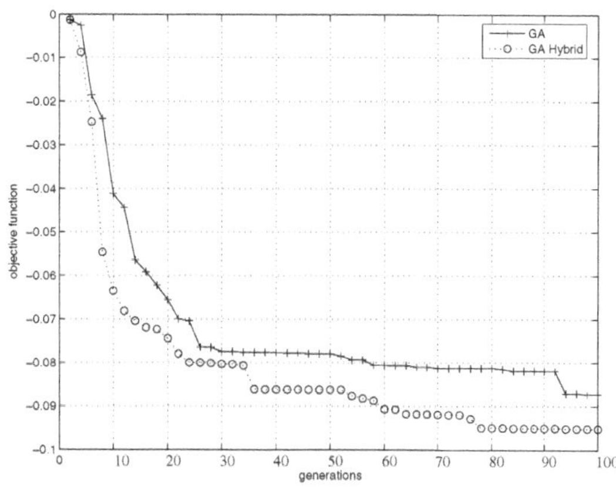

Fig. 7 Convergence Line for the Problem Test (T7) using the GAs

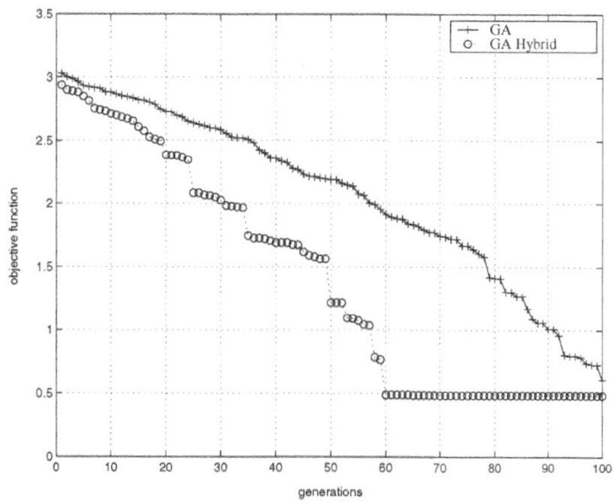

Fig. 8 Convergence Line for the Problem Test (T8) using the GAs

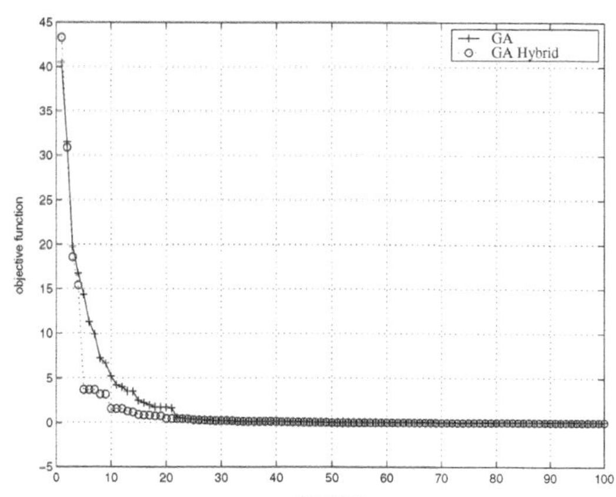

Fig. 9 Convergence Line for the Problem Test (T9) using the GAs

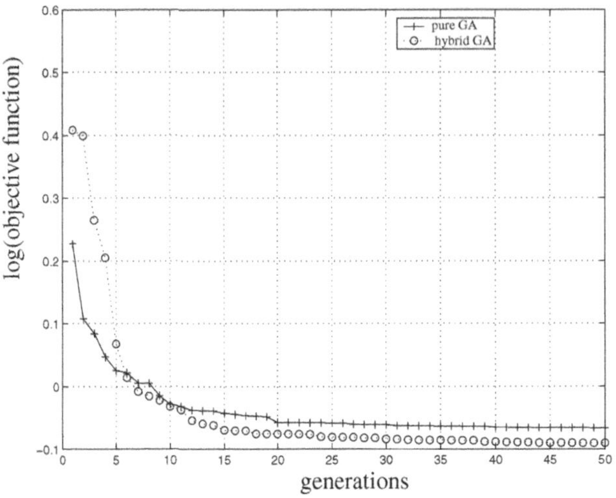

Fig. 10 Convergence line for the TEAM Benchmark Problem 22 using the pure and hybrid GA

This difference is due to the execution of the local search operator (both algorithms perform a similar number of objective function calls).

Figure 10 shows the mean convergence line which corresponds to the mean value of the best individual throughout the generations. In the graph, the x-axis represents the generation and the y-axis represents the base 10 logarithm of the objective function value of the best individual. The base 10 logarithm was used only to enforce the difference between the lines. We can observe that the quadratic local search operator can enhance the convergence speed (in terms of the number of function evaluations). Moreover, the hybrid algorithm is able to find a more accurate solution when compared to the pure algorithm.

Table 7 TEAM Benchmark Problem 22: Statistical Results for the value of $f(\mathbf{x})$ in feasible solutions

Algorithm	mean	stan. dev.
Pure GA	0.8580	0.1214
Hybrid GA	0.8125	0.0225

Table 8 TEAM Benchmark Problem 22: Time (in seconds) Required for running 50 generations with population of 40 individuals

Algorithm	mean	stan. dev.
Pure GA	6.1340×10^3	349.4566
Hybrid GA	6.2048×10^3	219.1207

Table 9 Best Solution Achieved by the Hybrid GA

Var	R_2	h_2	d_2	$f(x)$	E	B_{max}
Unit	(m)	(m)	(m)	(mT)	(MJ)	(T)
value	2.9709	0.6645	0.3026	0.7785	171.0	4.50

7 Conclusion

This paper has presented a new operator for dealing with inequality constraints within evolutionary algorithms. This operator is based on approximations of two types of the constraint functions: the locally convex constraints are approximated by quadratic functions, and the locally non-convex constraints are approximated by linear functions. The auxiliary problem defined by the approximated constraints and the approximated objective function is solved using an LMI (linear matrix inequality) formulation, and the approximated solution is re-inserted in the evolutionary algorithm population. It is noticeable that the proposed methodology does not need any additional function evaluation for being performed, since the function evaluations that are already performed by the evolutionary algorithm, in its normal operation, are re-used by the quadratic or linear approximation procedure for performing the least-squares function approximation.

The results that have been obtained show that the proposed methodology is competitive, leading to solutions that are equivalent or better than the solutions of recently published algorithms. The application of the proposed operator in a design problem with costly black-box objective function suggests that the proposed methodology can enhance the solutions in the case of problems of this class, without incurring in more computational effort.

Acknowledgements. The authors would like to thank CAPES, CNPq and Fapemig.

References

1. Alotto, P., Kuntsevitch, A.V., Magele, C., Molinari, G., Paul, C., Preis, K., Repetto, M., Richter, K.R.: Multiobjective optimization in magnetostatics: a proposal for benchmark problems. IEEE Transactions on Magnetics, Part 1 32(3), 1238–1241 (1996)
2. Back, T.: Proceedings of the Seventh International Conference on Genetic Algorithms. Morgan Kaufmann, San Mateo (1997)
3. Bazaraa, M.S., Sherali, H.D., Shetty, C.M.: Nonlinear Programming: Theory and Algorithms. John Wiley and sons, Inc., Chichester (1979)
4. Boyd, S.P., El Ghaoui, L., Feron, E., Balakrishnan, V.: Linear Matrix Inequalities in System and Control Theory. Siam Studies in Applied Mathematics, vol. 15. Society for Industrial & Applied Mathematics (1997)
5. Coello Coello, C.A.: Theoretical and numerical constraint-handling techniques used with evolutionary algorithms: A survey of the state of the art. In: Coello Coello, C.A. (ed.) Computer Methods in Applied Mechanics and Engineering, 2002, vol. 191, pp. 1245–1287 (2002)

6. Davis, L.: Handbook of genetic algorithms. Van Nostrand Reinhold (1991)
7. Dawkins, R.: The selfish gene. Oxford Press University, Oxford (1976)
8. Deb, K.: An efficient constraint handling method for genetic algorithms. Computer Methods in Applied Mechanics and Engineering 186, 311–338 (2000)
9. Deb, K., Pratap, A., Agarwal, S., Meyarivan, T.: A Fast and Elitist Multi-Objective Genetic Algorithm: NSGA–II, KanGAL Report No. 200001, Kanpur Genetic Algorithm Laboratory, Kanpur (2000)
10. Fonseca, C., Fleming, P.: Multiobjective optimization and multiple constraint handling with evolutionary algorithms I: a unified formulation. IEEE Transactions on Systems, Man and Cybernetics - Part A 28(1), 26–37 (1998)
11. Gen, M., Cheng, R.: Genetic Algorithms and Engineering Design. Wiley, New York (1997)
12. Goldberg, D.E.: Genetic algorithms in Search. In: Optimization and Machine Learning. Addison Wesley, Reading (1989)
13. den Hertog, D., de Klerk, E., Roos, K.: On convex quadratic approximation. Statistica Neerlandica 56(3), 376–385 (2002)
14. Himmelblau, D.M.: Applied Nonlinear Programming. McGraw-Hill, New York (1972)
15. Ishibuchi, H., Murata, T.: Multi-objective genetic local search algorithm. In: Proceedings of the IEEE International Conference on Evolutionary Computation, May 20-22, 1996, pp. 119–124 (1996)
16. Knowles, J.: Local-search and hybrid evolutionary algorithms for Pareto optimization. PhD thesis, Department of Computer Science, The University of Reading, UK (2002)
17. Knowles, J., Corne, D.: Recent Advances in Memetic Algorithms. In: Krasnogor, N., Smith, J.E., Hart, W.E. (eds.) Memetic algorithms for multiobjective optimization: issues, methods and prospects, pp. 313–352. Springer, Heidelberg (2004)
18. Lozano, M., Herrera, F., Krasnogor, N., Molina, D.: Real-coded memetic algorithms with crossover hill-climbing. Evolutionary Computation Journal 12(3), 273–302 (2004)
19. Mezura-Montes, E., Coello-Coello, C.A.: A Simple Multimembered Evolution Strategy to Solve Constrained Optimization Problem. IEEE Transactions on Evolutionary Computation 9(1), 1–17 (2005)
20. Michalewicz, Z.: Genetic Algorithms, Numerical Optimization and Constraints. In: Proceedings of the 6th International Conference on Genetic Algorithms, 1995, pp. 151–158. Morgan Kaufmann, San Francisco (1995)
21. Michalewicz, Z., Schoenauer, M.: Evolutionary algorithms for constrained parameter optimization problem. Evolutionary Computation 4(1), 1–32 (1996)
22. Moscato, P.: On Evolution, Search, Optimization, Genetic Algorithms and Martial Arts: Towards Memetic Algorithms. Caltech Concurrent Computation Program, C3P Report 826 (1989)
23. Moscato, P.: New Ideas in Optimization. In: Corne, D., Dorigo, M., Glover, F. (eds.) Memetic algorithms: a short introduction, pp. 219–234. McGraw-Hill, New York (1999)
24. Ramos, R.M., Saldanha, R.R., Takahashi, R.H.C., Moreira, F.J.S.: The real-biased multiobjective genetic algorithm and its application to design of wire antennas. IEEE Transactions on Magnetics 29(3), 1329–1332 (2003)
25. Richardson, J.T., Palmer, M.R., Liepins, G., Hilliard, M.: Some guidelines for genetic algorithm with penalty function. In: Proceedings of the Third International Conference on Genetic Algorithms, 1989, pp. 191–197. Morgan Kaufmann, San Francisco (1989)
26. Sturm, J.F.: Using SeDuMi 1.02, a MATLAB toolbox for optimization over symmetric cones. Optimization Methods and Software 11-12, 625–653 (1999)

27. Takahashi, R.H.C., Vasconcelos, J.A., Ramirez, J.A., Krahenbuhl, L.: A multiobjective methodology for evaluation genetic operators. IEEE Transactions on Magnetics 39, 1321–1324 (2003)
28. TEAM Workshop Problem 22: SMES Optimization Benchmark, `http://www.igte.tugraz.at/archive/team_new/description.php`
29. Wanner, E.F., Guimaraes, F.G., Saldanha, R.R., Takahashi, R.H.C., Fleming, P.J.: Constraint Quadratic Approximation Operator for Treating Equality Constraints with Genetic Algorithms. In: Proceedings of the IEEE Congress on Evolutionary Computation, 2005. IEEE Press, Los Alamitos (2005)
30. Wanner, E.F., Guimaraes, F.G., Takahashi, R.H.C., Fleming, P.F.: Local search with quadratic approximation into memetic algorithms for optimization with multiple criteria. In: Evolutionary Computation. MIT Press, Cambridge (to appear)

Constraint-Handling in Evolutionary Aerodynamic Design

Akira Oyama

Abstract. Constraint-handling techniques for evolutionary multiobjective aerodynamic and multidisciplinary designs are focused. Because number of evaluations is strictly limited in aerodynamic or multidisciplinary design optimization due to expensive computational fluid dynamics (CFD) simulations for aerodynamic evaluations, very efficient and robust constraint-handling technique is required for aerodynamic and multidisciplinary design optimizations. First, in Section 2, features of aerodynamic design optimization problems are discussed. Then, in Section 3 constraint-handling techniques used for aerodynamic and multidisciplinary designs are overviewed. Then, an efficient constraint-handling technique suitable to aerodynamic and multidisciplinary designs is introduced with real-world aerodynamic and multidisciplinary applications. Finally, an efficient geometry-constraint-handling technique commonly used for aerodynamic design optimizations is presented.

Keywords: real-world design optimization, aerodynamic design optimization, Pareto-based constraint handling.

1 Problem Statement

Without losing generality, constrained real-number optimization problems are written as:

Find $\mathbf{x}$ that minimizes

$$\mathbf{f}(\mathbf{x}) = (f_1(\mathbf{x}), \ldots, f_m(\mathbf{x}), \ldots, f_{m_{max}}(\mathbf{x})) \tag{1}$$

subject to

Akira Oyama
Institute of Space and Astronautical Science, Japan Aerospace Exploration Agency, 3-1-1 Yoshinodai, Sagamihara, Kanagawa, 229-8510, Japan
e-mail: oyama2@flab.isas.jaxa.jp

E. Mezura-Montes (Ed.): Constraint-Handling in Evolution. Optimi., SCI 198, pp. 219–236.
springerlink.com

$$g_1(\mathbf{x}) \le 0, \ldots, g_n(\mathbf{x}) \le 0, \ldots, g_{n_{max}}(\mathbf{x}) \le 0 \tag{2}$$

where $\mathbf{x} = (x_1, \ldots, x_l, \ldots, x_{l_{max}})$ is a vector of design parameters of the solution that minimizes the objective function(s) while satisfying the inequalities (2). l_{max} , m_{max} and n_{max} are numbers of design parameter(s), objective function(s) and constraint(s), respectively.

2 Features of Aerodynamic Design Optimization Problems

Most of real world aerodynamic or multidisciplinary design optimization problems are multiobjective and multi-constraint design optimization problems. For example, a typical transonic aircraft wing design involves maximization of drag divergence Mach number, minimization of mission block fuel, maximum take-off weight, and wing box weight while constraints on flutter speed, structural strength, manufacturing capability, fuel tank volume, etc. must be met. Another example is the supersonic transportation design presented in [33], which has four objectives (drag coefficients at transonic and supersonic cruise speeds, wing root bending moment and pitching moment) and constraints on lift coefficients at transonic and supersonic cruise speeds as well as wing thickness. Many other multiobjective and multi-constraint design optimization problems can be easily found, such as low-boom supersonic business jet design [5], expendable launcher design [10], and multistage compressor design [26].

A multiobjective optimization problem (MOP) simultaneously involves several competing objectives. While a single objective optimization problem may have a unique optimal solution, MOPs present a set of compromised solutions, largely known as the tradeoff surface, Pareto-optimal solutions or non-dominated solutions. The goal of MOPs is to find as many Pareto-optimal solutions as possible to provide useful information of the problem to the designers.

Other features of real-world aerodynamic or multidisciplinary design optimization problems are;

- Number of evaluations is severely limited because aerodynamic function evaluation using computational fluid dynamics (CFD) simulations are very expensive.
- Objective and constraint functions are highly multimodal due to nonlinearity of the flow governing equations.
- Design variables, objectives and constraints are typically real numbers.

For example, in the multidisciplinary design optimization of main wing of the regional jet that is under development in Japan (aimed entry to service is in 2013) [4], the objective and constraint function evaluations include; 1) aerodynamic evaluations, 2) aeroelasticity evaluation, 3) wing weight evaluation, and 4) flight envelope analysis (Fig.1), which required more than 100 hours of computational time for each design candidate evaluation even on a vector supercomputer (when one processing element is used). Therefore, in the example in [4], population size and number of generations are limited to 8 and 19, respectively. It should be noted that, in real world, evolutionary optimization with such small population size and number of

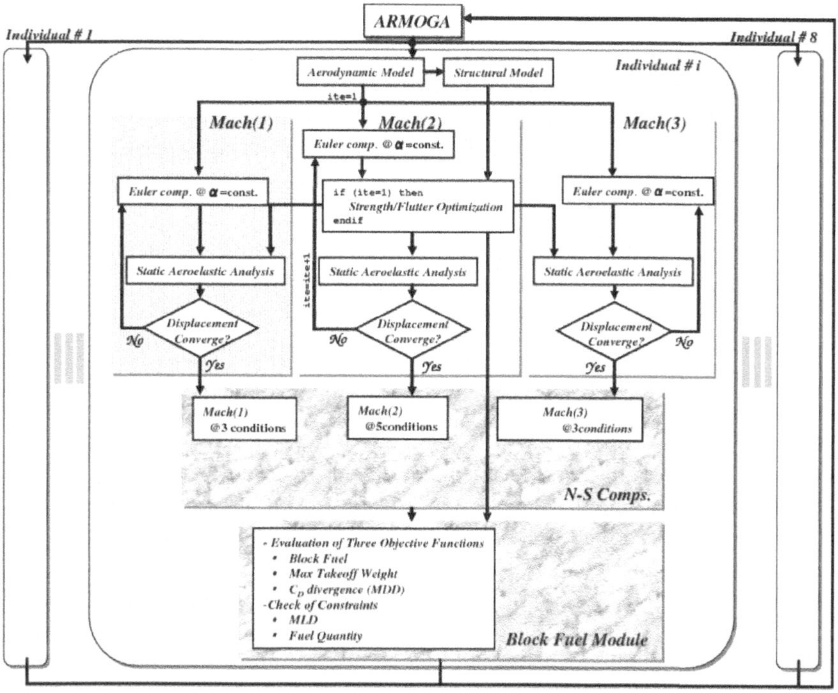

Fig. 1 Flowchart in the objective/constraint function evaluation module for the regional jet design [4]

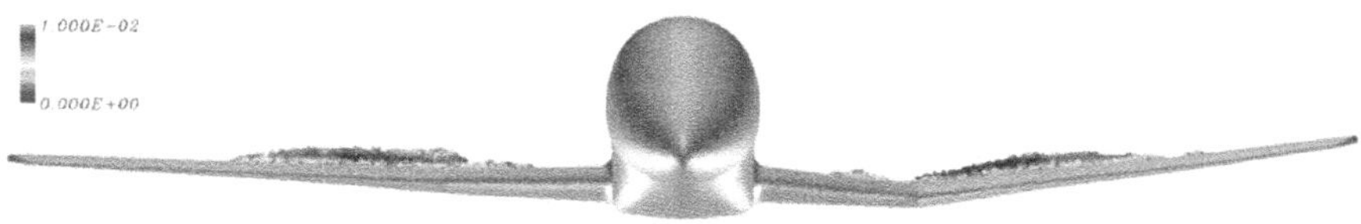

Fig. 2 Comparison of shock wave visualizations colored by entropy under the transonic cruising flight condition between the base design (left) and an optimized design (right) [4]. Weaker shock wave is observed for the optimized design

generations can give useful information to the designers. In fact, authors of [4] significantly improved block fuel (3.6% improvement) from the base design which was designed in conventional design manner (Fig.2).

3 Constraint-Handling Techniques Used for Aerodynamic Design Optimization

Real-world aerodynamic or multidisciplinary design optimization problems involve multiple objectives and multiple constraints. Among many design optimization approaches, therefore, evolutionary algorithms (EAs) are getting popular in aerodynamic and multidisciplinary design optimizations [3, 4, 10, 14–16, 18, 22, 24, 26–28, 31,33,35]. EAs are particularly suited for MOPs because they can uniformly sample various Pareto-optimal solutions in one optimization by maintaining a population of design candidates and using a fitness assignment based on the Pareto-optimality concept. In addition, EAs have other advantages such as robustness, efficiency, as well as suitability for parallel computing.

EAs, however, do not have any explicit mechanism to handle design constraints. A considerable amount of research on constraint handing techniques that incorporate objective function(s) and constraint(s) into the fitness function of design candidates has been carried out (good summaries are given in Coello [7] and Mezura-Montes [20]).

The simplest way to handle constraints is to remove infeasible design candidates out of optimization by applying fitness function of zero (for maximization problem) [15, 26]. However, this approach is not efficient because it wastes information that infeasible design candidates have, i.e., direction from infeasible region to the feasible region. As described in the previous section, number of design candidate evaluations is strictly limited in real-world aerodynamic or multidisciplinary design optimizations. Therefore this approach is not suitable to such design optimization problems.

Three constraint-handling approaches making use of information infeasible designs have been used for aerodynamic and multidisciplinary design optimizations as far as the author knows; penalty function approach (for example, see [9]), Deb's constraint-handling approach [8], and Oyama's constraint-handling approach [30].

Traditional and the most popular approach for handling design constraints in aerodynamic and multidisciplinary design optimizations is the penalty function method [9], where the fitness of a design candidate is determined based on scale function vector $\mathbf{F}(\mathbf{x}) = (F_1(\mathbf{x}), \ldots, F_m(\mathbf{x}), \ldots, F_{m_{max}}(\mathbf{x}))$, which is a weighted sum of the objective function value and the amount of design constraint violations. A typical scale function for minimization problem is given by

$$F_m(\mathbf{x}) = f_m(\mathbf{x}) + \sum_{n=nm1}^{nm2} \alpha_n \cdot max(g_n(\mathbf{x}), 0) \tag{3}$$

where α_n are the positive penalty function coefficients and constraints related to the objective function f_m is $(g_{nm1}, \ldots, g_{nm2})$. Though this approach is widely used in aerodynamic and multidisciplinary designs [14, 16, 18, 22, 31], this method requires careful tuning of the penalty function coefficients to obtain satisfactory designs. For example, if the penalty function coefficients are too small, the optimized designs would not satisfy the constraints. On the other hand, if the penalty function coefficients are too large, the optimized designs would not have satisfactory objective function values. In addition to the balance between the objective functions and the constraints, the balance among the constraints must also be carefully tuned so that the optimized designs satisfy all of the constraints.

Another constraint-handling approach used for aerodynamic evolutionary optimizations is Deb's constraint handling method [8]. This approach ranks design candidates using the following definition of domination between two design candidates,

Definition 1. a solution i is said to constrained-dominate a solution j if any of the following conditions is true,

1. Solutions i and j are feasible and solution i dominates solution j.
2. Solution i is feasible and solution j is not.
3. Solutions i and j are both infeasible, but solution i has a smaller constraint violation.

where

Definition 2. a solution i is said to dominate a solution j if both of the following conditions are true,

1. Solution i is no worse than solution j in all objectives, i.e.,

$$\forall f_m(\mathbf{x}_i) \leq f_m(\mathbf{x}_j) \tag{4}$$

2. Solution i is strictly better than solution j in at least one objective, i.e.,

$$\exists f_m(\mathbf{x}_i) < f_m(\mathbf{x}_j) \tag{5}$$

Flow chart of a procedure using this technique is presented in Fig.3. This approach does not require any penalty function coefficients to be tuned as long as the number of constraint is one. In this sense, this approach is very useful for EA-based design optimizations. In fact, Oyama et al obtained rotor blade designs that significantly outperform the baseline design using an EA coupled with Deb's constraint-handling technique [27, 28]. However, in [8], no approach for a problem with multiple constraints is not presented. Thus, this approach requires careful tuning of the weight

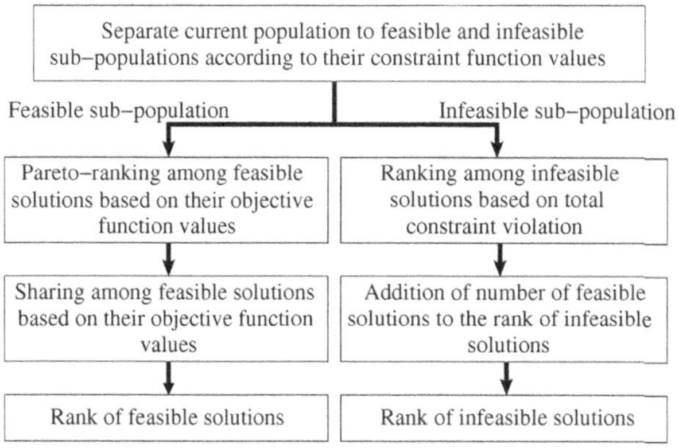

Fig. 3 Flow chart of a ranking procedure using Deb's constraint-handling technique

coefficients if weighted sum of the constraints is used to determine the constraint violation.

The last constraint-handling approach used in aerodynamic and multidisciplinary design optimization problems is Oyama's constraint handling method [30]. This approach is superior to the previous two approaches in the sense that no parameter tuning is required. This approach has been successfully applied to spaceplane conceptual design [30], aerodynamic compressor blade design [29], and aerodynamic airfoil shape optimization [35]. In the next section, this approach is described in detail and then two real-world applications are presented.

4 Oyama's Constraint-Handling Approach

4.1 Approach

Oyama's constraint-handling approach simply apply the idea of non-dominance and niching concepts in the objective function space to the constraint function space. This method bases on the following non-dominance concept.

Definition 3. Solution i is said to constrained-dominate solution j if any of the following conditions is true,

1. Solutions i and j are both feasible and solution i dominates solution j in the objective function space.
2. Solution i is feasible and solution j is not.
3. Solutions i and j are both infeasible, but solution i dominates solution j in the constraint space.

where dominance in the objective function space is defined as Definition 2 while dominance in the constraint space is defined as follows.

Definition 4. Solution i is said to dominate solution j in the constraint space if both of the following conditions are true,

1. Solutions i is no worse than solution j in all constraints, i.e.,

$$\forall G_n(\mathbf{x}_i) \leq G_n(\mathbf{x}_j) \tag{6}$$

2. Solution i is strictly better than solution j in at least one constraint, i.e.,

$$\exists G_n(\mathbf{x}_i) < G_n(\mathbf{x}_j) \tag{7}$$

where

$$G_n(\mathbf{x}) = max(0, g_n(\mathbf{x})) \tag{8}$$

Oyama's constraint-handling approach applies niching based on the amount of constraint violations to infeasible solutions. Here, a standard fitness sharing [13] is applied to the infeasible designs based on their constraint violations as

$$rank'(\mathbf{x}_i) = rank(\mathbf{x}_i) * penalty(\mathbf{x}_i) \tag{9}$$

$$penalty(\mathbf{x}_i) = 1 + \sum_{j=1, j \neq i}^{npop} sh_{ij} \tag{10}$$

$$sh_{ij} = \begin{cases} 1 - (d_{ij}/\sigma_{share})^{\alpha} & d_{ij} < \sigma_{share} \\ 0 & d_{ij} \geq \sigma_{share} \end{cases} \tag{11}$$

$$\sigma_{share} = \sum_{n=1}^{n_{max}} (gmax_n - gmin_n)/npop \tag{12}$$

$$d_{ij} = \sqrt{\sum_{n=1}^{n_{max}} (g_n(\mathbf{x}_i) - g_n(\mathbf{x}_j))^2} \tag{13}$$

$$gmax_n = max(g_n(\mathbf{x}_1), \ldots, g_n(\mathbf{x}_i), \ldots, g_n(\mathbf{x}_{npop})) \tag{14}$$

$$gmin_n = min(g_n(\mathbf{x}_1), \ldots, g_n(\mathbf{x}_i), \ldots, g_n(\mathbf{x}_{npop})) \tag{15}$$

where $npop$ is population size and α is set to 0.3. If the present approach is applied to a multiobjective optimization problem, a fitness sharing is used to the feasible designs based on their objective function values. Flow chart of a ranking procedure using this technique is presented in Fig. 4. Because this method simply uses the idea of non-dominance and niching concepts in the constraint function space, this idea can be coupled with most multiobjective EAs. For example, any ranking procedure can be used for ranking among feasible designs as well as infeasible designs. In addition, the use of stochastic ranking [32] may further improve efficiency and robustness.

The proposed method has a number of advantages.

- It does not require any coefficients to be tuned if a parameterless sharing method such as [12] is used. Even if a sharing method that has coefficients to be tuned is used, according to the author's experience, the parameter values used for sharing in the objective space can be used for sharing in the constraint function space.
- The number of objectives is not increased since non-dominance ranking is applied to feasible designs and infeasible designs separately. If the number of objectives is increased, it will be more difficult to obtain non-dominated solutions due to lower selection pressure.
- It is efficient and robust even when all individuals in the initial population are infeasible due to severe constraints because niching strategy is used in the constraint space. When all individuals are infeasible, the population could lose diversity in the next generation, if diversity in constraint space is not considered.
- It is efficient and robust even when the degree of violation of each constraint is very different because the total amount of constraint violation is not used. If total amount of constraint violation is considered and the degree of violation of each constraint is different, it is very difficult to obtain feasible solutions satisfying

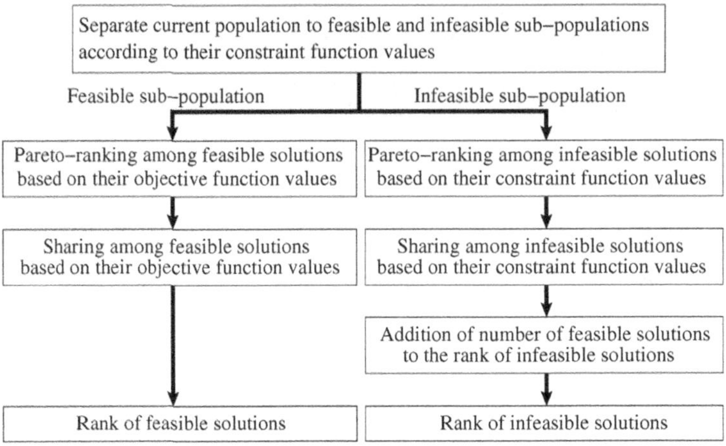

Fig. 4 Flow chart of ranking procedure using Oyama's constraint-handling technique

constraints that has smaller violation in average. There are some approaches that use total amount of constraint violation with dynamically tuned weights of constraints. However, such methods may lose diversity in the population when most of the population are infeasible because niching in the constraint space is not applicable.

- Implementation is easy because Pareto-ranking and sharing method based on the objective functions can be usually applicable to Pareto-ranking and sharing based on the constraint function.

Though this method may increase the computational time required for an EA, the increase is usually negligible in real-world aerodynamic design optimization problems where the computational time required for objective and constraint function evaluations is very large. In the next two subsections, real-world design optimizations using an EA coupled with this constraint-handling approach are presented.

4.2 Conceptual Design Optimization of a Two-Stage-To-Orbit Spaceplane

In this subsection, conceptual design optimization of a two-stage-to-orbit (TSTO) spaceplane is presented. The TSTO spaceplane consists of a booster with air-breathing engines and an orbiter with rocket engines. The orbiter is separated from the booster at a certain altitude to reach low-earth-orbit (LEO) for delivering the payload (Fig. 5).

4.2.1 Formulation of the Design Optimization Problem

The present TSTO mission is to put a payload of 10t into an equatorial orbit at the altitude of 400km. For simplicity, the take-off and landing sites are assumed to be on the equator. The engine of the booster is an air-turbo-ramjet engine with expander cycle (ATREX) [36], which is under development in Japan. The objective is to minimize the gross take-off weight of the spaceplane. The separation time is constrained to be smaller than 550 seconds. The maximum thrust of the booster is also constrained to be smaller than 2.5 MN. The gross take-off weight, separation time and maximum thrust of the booster are iteratively computed from the propulsion, aerodynamics, trajectory and structure modules [19, 34] as shown in Fig. 6. Propulsion, trajectory and airframe configuration parameters (ten parameters in total) are considered as design variables.

4.2.2 Optimization Approach

The present EA uses floating-point representation [21] to represent design parameters of design candidates where an individual is characterized by a vector of real numbers. Random parental selection and the best-N selection [37], where the best N individuals are selected for the next generation among N parents and N children based on Pareto-optimality, are used. The blended crossover (BLX-0.5) [11] is used with crossover rate of 1 for reproduction. Since strong elitism is used, a high mutation rate of 0.2 is applied and a random disturbance is added to the parameter in the amount up to $\pm 20\%$ of the design space. The initial population is generated randomly over the entire design space. The capability of the present EA to find quasi-optimal solutions has been well validated [23, 25].

The rank of each design candidate is defined according to Definition 3. Fonseca and Fleming's Pareto-based ranking [12] is used to rank feasible designs as well as infeasible designs. The population size and number of generations are set to 50.

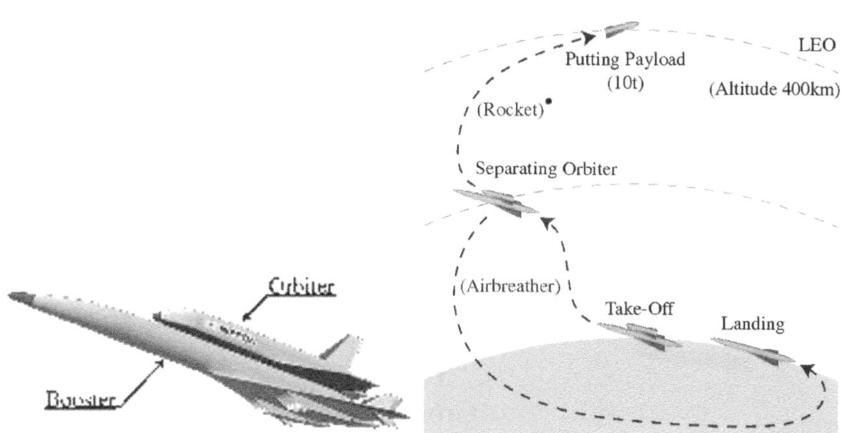

Fig. 5 Image of the TSTO spaceplane (left) and its mission (right)

4.2.3 Results

Optimization is repeated one hundred times with different initial populations. The present EA coupled with Oyama's constraint-handling approach found feasible designs in each run. Average weight of the optimized designs, weight of the best optimum and standard deviation of the optimum weight were 371.19 kt, 369.00 kt, and 1.5787 kt, respectively.

For comparison of constraint-handling techniques, the result is compared with that obtained with the same EA with different constraint-handling techniques; Deb's approach [8], Coello's approach [6], and dynamic penalty approach [17]. To handle multiple constraints with the Deb's approach, the constraints are combined into one constraint violation function where all weights are 1. For the dynamic penalty approach, all weights in equation (3) are 1. The parameter values used in the dynamic penalty approach are C=0.2, α=2, and β=2. The result is summarized in Table 1. The dynamic penalty approach and Deb's approach failed to find feasible designs in 100 optimizations for this design problem. The reason is probably that both methods adopt simple sum of the amounts of constraint violation of different order of magnitude. On the other hand, Oyama's approach and Coello's approach got good

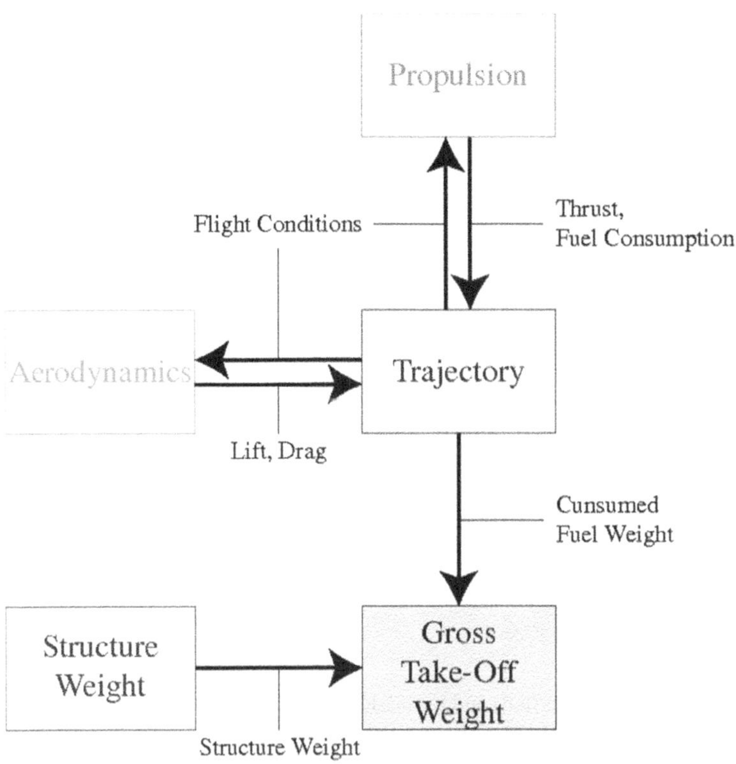

Fig. 6 Multidisciplinary TSTO evaluation

Table 1 Comparison between the constraint-handling methods

Approach	Number of successes	Average weight [kt]	Weight of the best design [kt]	Standard Deviation [kt]
Oyama's approach	**100**	**371.19**	**369.00**	**1.5787**
Coello's approach	99	371.29	369.04	1.6239
Deb's approach		No feasible design is found		
Dynamic penalty		No feasible design is found		

scores, while the proposed method was slightly better than Coello's approach in every measure.

4.3 High-Fidelity Aerodynamic Design Optimization of an Axial Compressor Blade

4.3.1 Formulation of the Design Optimization Problem

The next optimization problem is to seek a redesign of NASA rotor67 [38], which is a low-aspect-ratio transonic axial-flow fan rotor and is the first-stage rotor of a two-stage fan. The fan was designed and tested to help provide the technology to develop efficient, lightweight engines for short-haul aircraft in 1970s. The rotor 67 was designed by using a streamline-analysis computational procedure, which provides an axisymmetric, compressible-flow solution to the continuity, energy, and radial equilibrium equations.

The rotor design pressure ratio is 1.63 at a mass flow of 33.25 kg/sec. The design rotational speed is 16043 rpm, which yields a tip speed of 429 m/sec and an inlet tip relative Mach number of 1.38. The rotor has 22 blades and aspect ratio of 1.56 (based on average span/root axial chord). The rotor solidity varies from 3.11 at the hub to 1.29 at the tip. The inlet and exit hub/tip radius ratios are 0.375 and 0.478, respectively. Reynolds number is 1,797,000 based on the blade axial chord at the hub.

The objective of the aerodynamic rotor shape design optimization problem is to minimize the flow loss manifested via entropy generation. Here, mass-averaged entropy production from inlet to exit at the design point of rotor67 is considered as the objective function to be minimized. Because an optimized rotor design should meet the required mass flow rate and pressure ratio, they are maintained by specifying constraints on them:

$$\left|\frac{massflowrate_{design} - massflowrate_{rotor67}}{massflowrate_{rotor67}}\right| \leq 0.005 \quad (16)$$

$$\left|\frac{pressureratio_{design} - pressureratio_{rotor67}}{pressureratio_{rotor67}}\right| \leq 0.010 \quad (17)$$

In addition, thickness of the optimized design is constrained to be equal to or larger than that of the rotor 67;

$$\sum max(0, thickness_{rotor67} - thickness_{design}) \leq 0 \quad (18)$$

where thickness of the designs and rotor 67 is measured at 10%, 20%, ..., 90% chord positions on 57 blade profiles from root to tip.

4.3.2 Blade Shape Parameterization

Here a rotor blade shape is represented by four blade profiles, respectively at 0%, 31%, 62%, and 100% spanwise stations (all spanwise locations discussed here are measured from the hub), the spanwise twist angle distribution, and the stacking line. Each of these sectional profiles can be uniquely defined by using a mean camber line and a thickness distribution. Here, they are parameterized by the third-order B-Spline curves and positions of control points of the B-Spline curves are considered as the design parameters. As illustrated in Fig. 7, five control points are used for the mean camber line. For the thickness distribution, two control points are added at the leading edge and the trailing edge so that these points represent leading edge and trailing edge radii, respectively. Chordwise locations of the control points at leading edge and trailing edge are frozen to zero and one, respectively. The thickness control points at the leading and trailing edges are defined so that the leading and trailing radii of the designs are identical to those of the rotor 67. These profiles are linearly interpolated from hub to tip. Stagger angles are defined at 0%, 33%, 67%, and 100% spanwise stations and linearly interpolated. Spanwise chord length distribution remains identical to that of the rotor 67. Final blade shape is defined by stacking the blade profiles around the center of gravity of each profile. Here, streamwise and circumferential the stacking lines are defined by B-Spline curves as shown in Fig. 7, respectively. As a result, each blade shape is represented with 49 design parameters.

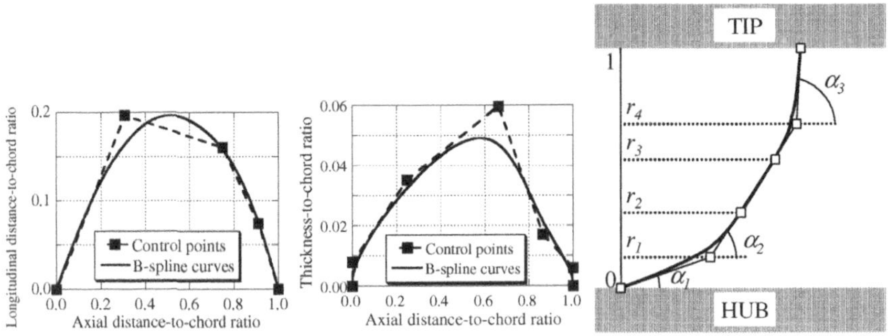

Fig. 7 B-Spline curves for mean camber line (left) and thickness (middle) distributions and stacking line definition (right)

4.3.3 Optimization Approach

The same EA described in 4.2.2 is used. The parameter values used in the EA are also same except for the population size. The population size is increased from 50 to 64 because number of the design parameters is increased. To handle constraints on mass flow rate and pressure ratio, Oyama's constraint-handling approach is used. For blade thickness constraints, an approach for geometry constraints described in section 5 is used.

4.3.4 Aerodynamic Evaluation

The three-dimensional Navier-Stokes (N-S) code used in the present research is TRAF3D [1,2]. Capability of the present code has been validated by comparing the computed results to some experiments such as the Goldman annular vane with and without end wall contouring, the low speed Langston linear cascade [1] as well as the NASA rotor67 [2]. Detail of this code is described in [27].

The three-dimensional grids are obtained by stacking two-dimensional grids generated on the blade-to-blade surface. These two-dimensional grids are of C-type and are elliptically generated, with controlled grid spacing and orientation at the wall. The number of the grid points is 201 chordwise $\times$ 53 tangential $\times$ 57 spanwise. The computational grid for the NASA rotor 67 is shown in Fig. 8. All computations were performed on the NEC SX-6 supercomputer consisting of 128 vector processing elements (PEs) located at JAXA Institute of Space and Astronautics Science in Japan. Aerodynamic evaluations of design candidates at each generation is parallelized using the simple master-slave concept; the grid generations and the flow calculations associated to the design candidates of a generation are distributed into 32 PEs of the NEC SX-6 machine. Aerodynamic function evaluation of each design candidate took about 40 minutes on one PE of the NEX SX-6 machine (For 50 generations, it took more than 66 hours of computational time on 32 vector PEs of the supercomputer).

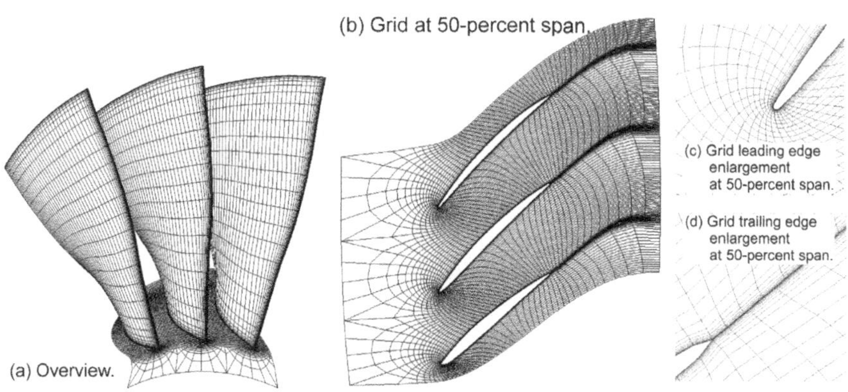

Fig. 8 Computational grid over NASA rotor67. Every other line is shown

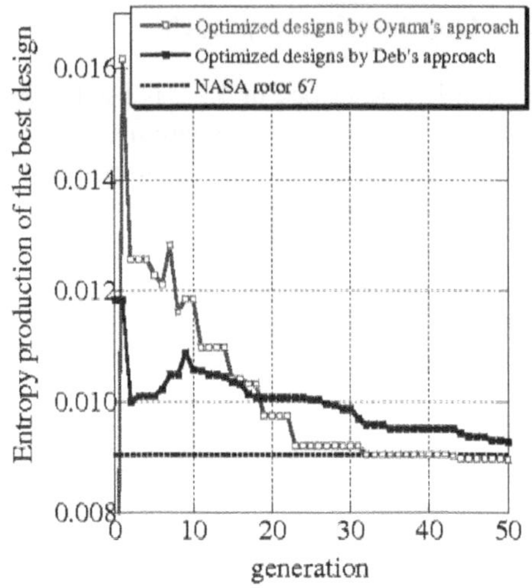

Fig. 9 Comparison of the optimization histories

4.3.5 Result

Figure 9 presents optimization history in terms of the objective function (entropy production) compared with the NASA rotor 67. In the same figure, optimization history of the same EA coupled with Deb's constraint-handling approach is presented for comparison purpose where constraint violation *CV* is defined as

$$CV = 2 \cdot CV_{massflowrate} + CV_{pressureratio} \tag{19}$$

For both cases, the optimized designs obtained after the eighth generation satisfied all the constraints. However, the final design obtained by Oyama's approach has a smaller entropy production than the NASA rotor 67 while the optimized design by Deb's approach could not improve this result in 50 generations. It may be because diversity in the population is lost before a feasible solution is found at the ninth

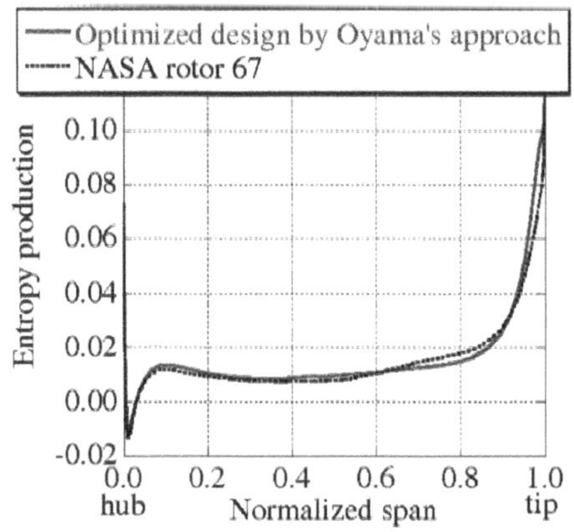

Fig. 10 Comparison of the spanwise entropy distributions

generation when Deb's approach is used. Spanwise entropy distributions of the optimized design and the NASA rotor 67 are compared in Fig. 10. The figure shows that the entropy production can be reduced mainly between 60% to 90% span while it is increased near the tip.

5 Geometry-Constraint-Handling Used for Aerodynamic Design Optimization

In general, aerodynamic design optimization problem involves aerodynamic constraints and other constraints which don't require CFD simulation for function evaluation. A typical example is geometry constraint. For example, aerodynamic drag minimization of a transonic wing without any geometric constraints would result in very thin wing shape. However, such wing shape does not have enough structural strength to withstand the bending moment due to the lift force on the wing. In many cases, structural strength is guaranteed by constraint on minimum wing thickness or minimum wing profile area. In such cases, the constraint function can be evaluated without CFD evaluation of candidate wings. Therefore, in some aerodynamic design optimizations, geometry constraints are evaluated as soon as a new design candidate is generated and if the design candidate does not satisfy the geometry constraints, it is discarded and another design candidate is generated until the new design candidate satisfies all geometry constraints. By doing this procedure, all design candidates satisfy geometry constraints and thus, expensive aerodynamic evaluations can be significantly saved [3,4,24,29].

6 Conclusions

In this chapter, features of aerodynamic design optimization problems were presented and constraint-handling techniques for evolutionary multiobjective aerodynamic and multidisciplinary designs were overviewed. Because number of evaluations is limited in aerodynamic and multidisciplinary design optimizations, a very efficient and robust constraint-handling technique is required for aerodynamic and multidisciplinary design optimizations. Oyama's constraint-handling approach is suitable to aerodynamic and multidisciplinary design optimizations in this sense. Conceptual design of TSTO spaceplane and high-fidelity aerodynamic rotor blade design optimization demonstrated that Oyama's approach is better than traditional constraint-handling methods for real-world aerodynamic and multidisciplinary design optimization problems.

References

1. Arnone, A., Liou, M.S., Povinelli, L.A.: Multigrid calculation of three-dimensional viscous cascade flows. NASA. TM–105257 (1991) (accessed March 1, 2008), http://ntrs.nasa.gov

2. Arnone, A.: Viscous analysis of three-dimensional rotor flow using a multigrid method. ASME J. Turbomach 116, 435–445 (1994)
3. Beume, N., Naujoks, B., Emmerich, M.: SMS-EMOA: Multi-objective selection based on dominated hypervolume. European J. Operational Res. 181, 1653–1669 (2007)
4. Chiba, K., Oyama, A., Obayashi, S., Nakahashi, K., Morino, H.: Multidisciplinary design optimization and data mining for transonic regional-jet wing. J. Aircr. 44(4), 1100–1112 (2007)
5. Choi, S., Alonso, J.J., Kroo, I.M., Wintzer, M.: Multi-fidelity design optimization of low-boom supersonic business jets. American Institute of Aeronautics and Astronautics. AIAA-2004-4371(2004) (accessed March 1, 2008), `http://www.aiaa.org`
6. Coello, C.A.C.: Constraint-handling using an evolutionary multiobjective optimization technique. Civil Engrg. Environ. Syst. 17, 319–346 (2000)
7. Coello, C.A.C.: Theoretical and numerical constraint-handling techniques used with evolutionary algorithms: A survey of the state of the art. Comput Methods Appl. Mech. Engrg. 191, 1245–1287 (2002)
8. Deb, K.: An Efficient constraint-handling method for genetic algorithms. Comput. Methods Appl. Mech. Engrg. 186, 311–338 (2000)
9. Deb, K.: Multi-objective optimization using evolutionary algorithms. John Wiley & Sons, Chichester (2001)
10. Duranté, N., Dufor, A., Pain, V., Baudrillard, G., Schoenauer, M.: Multidisciplinary analysis and optimization approach for the design of expendable launchers. American Institute of Aeronautics and Astronautics. AIAA-2004-4441 (2004) (accessed March 1, 2008), `http://www.aiaa.org`
11. Eshelman, L.J., Schaffer, J.D.: Real-coded genetic algorithms and interval schemata. In: Whitley, L.D. (ed.) Foundations of genetic algorithms, vol. 2. Morgan Kaufmann Publishers, Inc., San Mateo (1993)
12. Fonseca, C.M., Fleming, P.J.: Genetic algorithms for multiobjective optimization: Formulation, discussion and generalization. In: Proceedings of the fifth international conference on genetic algorithms. Morgan Kaufmann Publishers, Inc., San Mateo (1993)
13. Goldberg, D.E., Richardson, J.: Genetic algorithms with sharing for multimodal function optimization. In: Proceedings of the second international conference on genetic algorithms. Lawrence Erlbaum Associates, Inc., Mahwah (1987)
14. Gonzalez, L.F., Periaux, J., Srinivas, K., Whitney, E.J.: A generic framework for the design optimization of multidisciplinary UAV intelligent systems using evolutionary computing. American Institute of Aeronautics and Astronautics. AIAA-2006-1475 (2006) (accessed March 1, 2008), `http://www.aiaa.org`
15. Holst, T.L.: Genetic algorithms applied to multi-objective aerospace shape optimization. J. Aerosp. Comput. Inf. Commun. 2(4), 217–235 (2005)
16. Jeong, S., Yamamoto, K., Obayashi, S.: Kriging-based probabilistic method for constrained multi-objective optimization problem. American Institute of Aeronautics and Astronautics. AIAA-2004-6437 (2004) (accessed March 1, 2008), `http://www.aiaa.org`
17. Joines, J.A., Houck, C.R.: On the use of non-stationary penalty functions to solve nonlinear constrained optimization problems with gas. In: Proceedings of the first IEEE conference on evolutionary computation. IEEE Press, Orlando (1994)
18. Kampolis, I.C., Giannakoglou, K.C.: A multilevel approach to single- and multiobjective aerodynamic optimization. Comput. Methods Appl. Mech. Engrg. (2008) doi: 10.1016/j.cma.2008.01.015

19. Kobayashi, H., Tanatsugu, N.: Optimization method on TSTO spaceplane system powered by airbreather. American Institute of Aeronautics and Astronautics. AIAA-2001-3965 (2001) (accessed March 1, 2008), http://www.aiaa.org
20. Mezura-Montes, E., Coello Coello, C.A.: Constrained Optimization via Multiobjective Evolutionary Algorithms. In: Knowles, J., Corne, D., Deb, K. (eds.) Multi-Objective Problem Solving from Nature: From Concepts to Applications. Springer, Germany (2008)
21. Michalewicz, Z.: Genetic Algorithms + Data Structures = Evolution Programs, 3rd Revised and Extended edn. Springer, New York (1996)
22. Nelson, A., Nemec, M., Aftosmis, M.J., Pulliam, T.H.: Aerodynamic optimization of rocket control surfaces using Cartesian methods and CAD geometry. American Institute of Aeronautics and Astronautics. AIAA-2005-4836 (2005) (accessed March 1, 2008), http://www.aiaa.org
23. Obayashi, S., Sasaki, D., Oyama, A.: Finding tradeoffs by using multiobjective optimization algorithms. Trans. Jpn. Soc. Aeronaut. Space Sci. 47, 51–58 (2004)
24. Oyama, A., Obayashi, S., Nakahashi, K., Nakamura, T.: Euler/Navier-Stokes optimization of supersonic wing design based on evolutionary algorithm. AIAA J. 37(10), 1327–1329 (1999)
25. Oyama, A., Liou, M.S.: Multiobjective optimization of rocket engine pumps using evolutionary algorithm. AIAA J. Propuls Power 18(3), 528–535 (2002)
26. Oyama, A., Liou, M.S.: Multiobjective optimization of a multi-stage compressor using evolutionary algorithm. American Institute of Aeronautics and Astronautics. AIAA-2002-3535 (2002) (accessed March 1, 2008), http://www.aiaa.org
27. Oyama, A., Liou, M.S., Obayashi, S.: High-fidelity swept and leaned rotor blade design optimization using evolutionary algorithm. American Institute of Aeronautics and Astronautics. AIAA-2003-4091 (2003) (accessed March 1, 2008), http://www.aiaa.org
28. Oyama, A., Liou, M.S., Obayashi, S.: Transonic axial-flow blade shape optimization using evolutionary algorithm and three-dimensional Navier-Stoke solver. AIAA J. Propuls Power 20(4), 612–619 (2004)
29. Oyama, A., Fujii, K., Shimoyama, K., Liou, M.S.: Pareto-optimality-based constraint-handling technique and its application to compressor design. American Institute of Aeronautics and Astronautics. AIAA-2005-4983 (2005) (accessed March 1, 2008), http://www.aiaa.org
30. Oyama, A., Shimoyama, K., Fujii, K.: New constraint-handling method for multi-objective and multi-constraint evolutionary optimization. Trans. Jpn. Soc. Aeronaut. Space Sci. 50(167), 56–62 (2007)
31. Pediroda, V., Poloni, C., Clarich, A.: A fast and robust adaptive methodology for airfoil design under uncertainties based on game theory and self-organizing-map theory. American Institute of Aeronautics and Astronautics. AIAA-2006-1472 (2006) (accessed March 1, 2008), http://www.aiaa.org
32. Runarsson, T.P., Yao, X.: Stochastic ranking for constrained evolutionary optimization. IEEE Trans. Evol. Comput. 4, 284–294 (2000)
33. Sasaki, D., Obayashi, S., Nakahashi, K.: Navier-Stokes optimization of supersonic wings with four objectives using evolutionary algorithm. J. Aircr. 39(4), 621–629 (2002)
34. Shimoyama, K., Fujii, K., Kobayashi, H.: Improvement of the optimization method of the TSTO configuration - Application of accurate aerodynamics. In: Groth, C., Zingg, D.W. (eds.) Proc. third international conference on computational fluid dynamics (2004)

35. Shimoyama, K., Oyama, A., Fujii, K.: Multi-objective six sigma approach applied to robust airfoil design for Mars airplane. American Institute of Aeronautics and Astronautics. AIAA-2007-1966 (2007) (accessed March 1, 2008), http://www.aiaa.org
36. Tanatsugu, N., Carrick, P.: Earth-to-orbit combined rocket/airbreathing propulsion. American Institute of Aeronautics and Astronautics. AIAA-2003-2586 (2003) (accessed March 1, 2008), http://www.aiaa.org
37. Tsutsui, S., Fujimoto, Y.: Forking genetic algorithms with blocking and shrinking modes (fGA). In: Proc. the fifth international conference on genetic algorithms. Morgan Kaufmann Publishers, Inc., San Mateo (1993)
38. Walter, S.C., William, S., Donald, C.U.: Design and performance of a 427-meter-per-second-tp-speed two-stage fan having a 2.40 pressure ratio. NASA. TP–1314 (1978) (accessed March 1, 2008), http://ntrs.nasa.gov

Handling Constraints in Global Optimization Using Artificial Immune Systems: A Survey

Nareli Cruz-Cortés

Abstract. Artificial Immune Systems (AIS) are computational intelligent systems inspired by some processes or theories observed in the biological immune system. They have been applied to solve a wide range of machine learning and optimization problems. In this chapter the main AIS-based proposals for solving constrained numerical optimization problems are shown. Although the first works were hybrid solutions partially based on Genetic Algorithms, the most recent proposals are algorithms completely based on immune features. We show that these algorithms represent viable alternatives to the penalty functions and other similar mechanisms to handle constraints in numerical optimization problems.

Keywords: Artificial Immune Systems, Genetic Algorithms, Constrained Numerical Optimization.

1 Introduction

Artificial Immune Systems (AIS) are computational intelligent systems whose design is inspired by the biological immune system. This paradigm has been successfully used to solve complex problems in domains such as machine learning, global optimization, and information security, among others.

It is worth to stress that when solving optimization problems, AIS utilize oversimplifications of the natural immune system. This is because the main goal of this paradigm is to obtain algorithms capable of solving specific well-determined problems. Hence, it is generally considered not crucial to mimic the immune principles in a strict manner.

Nareli Cruz-Cortés
Center for Computing Research, National Polytechnic Institute (CIC-IPN),
07738 Mexico City, Mexico
e-mail: nareli@cic.ipn.mx

E. Mezura-Montes (Ed.): Constraint-Handling in Evolution. Optimi., SCI 198, pp. 237–262.
springerlink.com

As many bio-inspired algorithms (particularly Evolutionary Algorithms), AIS have been very successful in the solution of a wide variety of optimization problems, such as global and multimodal optimization [7, 14, 20, 24] and multiobjective optimization [10, 15, 19, 21, 28].

Evolutionary Techniques are based on unconstrained search spaces. As consequence, these techniques need additional mechanisms to be able of handling the constraints which are typically presented in optimization problems. The most popular mechanism adopted by researchers is probably the so-called *penalty functions* that allow to incorporate constraints into the fitness functions. The crucial idea of a penalty function is to "punish" an infeasible solution by decreasing its fitness value (assuming a maximization problem). Despite their popularity, penalty functions have several disadvantages, being the main one the relatively high difficulty to define accurate penalty factors.

Until date, several AIS used for solving constraint optimization problems have tried to overcome the main disadvantages of the penalty functions. In some cases, this has been done by reducing the parameters defined by the user, thus obtaining highly competitive results when compared against the state-of-the-art evolutionary techniques.

Formally, the problem we are interested to solve is the general nonlinear programming problem which is defined as follows.
Find $\overrightarrow{X}$ which optimizes $f(\overrightarrow{X})$ subject to:

$$g_i(\overrightarrow{X}) \leq 0, i = 1, \ldots, m \tag{1}$$

$$h_j(\overrightarrow{X}) = 0, j = 1, \ldots, p \tag{2}$$

where $\overrightarrow{X} \in \Re^n$ is the vector of decision variables $\overrightarrow{X} = [x_1, x_2, \ldots, x_n]$ where each x_i, $i = 1, \ldots, n$ is bounded by a lower and upper limits $L_i \leq x_i \leq U_i$ defining the search space S, and F is the feasible region $F \subseteq S$; m is the number of inequality constraints and p is the number of equality constraints, both kind of constraints could be linear or nonlinear.

It is customary to transform the equality constraints into inequality constraints of the form:

$$g_j(\overrightarrow{X}) = |h_j(\overrightarrow{X})| - \varepsilon \leq 0, \; j = m+1, m+2, \ldots, m+p \tag{3}$$

where ε is the tolerance allowed (which is a very small value).

In this chapter, we review the most relevant AIS proposals designed to handle constraints published until the present. We classify them into two groups: (1) AIS based on Evolutionary Algorithms (EA)[1], and (2) AIS that are not based on any Evolutionary Algorithms (EA).

This document is organized as follows: In Section 2 we will describe the biological immune principles in which most of the AIS are based. Then, in Section 3 we will expose the AIS based on EA. After that, in Section 4, the AIS that are not

[1] Specifically we refer to Genetic Algorithms.

based on any EA will be presented. In Section 5 we will show the statistical results from some selected algorithms applied to solve several test functions, and some final remarks will be presented in Section 6.

For each AIS presented in this chapter, we will show the components of the system, pseudo-code, its parameters, the manner that the algorithm was validated and some remarks we consider important to point out.

2 Biological Immune System

In this section we will discuss some processes and theories observed in the biological immune system that are useful to understand the algorithms to be presented in the remaining of this chapter. The immune system is a very complex system whose complexity is only comparable with that of the brain [12], this is why we stress that the explanations given in this section are very rough and over simplifications of what really happens in the biological immune system. We start our description by mentioning that the *antigens* are cells or molecules capable of induce an immune response. There are two levels of the immune response: the innate and the adaptive.

The innate response is mainly represented by the cells called *granulocytes* and *macrophages*, this response presents the same response to any type of antigen; i.e., it is unspecific.

On the other hand, the adaptive response is specific to the type of antigen presented to the system; it is mainly executed by the cells named *lymphocytes*. The adaptive immune system is capable of learning and memorizing encounters with the antigens.

Lymphocytes have associated molecules that have the ability to recognize special types of antigens, i.e., lymphocytes' detectors can recognize antigens with different characteristics. There are two types of lymphocytes: the B lymphocytes (also called B-cells) and T lymphocytes (or T-cells). B and T lymphocytes develop in the bone marrow. B-cells mature into the bone marrow, whilst the T-cells travel to the thymus to mature. The B-cell receptors interact with antigens free in solution, while the T-cell receptors recognize antigens processed and bound to a surface molecule called *major histocompatibility complex*. The B cells have expressed on their

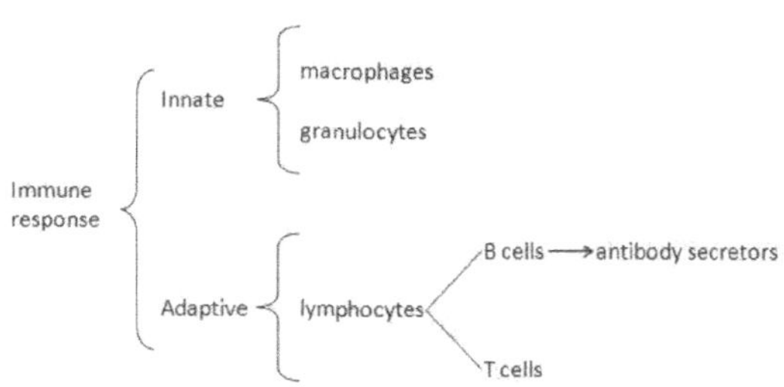

Fig. 1 Main immune system cells

surfaces molecules called *antibodies* that are the main actors of the immune response induced by B cells. The genetic information for an antibody is encoded into gene segments which compose a so-called *library of gene segments*. The main immune system cells are illustrated in Figure 1.

The *antibody* molecule contains a specialized region able to recognize other molecules called *paratope*. The shape of the paratope is determined by the type of aminoacids from which it is composed, and this shape is used to identify antigens. If a match between an antibody and an antigen occurs, the antibody attaches to the antigen and leads to its elimination.

2.1 Clonal Selection Principle

The Clonal Selection Principle is a theory that tries to explain the process by which the immune system finds and neutralizes antigens. It was proposed by F. M. Burnet [5]. Its main idea is that only those lymphocytes with the highest affinity with respect to the antigen are stimulated to be cloned.

Clonal Selection operates on both T-cells and B-cells, however there are some differences when operating on each of them. On the one hand, the B-cells suffer *somatic mutation*[2] during reproduction and are active *antibody* secreting cells. On the other hand, T-cells do not suffer *somatic mutation* during reproduction and they are active lymphokine secretors or T-killer cells [13]. Due to the fact that the B-cells suffer mutation and adaptation capabilities, we will focus on the Clonal Selection of the B-cells.

The B-cells receptors that match an antigen are stimulated to proliferate by cloning. Then the new clones suffer mutation with high rates. In addition to that, some B-cells can differentiate into long-lived memory cells. Memory cells circulate through the blood, lymph and tissues and, if in future they are exposed to the same

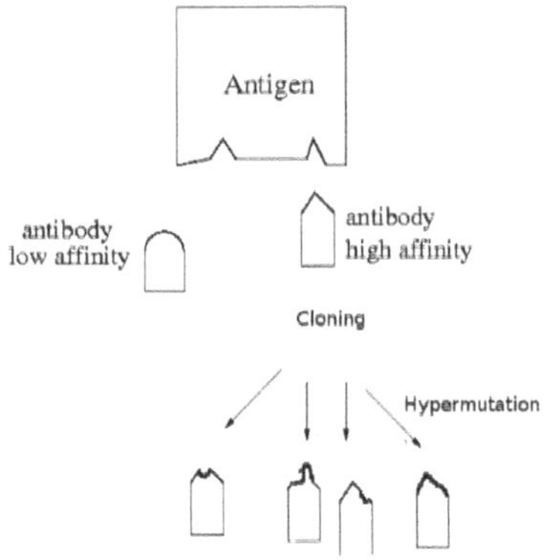

Fig. 2 The Clonal Selection Principle

[2] Mutation with high rate or hypermutation.

antigen (or a similar one) they rapidly start differentiating into plasma cells capable of producing high affinity antibodies. This description is illustrated in Figure 2.

2.2 *Affinity Maturation*

When an antibody is stimulated by a specific antigen for the first time, its response is called *primary immune response*. If the same antibody encounters the same antigen (or a similar one) the immune system presents a quicker and more efficient response. This is called *second response*. The antibodies present in a second response have on average higher affinity to the antigen than those of the early primary response [13], this phenomenon is referred to as the *maturation of the immune response*. This maturation requires that the antibody's receptor in the matured response should be structurally different from those present in the primary response. The repertory of an antibody's receptor is mainly diversified by two mechanisms: *hypermutation* and *receptor editing*. Hypermutation refers to mutation with high rates. Due to the random nature of the mutation process, some cells could result with low affinity receptors or self-reactive cells, thus they must be eliminated through a process called *apoptosis*. Occasionally antibodies with self-reactive detectors suffer *receptor editing* instead of elimination, receptor editing deletes the self-reactive (or low affinity) receptor and develops entirely new receptors.

2.3 *Immune Network Theory*

There exist some alternative theories to the Clonal Selection Principle, one of these is the theory termed *Immune Network Theory* (or Idiotypic Network Theory) proposed by N. K. Jerne [18], which contradicts the Clonal Selection Principle explanation of the immune system. Until date the ideas exposed by Jerne are subject of intense debates among the immunologist community.

The Immune Network Theory proposes that the immune system is composed of a set of regulated molecules' networks and cells that recognize one another even in the absence of antigens. The segment of an antigen that can be recognized by an antibody is named *epitope*. The portion of an antibody responsible for recognizing an epitope is called *paratope*. An *idiotype* is defined as a set of epitopes displayed by a set of antibodies. According to this theory the immune system is a huge and complex network of paratopes that recognize sets of idiotopes and of idiotopes that are recognized by sets of paratopes. These interactions lead to the establishment of a network. After an antibody recognizes an epitope or an idiotope, it can respond either positively or negatively to this recognition signal. A positive response would result into cell activation, cell proliferation and antibody secretion, whereas a negative response would lead to tolerance and supression [13].

The most important AIS to handle constraints will be reviewed in the following sections. They used as a source of inspiration one or several immune mechanisms described before.

3 Handling Constraints with AIS Based on Genetic Algorithms

In this section we describe hybridized mechanisms able to handle constraints in optimization problems through AIS modeled with Genetic Algorithms (GA).

The first work that modeled an AIS to handle constraints was based on a GA [16,17]. In this section we will talk about this seminal work and posterior extensions and modifications to it.

The idea is to use a standard GA that optimizes the objective function without considering the problem's constraints (we will call it the outer GA), and an AIS inserted into it to handle the constraints (we will call it the inner AIS). The premise is that the inner AIS will reduce the amount of constraint violation of the infeasible individuals.

A general model for the AIS based on a GA is shown in the Algorithm 1. All the AIS presented in this section are based on this model, the main difference among all the algorithms [3,4,8,9,17] is the manner of defining the inner AIS (Step 5 from Algorithm 1). Then, for each algorithm presented, we will focus only on that inner AIS assuming that the outer GA is the one presented in Algorithm 1.

Algorithm 1. General model for an AIS based on the GA

```
1: BEGIN outer GA
2: Generate the initial population randomly with binary representation.
3: repeat
4:     For each individual determine if it is feasible or infeasible.
5:     Execute the AIS.
6:     The current AG's population is composed by the antibodies and antigens received from the AIS executed in the
   previous step.
7:     For each individual compute the fitness function value (unconstrained).
8:     Apply the conventional GA's operators (parents selecion, crossover and mutation).
9:     The offsprings replace the current population.
10: until a predetermined number of times or until a determined convergence condition is reached
11: OUTPUT: Best individual
12: END outer GA
```

3.1 *Hajela's Algorithm*

P. Hajela et al. [16, 17] proposed an AIS that was able to handle constraints into a GA. It is based on the process in which if an antigen is detected by the immune system, the antibodies should *learn* which are the correct antibodies able to neutralize that antigen.

The goal is that the antibodies (infeasible individuals) increase their similarity (affinity) with respect to the antigens (feasible individuals), that is, the infeasible solutions will evolve towards feasible solutions.

Then, we have two GAs. The outer GA (Algorithm 1) that optimizes the objective function (unconstrained). The other GA is called AIS or inner GA which is inserted into the outer GA. The AIS handle the problem's constraints. This AIS represents the Step 5 from Algorithm 1.

This proposal will be explained in detail next.

Components of the system

The main components of the inner GA (that is the AIS) are:

- Antigens → Feasible individuals.
- Antibodies → Infeasible individuals.

Both antigens and antibodies are represented by binary strings with the same length codifying the problem's variables.

Algorithm

This algorithm consists of two GAs, the first is the conventional GA (Algorithm 1) which optimizes the objective function without constraints, the second one is another GA inserted into the first one. This inner GA (Algorithm 2) is the AIS composed by *antigens* and *antibodies* handling the constraints. The main idea is that the population of antibodies will be evolved towards the antigens through the GA. An antibody's fitness value (called Z) is measured by means of how similar it is with respect to an antigen. Due to the fact that the representation utilized in this work is binary, authors proposed to compute the antibodies' fitness by counting the coincidences along both binary strings (antigen-antibody). Higher values of Z mean higher degree of match between the two strings.

The pseudo-code for this AIS is illustrated in Algorithm 2, which is inserted into the outer GA defined in Algorithm 1 Step 5.

Algorithm 2. Inner AIS proposed by Hajela et al. in [17]

```
1: BEGIN AIS-Hajela
2: INPUT: it receives feasible and infeasible individuals from the outer GA (Algorithm 1).
3: The feasible and infeasible individuals are separated and each group ranked in an order which places the best
   objective function value at the top of the scale. A fraction of the feasible individuals are selected and denoted as the
   antigen population. The infeasible individuals are called antibodies.
4: Initialize the fitness of all antibodies to zero.
5: Compute the fitness of the antibody pool based on similarity to the antigens; this requires the following specific
   steps:
6: repeat
7:     An antigen is selected at random.
8:     A sample of antibodies of size μ is selected from the antibody pool without replacement.
9:     The match score of each antibody is computed by comparing against the selected antigen, and the antibody with
   the highest score has the match score added to its fitness value; the fitness of the other antibodies is unchanged.
10:     the antibodies are then returned to the antibody population
11: until a determined number of times is reached (typically two to three times the antibody population size)
12: Based on the fitness computed in Steps 5-10, a GA simulation is conducted (parent selection, crossover, mutation,
   population replacement)
13: The process is repeated from Step 4 a predetermined number of iterations.
14: OUTPUT: Both antigen and antibodies populations are returned to the outer GA.
15: END AIS-Hajela
```

Parameters

The parameters introduced by this algorithm are:

- All the necessary parameters to define the inner and outer GAs, i.e., type of parents selection, crossover type and rate, mutation rate, number of generations and population size for both algorithms.
- The value of μ which determines the size of the antibodies' sample in Step 7 from Algorithm 2.

Algorithm's validation

This algorithm was originally validated on two engineering designs problems [16]. The first one optimizes the size of a truss structure and the second one is focused on the minimal thickness design of simply-sported composite plate. They were compared against a plain GA. For the first problem this algorithm obtained a very superior result than the plain GA, and very similar results with both approaches for the second problem.

Rajasekaran and Lavanya [26] applied this algorithm on large dimensionality design problems where the computational efficiently is a major concern. They used one well-known optimization problem of 25-bar truss and two practical examples: a braced barrel vault (static load) and a double-layer grid (earthquake loading). They obtained better results than the ones achieved through the usage of other techniques previously reported.

Remarks

This algorithm represents the first proposal using an AIS to handle constraints. One of its main advantages is that it does not require the definition of any penalization function and its conceptual simpleness. It is important to remark that, despite that this scheme consists of a GA inside another GA, the fitness functions evaluations are not increased because only the outer GA evaluates the objective function. The inner GA, instead, assigns the antibodies's fitness just by counting the coincidences between two binary strings.

The authors assume that, in a randomly generated initial population, there are feasible and infeasible individuals, however this assumption is not necessarily true, especially when faced with constrained problems whose feasible's region is small compared with the complete search space.

3.2 Coello's Algorithm

Coello et al., [8, 9] published a work based on Hajela's algorithm giving some ideas on how to improve it. Furthermore they proposed a parallel version of the algorithm obtaining promising results.

Components of the system

The algorithm's components are the same than the ones in Hajela's work (Section 3.1), i.e., *antigens* are the feasible individuals and *antibodies* are the infeasible ones representing potential solutions to the problem through binary strings with the same length.

Algorithm

In [9] authors argue that Hajela's algorithm needs to sort the population in the AIS which is a task that increases the execution time, and it does not indicate how to proceed when no feasible individuals are present. Then this proposal tries to overcome these inconveniences, see pseudo-code in Algorithm 3.

Algorithm 3. AIS proposed by Coello et al., [9]

```
BEGIN AIS
INPUT: it receives feasible and infeasible individuals from the outer GA (Algorithm 1).
if the population contains feasible and infeasible individuals then
    the population is divided into antibodies (infeasible individuals) and the antigens (feasible individuals)
else
    the best individual in the population is used as "antigen", this is the individual with the lowest amount of constraint violation.
end if
repeat
    Select randomly a sample of antibodies of size μ
    The fitness of the antibodies is computed:
    An antigen is randomly selected. Each antibody in the sample is compared against the antigen selected, and the result of the comparison is computed (called Z): Z = Σ_{i=1}^{L} t_i where t_i = 1 if there is a matching at position i = 1,...,L (L is the length of the chromosome)
    Based in the fitness computed in the previous step, the population of antibodies is reproduced in a traditional GA (using parent selection, crossover and mutation)
until a predetermined number of iterations is reached.
OUTPUT: All the individuals are returned to the outer GA.
END AIS.
```

Furthermore, authors suggested some changes into the Algorithm 1 as follows: The parent selection from the outer GA (Step 8) is a binary tournament following the next rules:
- If one individual is infeasible and the other is feasible, then the feasible individual wins.
- If both individuals are feasible, then the one with the highest fitness value is the winner.
- If both individuals are infeasible, then the winner is the one with lowest constraint violation value.

Parallel version of Coello's algorithm

A parallel version of this algorithm [8, 9] was proposed by using an island model or *multi-deme* where there are a number of sub-populations or *demes* evolving separately but interchanging individuals at certain intervals (number of generations called *epoch*). This model is called multi-population GA or coarse-grained GA [6]. In this work each deme has both, the inner and outer GAs.

Parameters

The parameters utilized in this algorithm are the following:

- All the necessary parameters to define both the inner and outer GAs, i.e., type of selection, crossover type and rate, mutation rate, number of generations and population size.
- The value of μ which determines the size of the antibodies' sample in Step 8 from Algorithm 3.

Furthermore, the parallel version requires to define:

- The number of demes (sub-populations), the size of the epoch, the value of r which represents the number of antibodies selected to migrate, the demes' interconnection topology, and the migration's policy.

Algorithm's validation

This algorithm is tested on a well known benchmark [23, 27] and compared against other AIS. Details can be found in Section 5.

Remarks

This algorithm is very similar to Hajela's. Authors argue that their proposal has a lower execution time by introducing only some small changes.

3.3 Bernardino's Algorithm

Bernardino et al., [3,4] proposed a hybrid GA combined with an AIS. This approach is based on the previous mentioned Hajela's algorithm [16, 17] but the inner GA is substituted by an AIS which is not based on a GA. This AIS is a simplification of the Clonal Selection Algorithm (CLONALG) originally proposed by De Castro y Von Zuben [14] for pattern recognition tasks and multimodal optimization.

The biological immune process in which this AIS is based is the Clonal Selection Theory of the B-cells, whose main idea is that those B-cells with the highest affinity with respect to the antigen are stimulated to be cloned. The stimulated B-cells suffer a cloning process, then the new clones are mutated. Some cells would increase their affinity after these processes, but others would decrease it. The best cells are allowed to survive and the worst are eliminated from the system.

Components of the system

The system's components defined for this inner AIS are antigens and antibodies:

- Antigens → feasible individuals.
- Antibodies [3] → infeasible individuals.

represented by fixed length binary strings with Gray codes codifying the problem's variables.

Algorithm

The AIS's general idea is that the best antibodies are cloned. After that the new clones are mutated. Next, the distances (affinities) between antibodies and antigens are calculated. Then, the antibodies with the highest affinity values are selected to survive to the next algorithm's iteration. This algorithm utilizes binary representation, then authors suggest to compute the antibodies affinities through Hamming distances with respect to antigens.

This algorithm is composed by the outer GA shown in Algorithm 1 and the inner AIS which is illustrated in Algorithm 4.

Algorithm 4. Inner AIS proposed by Bernardino et al. [4]

BEGIN AIS
INPUT: it receives feasible and infeasible individuals from the outer GA (Algorithm 1).
if the population contains feasible and infeasible individuals **then**
    the population is divided into antibodies (infeasible individuals) and the antigens (feasible individuals)
else
    the two best individuals in the population are used as "antigens", these are the individuals with the lowest amount of constraint violation.
end if
repeat
    Clone the antibodies.
    Mutate the new clones.
    Compute the affinity between antigens and antibodies.
    Based on their affinity values, select the best antibodies to survive .
until a predetermined number of iterations
OUTPUT: All the antigens and antibodies are returned to the outer GA.
END AIS

Furthermore, authors suggested some changes into the outer GA (Algorithm 1):

- To use a binary tournament to select the parents (Step 8) as follows: (1) between a feasible and an infeasible individual, the feasible individual wins, (2) between two feasible individuals, the one with the higher fitness wins, and (3) between two infeasible individuals the one with the smaller constraint violation wins.
- To apply the crossover operator only to similar individuals (Step 8).
- In Step 9 the individuals that will form the population for the next generation are selected. This is done by joining the current population and the new individuals,

[3] Note that we do not make difference between B-cells and antibodies.

then the *clearing operator* is applied to them. The *clearing operator* was proposed by Petrowski [25] for multimodal problems as a niching mechanism. The rationale of this operator is inspired in the principle of sharing limited resources within sub-populations of individuals.

Parameters

The parameters introduced by this algorithm are the following:

- Parameters needed by the outer GA: Population size, number of generations, mutation rate, crossover rate, and the Critical distance to compute the clearing process.
- Parameters needed by the inner AIS:
  - Number of clones.
  - Number of antibodies to be cloned.
  - Mutation rate and type.
  - Number of iterations.

Algorithm's validation

The algorithm was tested on six mechanical engineering optimization problems [4] and compared against the EA presented in [22]. Authors obtained very competitive results, concluding that their algorithm performed well in problems with continuous design variables. At the same time, they observed a lower performance for problems with discrete design variables.

Remarks

This algorithm combines a GA with a pure AIS. It seems to be that the cloning process into the AIS accelerates the antibodies's convergence towards the antigens, i.e., the antibodies will be very similar to antigens very soon. This could be helpful for certain type of problems, however for other cases, it could provoke premature convergence. Perhaps authors tried to avoid premature convergence by using the clearing operator.

There are two features that seem to be very important for this algorithm, the mutation operator used into the AIS and the critical distance used into the clearing operator. Authors do not give any clue about their effect to algorithm's performance.

4 Handling Constraints with AIS

In this section we describe AIS that are exclusively based on processes and theories directly extracted from the biological immune system, which we call *pure schemes* because they do not use any EA.

The algorithms presented in this section [1, 2, 11, 29] are based on three different immune processes: Firstly algorithms based on the Clonal Selection Algorithm

(CLONALG). Secondly an algorithm based on a process suffered by the T-cells; and thirdly an algorithm combining the Clonal Selection and the Immune Network Theories.

4.1 Handling Constraints with AIS Based on the Clonal Selection Principle

The Clonal Selection Principle explains how the adaptive immune system deals with antigens. Briefly, when a B-cell recognizes an antigen with certain affinity, it is stimulated to produce antibodies by cloning itself in large quantities. During this cloning process, the new cells (clones) suffer mutation with high rates (hypermutation) [13].

The Clonal Selection Algorithm (named CLONALG) was proposed by De Castro and Von Zuben [14]. It is a learning algorithm well-suited for solving pattern recognition and multimodal optimization problems. This is a population based algorithm whose only variation operator is *mutation* (or hypermutation).

The CLONALG to optimization problems has two main components: antibodies represented by the problem's variables and antigen which is the objective function. The antibodies affinity is their evaluation into the objective function.

A general model for the Clonal Selection Algorithm (CLONALG) [14] to numerical optimization is presented in Algorithm 5. The main features observed in this algorithm are the following:

- The cloning rate of each antibody is proportional to its affinity with respect to the antigen, i.e., the higher the affinity is, the higher becomes the number of clones generated for each antibody and viceversa.
- The mutation suffered by each clone during reproduction is inversely proportional to its affinity with the antigen, i.e., the higher the affinity, the smaller the mutation rate and vice versa.

Algorithm 5. A general model for CLONALG to numerical optimization [14]

BEGIN CLONALG
INPUT: Parameters values
Generate j antibodies randomly.
repeat
    Determine the affinity of each antibody (Ab). This affinity corresponds to the evaluation of the objective function.
    Sort the antibodies by using their affinity values, from the highest to the lowest.
    The antibodies are cloned. The number of clones for the antibody i (NC_i) is calculated with:
$NC_i = \Sigma_{i=1}^{j}(round)(\frac{\beta * j}{i})$ where β is a multiplier factor.
    All the clones are subjected to a hypermutation process inversely proportional to their antigenic affinity.
    Determine the affinity of the mutated clones.
    From this set of clones and antibodies (Ab) select the j highest affinity clones to compose the new antibodies' population.
    Replace the d lowest affinity antibodies by new individuals generated at random.
until a predetermined number of times is reached
OUTPUT: Best antibody.
END CLONALG

The authors proposed to use self-adapting step size to mutate individuals as in Evolution Strategies.

The next two algorithms (Constrained-CLONALG [11] and AIS_{const} [1]) are based on this schema, they are only variations on it.

4.1.1 Constrained-CLONALG [11]

Cruz-Cortés et al. [11] adapted the CLONALG shown in Algorithm 5 to handle constraints in the following manner:

Components of the system

The system's components are the following:

- *Antigens* are represented by the objective function we want to optimize and the problem's constraints.
- *Antibodies* are represented by the variables of the problem with real numbers. Antibodies could be feasible or infeasible depending if they satisfy or not the problem's constraints.

Algorithm

This algorithm is based on the one presented in Algorithm 5. The changes suggested by the authors are:

- The manner to determine the antibodies affinity values (Steps 4 and 8).
- The hypermutation operator (Step 7). Authors suggest to define a step size which is in function of the range of each decision variable, the size of antibodies' population and their affinity value (instead of the self-adapting mechanism).
- The antibodies that are allowed to survive (Step 10).

These changes are presented in the pseudo-code in Algorithm 6.

Parameters

The parameters introduced by this algorithm are the following:

- Number of antibodies.
- Number of algorithm's iterations.
- d that determines the number of the lowest affinity antibodies to be replaced.
- β which is used to compute the number of clones.
- q that determines the number of infeasible antibodies allowed to survive for the next iteration.

Algorithm's validation

This algorithm is compared against other AIS presented in this chapter. Details are shown in Section 5.

Algorithm 6. Constrained-CLONALG [11]

1: BEGIN Constrained-CLONALG
2: INPUT: Parameters values.
3: Generate j antibodies randomly.
4: **repeat**
5: Determine the affinity of each antibody (Ab). This affinity corresponds to the evaluation of the objective function and feasibility. The feasible antibodies are in top of the list (the best), followed by the infeasible ones. Into the feasible group the best are those with highest objective function values. Into the infeasible antibodies, those with less constraint violation value are the best.
6: Sort the antibodies using their affinity values from the highest to the lowest.
7: The antibodies are cloned. The number of clones for the antibody i (NC_i) is calculated with:
8: $NC_i = \Sigma_{i=1}^{j}(int)(\frac{\beta * j}{i})$ where β is a multiplier factor.
9: The clones are subjected to a hypermutation as follows:
- For each decision variable x_k, compute $R_k = UB - LB$, where UB and LB are the upper and lower bounds of that variable, respectively, and $R - k$ is the search space size of the k-th variable.
- Compute $\Delta_k = R_k / j$ where j is the number of antibodies in the population.
- The clone population is sorted by affinity values in descending order.
- The mutation operator is applied to each size-g clone group coming from the same parent
- - For each variable k, compute $\delta_k = \Delta_k / g$
- - Apply mutation to each variable x_k by using $x_k^{new} = x_k + U(0,)$ where U is a random number in the range from 0 to δ_k with uniform distribution.

10: Determine the affinity of the mutated clones. (Similar to Step 4).
11: From the union of clones and antibodies select the j higher affinity clones to compose the new antibodies' population.
12: Replace the d lowest affinity antibodies by new individuals generated at random. Allow at least q infeasible antibodies to survive to the next iteration.
13: **until** a predetermined number of times is reached
14: OUTPUT: Best antibody.
15: END CLONALG

Remarks

This proposal represents the first attempt to use a pure AIS which is not based on any EA. The algorithm promotes the search into the boundary between feasible and infeasible region. Authors conclude that the antibodies representation and type of mutation is crucial. They obtained an algorithm's improvement when using real numbers representation and controlled and uniform mutation. One of the main advantages of this algorithm is that it avoids the use of any penalty function and it requires only few parameters to be defined however, its authors did not give any clue on how important are those parameters for the algorithm's performance.

4.1.2 AIS$_{const}$ [1]

Aragón et al. [1] suggested an algorithm based on the CLONALG (Algorithm 5) scheme called AIS$_{cont}$. The main difference between both works is the mutation operator, which performs differently when applied to feasible or infeasible antibodies. In fact, AIS_{const} proposes two different mutation operators: one specially designed to feasible antibodies, and the other one to infeasible antibodies.

Components of the system

The system's components are the following:

- *Antigens* are represented by the objective function we want to optimize and the problem's constraints.
- *Antibodies* are represented by the variables of the problem which are potential solutions to the problem by using real numbers.

Algorithm

Aragón et al. scheme is based on the one shown in Algorithm 5. The following changes are proposed by the authors:

- The way to determine the highest affinity antibodies (Steps 4 and 8).
- The hypermutation operator (Step 7) which depends on the antibodies feasibility: If the antibodies are feasible then the mutation is small. However, for infeasible antibodies, it randomly apply a small or a large mutation.

These changes are illustrated in pseudo-code presented in Algorithm 7.

Algorithm 7. AIS_{const} [1]

1: BEGIN AIS_{const}
2: INPUT: Parameters values.
3: Generate j antibodies randomly.
4: **repeat**
5: Determine the affinity of each antibody (Ab). This corresponds to the evaluation of the objective function and feasibility. The feasible antibodies are in top of the list, followed by the infeasible ones. Into the feasible group the best are those with highest objective function values. Into the infeasible antibodies, those with less constraint violation value are the best.
6: Sort the antibodies using their affinity values from the highest to the lowest.
7: The antibodies are cloned. The number of clones for the antibody i (NC_i) is calculated with:
8: $NC_i = \Sigma_{i=1}^{j}(int)(\frac{\beta * j}{i})$ where β is a multiplier factor.
9: The clones are subjected to a hypermutation process depending on the clones feasibility:

- If a cell is a feasible solution then only a single position of the string is changed according to a randomly chosen value.
- If the clone is an infeasible solution then for each decision variable x_i
  - Assign a binary random number to r:
  - if $r = 0$ mutate using the equation:
  - $x_i = x_i \pm rand(0,1) * \frac{range(x_i)}{generation} * NC$
  -
  - else use the equation:
  - $x_i = x_i \pm rand(0,1) * \frac{range(x_i)}{generation * NC}$
  -

where $rand(0,1)$ is a random number with uniform distribution between 0 and 1, $range(x_i)$ is a random number in the allowable range of x_i, *generation* is the current generation and NC is the number of clones.

10: Determine the affinity of the mutated clones (similar to Step 4).
11: From this set of clones and antibodies, select the j higher affinity clones to compose the new antibodies' population.
12: Replace the d lowest affinity antibodies by new individuals generated at random.
13: **until** a predetermined number of times is reached
14: OUTPUT: Best antibody.
15: END AIS_{const}

Parameters

The parameters utilized in this algorithm are the following:

- Number of antibodies.
- Number of algorithm's iterations.
- d that determines the number of the lowest affinity antibodies to be replaced.
- β which is used to compute the number of clones.

Algorithm's validation

This algorithm is tested by using a benchmark [23, 27] and compared against other AIS found in this chapter. The results are presented in Section 5.

Remarks

The main difference among this work and the previously reported approaches is the mutation operator that makes one distinction between feasible and infeasible antibodies. For the infeasible antibodies case, the mutation operator considers the range of each decision variable, the current generation number and the number of clones, trying to reduce the step size on infeasible antibodies. This algorithm does not require any penalty function and has only few parameters to define.

4.2 AIS Based on a T-Cell Model [2]

A novel approach to handle constraints was proposed recently by Aragón et al. [2] based on the manner that the T cells go through different phases.

The main idea behind this model is a three-population architecture which emulates the T cells going from Virgin Cells (VC), to Effector Cells (EC). Finally some of them are converted into Memory Cells (MC). Furthermore, the *apoptosis* process is modeled. This process means that the useless cells are eliminated. The cells go to another population when they "react", i.e., when they change by means of a mutation operator.

The Virgin Cells (VC) provide the diversity, the Effector Cells (EC) explore the region close to the bound between feasible and infeasible regions, whereas the Memory Cells (MC) explore the neighborhood of the best solutions found so far. The authors suggested the use of a Dynamic Tolerance Factor (DTF) into the virgin and effector cells populations. The goal of adding this factor is to avoid that the search falls into a local optimum. DTF is computed by adding the constraint violation of each cell and dividing it by the number of VC cells or three times the number of ECs for the virgin or effector cells, respectively. DTF relaxes the tolerance factor making easy to generate solutions considered "feasible" although they may become infeasible if evaluated with the actual required precision.

Components of the System

The system's components are the following:

- T-Cells $\rightarrow$ potential solutions to the problem represented with real numbers.

Algorithm

The pseudo-code is presented in Algorithm 8. Here, "react" (Steps 8 and 14) means to apply the mutation operator. The algorithm applies three different mutation operators, two for the Effector cells (feasible and infeasible ones) and one more for the Memory cells. They are explained next.

- **The mutation operator applied to the infeasible Effector Cells ($\mathbf{EC}_{inf}$)** is defined in the following manner:
  First it identifies the most violated constraint (c). If this constraint c value is larger than the sum of the constraints violation divided by the total number of constraints then: Each bit from each decision variable involved in c is changed with probability 0.05 Otherwise, each bit from one decision variable involved in c is changed with probability 0.05
- **The mutation operator for the feasible Effector Cells ($\mathbf{EF}_f$).** This operator generates two mutated cells and the best of them survive to the following iteration:

  1. It identifies the constraint with the most negative value and changes each bit from each decision variable involved in that constraint with probability 0.05
  2. It changes each bit from all the decision variables with probability 0.5

- **The mutation operator for Memory Cells** is defined by the following equation:

$$x' = x \pm \left(\frac{U(0,1)lu - ll}{1000000gen|const||dev|}\right)^{U(0,2)} \tag{4}$$

  where x and x' are the original mutated decision variables, respectively. U(0,1) and U(0,2) are randomly generated numbers with uniform distribution between (0,1) and (0,2), respectively. lu and ll are the upper and lower bounds of x, $|const|$ is the number of constraints in the problem, $|dev|$ are the number of decision variables of the problem and gen is the current number generation.

Parameters

The parameters in this algorithm are the following:

- Population size for EC, MC, and VC.
- Replacement policy for the cells in EC and MC
- Number of iterations for each of the three loops.

Algorithm's validation

This algorithm is tested on a well known benchmark [23, 27] and compared against other AIS presented in this survey. The details are shown in Section 5.

Algorithm 8. T-Cell model to handle constraints [2]

```
BEGIN T-CELL
INPUT: Parameters values.
repeat
   Randomly generate Virgin Cells
   Calculate DTF's VC
   Evaluate VC with its own DTF
   Insert a percentage of Virgin Cells into the Effector Cells population
   repeat
      Make the Effector Cells React
         Calculate DTF's ECs
         Evaluate ECs with its own DTF
   until a predetermined number of times is reached
   Insert a Percentage of Effector Cells in Memory Cells population
   repeat
      Make the Memory Cells React
      Evaluate MC
   until a predetermined number of times is reached
until a predetermined number of times is reached
OUTPUT: Best antibody from Memory Cells (MC)
END T-CELL
```

Remarks

This proposal shows a novel algorithm which is not similar to the ones presented before. It proposes a new mutation operator that incorporates knowledge of the problem by modifying the decision variables involved in the most violated constrained. This algorithm shows an efficient local search which allows the model to improve the feasible solutions found. It requires to define only few parameters and does not require any penalty function.

4.3 AIS Based on the Clonal Selection and Idiotypic Network Theories

This algorithm was proposed by Wu [29]. Even though the Clonal Selection and the Idiotypic Network theories are in conflict to each other (see Section 2) they are combined in this paper to handle constraints in an optimization problem.

Components of the System

The system's components are the following:

- Antigen (ag) → represented by the objective and constraints.
- Antibodies (ab) are constructed by the combination of paratope and idiotope: paratope → problem's decision variables represented with real numbers idiotope → is responsible for the antibody-antibody recognition.

Algorithm

This proposal is shown in Algorithm 9. In this algorithm (Step 4) the antibody-antigen affinity (that is the antibody affinity value) is measured by using an adaptive penalty function defined as follows,

Maximize:

$$f_{pseudo} = -1 * \{f(x) + \rho_g\{\sum_{m=1}^{M}\{max[0, g_m(x)]\}^2\}\} \tag{5}$$

where ρ_g is a penalty parameter in the current generation g. The parameter ρ_g can be adaptively modified.

The *Clonal Selection* applied from Step 5 to 13, utilizes the so-called *idiotypic network selection* to control the number of antibodies to be cloned by computing their probability p_{rj} as follows:

$$p_{rj} = \frac{1}{N}\sum_{n=1}^{N}\frac{1}{e^{d_{nj}}} \tag{6}$$

$$d_{nj} = |\frac{x_n^* - x_{nj}}{x_n^*}|, j = 1,2,\ldots,rs, n = 1,2,\ldots,N \tag{7}$$

where p_{rj} is the probability of the antibody j to recognize the ab* (ab with the highest affinity) defined by the decision variables $x*_{nj}$.

The *Affinity Maturation* (from Step 14 to 22) is composed by the next two mechanisms:

- The *somatic hypermutation* applies a multi-non-uniform mutation with perturbations generated with normal distributed random numbers, which promotes uniform search and local fine-tuning (used in Step 18).
- The *receptor editing* process is modeled by applying perturbations with random numbers with a Cauchy distribution, which allows large jumps in the search space (used in Step 20).

The *bone marrow operator* applied in Step 23, is used to introduce diversity into the antibodies population by interchanging gene segments taken randomly from pairs of antibodies. Furthermore, these segments are perturbed with random number with normal distribution in the rank $[0, 1]$.

Parameters

The parameters for this algorithm are the following:

- rs: defines the repertory (population) size.
- p_{rt}: threshold degree of antibody-antibody recognition.
- $gmax$, maximum number of iterations.
- Number of clones.
- ρ: penalty factor.

Algorithm's validation

The work published in [29] validates the algorithm by using four benchmark functions from [23]. The results presented seem to be competitive, however, authors

Algorithm 9. AIS based on the Clonal Selection and Idiotypic Network Theories [29]

```
BEGIN AIS
INPUT: parameters values
The initial population of rs antibodies repertoire (ab) is created randomly
repeat
    For each antibody in the population compute the antigen-antibody affinity using the Equation 5
    Clonal Selection:
    for Each antibody do
        Computes its probability p_rj to be cloned with Equations 6 and 7.
        if p_rj ≥ p_rt then
            the antibody is cloned to create the intermediate antibody repertory,
        else
            it is suppressed.
        end if
    end for
    Affinity Maturation:
    for Each antibody in the intermediate repertory do
        Generate a random number U with uniform distribution in the rank (0,1)
        if U ≤ 0.5 then
            Apply somatic hypermutation.
        else
            Apply receptor editing.
        end if
    end for
    Select two antibodies randomly, then apply to them the bone marrow operator.
    Update the antibody repertory.
until a predetermined number of iterations is reached
OUTPUT: Best antibody.
END AIS
```

do not specify how many objective function evaluations were performed to obtain these results, which makes difficult to determine the algorithm's robustness.

Remarks

This algorithm tries to emulate several mechanisms from the immune system, which is clearly an interesting feature that the previous algorithms did not present, however it increases the algorithm's complexity.

Note that computing the antibodies affinity values require to define a penalty factor, and it is difficult to assess its impact on algorithm's performance.

5 Comparative Results

In this section we present the results given by four AIS applied to a well known benchmark proposed by Michalewicz and Runarsson [23, 27]. This benchmark is composed by thirteen functions. The selected algorithms are the following:

- Coello's algorithm (Algorithm 3 presented in Section 3.2.)
- Constrained-CLONALG (Algorithm 6 presented in Section 4.1.1.)
- AIS_{const} (Algorithm 7 presented in Section 4.1.2)
- T-Cell (Algorithm 8 presented in Section 4.2)

The parameters used for each algorithm are the following:

- Coello's algorithm: The following parameters were applied to both, the inner and outer GAs: Population Size=30; Two Points Crossover; Crossover Rate=0.8; Dynamic Mutation Rate: begin=0.4 end=$1/L$ where L is the individual's length. Furthermore, the generations of the inner GA are 30 and the parameter σ is equal to all the infeasible individuals into the population.
- Constrained-CLONALG: Number of antibodies=20, $d = 20\%$, $\beta = 1$.
- AIS_{const}: Replacement = 20%, $\beta = 1$.
- T-Cell: EC and MC Size= 20 cells each; VC Size=100 cells. Replacement policy for cells in EC=100%, MC=50%.

All the presented algorithms evaluated 350,000 times the objective function. The statistical results obtained from 30 independent runs are shown in Tables 1, 2, 3, and 4.

Table 1 Results obtained by the sequential version of Coello's algorithm [9] (shown in Sec. 3.2, Algorithm 3). The asterisk (*) indicates cases in which only 75% of the runs converged to a feasible solution

Test Funcion	Optimal	Best	Mean	Worst	Std. Dev.
g01	-15	-15.0	-15.0	-15.0	0.0
g02	0.803619	0.770337	0.704021	0.59632	0.0449
g03	1.0005	1.0	0.997	0.981	0.006
g04	-30665.5386	-30665.0815	-30648.175046	-30613.442569	14.0335
g05*	5126.4967	5126.686105	5307.201594	5927.367117	249.6318
g06	-6961.81387	-6961.179206	-6959.550307	-6955.603388	1.2393
g07	24.3062	24.332397	31.514070	47.214202	4.3983
g08	0.095825	0.095825	0.095825	0.095825	0.00
g09	680.63	680.831650	682.193733	688.603687	1.4476
g10	70.49.33	7133.1280	8158.9658	9493.8894	677.29
g11	0.7499	0.75	0.7505	0.7516	0.0004
g12	1.0	1.0	1.0	1.0	0.0
g13	0.05394	0.06716	1.29267	14.71544	2.57983

Table 2 Results obtained by Constrained-CLONALG [11] (shown in Sec. 4.1.1 Algorithm 6). The asterisk (*) indicates a case in which only 90% of the runs converged to a feasible solution

Test Funcion	Optimal	Best	Mean	Worst	Std. Dev.
g01	-15	-14.9874	-14.7264	-12.9171	0.6070
g02	0.803619	0.8017	0.7434	0.6268	0.0414
g03	1.0005	1.0	1.0	1.0	0.0
g04	-30665.5386	-30665.5387	-30665.5386	-30665.5386	0.0
g05*	5126.4967	5126.999	5436.1278	6111.1714	300.8854
g06	-6961.81387	-6961.8105	-6961.8065	-6961.7981	0.0027
g07	24.3062	24.5059	25.4167	26.4223	0.4637
g08	0.095825	0.095825	0.095825	0.095825	0.0
g09	680.63	680.6309	680.6521	680.6965	0.0176
g10	70.49.33	7127.9502	8453.7902	12155.1358	1231.3762
g11	0.7499	0.75	0.75	0.75	0.0
g12	1.0	1.0	1.0	1.0	0.0
g13	0.05394	0.05466	0.45782	1.49449	0.37900

Table 3 Results obtained by AIS_{const} [1] (shown in Sec. 4.1.2 Algorithm 7). The asterisk (*) indicates a case in wich only 75% of the runs converged to the feasible solution

Test Funcion	Optimal	Best	Mean	Worst	Std. Dev.
g01	-15	-14.993	-14.989	-14.982	0.002982
g02	0.803619	0.7821	0.7573	0.7230	0.014765
g03	1.0005	1.0	0.9880	0.9108	0.025057
g04	-30665.5386	-30665.1117	-30645.9122	-30533.7827	31.929167
g05*	5126.4967	5126.660	5468.743	6112.072	339.183
g06	-6961.81387	-6961.7940	-6960.3768	-6956.7421	1.183813
g07	24.3062	24.531708	25.644893	27.056295	0.667470
g08	0.095825	0.095825	0.095825	0.095825	0.0
g09	680.63	680.6519	680.8343	681.1474	0.134034
g10	7049.33	7058.45	8344.69	15787.89	1793.850342
g11	0.7499	0.7499	0.7499	0.7499	0.000001
g12	1.0	1.0	1.0	1.0	0.0
g13	0.05394	0.05820	1.37142	16.43139	2.904695

Table 4 Results obtained by T-cell algorithm [2] (shown in Sec. 4.2 Algorithm 8). The asterisk (*) indicates cases in which only 68% of the runs converged to a feasible solution

Test Funcion	Optimal	Best	Mean	Worst	Std. Dev.
g01	-15	-15.0	-15.0	-15.0	0.0
g02	0.803619	0.803102	0.783593	0.752690	0.013761
g03	1.0005	1.00041	0.998627	0.984513	0.004208
g04	-30665.5386	-30665.5386	-30665.5386	-30665.5386	0.00
g05*	5126.4967	5126.4982	5231.7186	5572.0024	143.0598
g06	-6961.81387	-6961.81387	-6961.81387	-6961.81387	0.0
g07	24.3062	24.3503	25.3877	28.8553	1.2839
g08	0.095825	0.095825	0.095825	0.095825	0.0
g09	680.63	680.63701	680.74652	680.94299	0.078017
g10	7049.24	7086.7891	7955.0428	9592.7752	766.493969
g11	0.7499	0.7499	0.7553	0.7983	0.010717
g12	1.0	1.0	1.0	1.0	0.0
g13	0.05394	0.054448	0.2232	0.94019	0.25325

We can observe that, in general the four algorithms show very good approximations to the optimal values. Note that the results given by Coello, Constrained-CLONALG and AIS_{const} (Algorithms 3, 6, and 7) are very similar to each other. However, the T-Cell (Algorithm 8) seems to be the best of all of them because it obtained the optimal value (or a very close value) for more functions, and in general shows the best mean values.

6 Closing Remarks

Artificial Immune Systems are computational tools to solve engineering and machine-learning problems that use a process or theory observed in the biological immune system. This is of course, a rough simplification of what really happens into the biological immune system, however it seems to be enough for solving most global optimization problems.

We have reviewed the Artificial Immune Systems that handle constraints in numerical optimization problems. The works published in this domain are relatively few in spite that the results obtained for some of them are very competitive compared with the state-of-the-art Evolutionary Computation schemes handling constraints.

We can observe that the early works were based on Genetic Algorithms, however the most recent works are pure Artificial Immune Systems, that is, AIS without any EA mechanism.

The reviewed proposals can be classified into two groups: (1) AIS based on GA, (2) AIS not based on GAs. Into this second group we can find algorithms based on: the Clonal Selection Theory, the T-cells, and on both the Clonal Selection and Immune Network theories.

We consider that the immune system is a realistic option to advance on designing efficient algorithms capable of handling constraints in numerical optimization problems.

References

1. Aragón, V.S., Esquivel, S.C., Coello, C.A.: Artificial Immune System for Solving Constrained Optimization Problems. Revista Iberoamericana de Inteligencia Artificial 35, 55–66 (2007)
2. Aragón, V.S., Esquivel, S.C., Coello, C.A.: A Novel Model of Artificial Immune System for Solving Constrained Optimization Problems with Dynamic Tolerance Factor. In: Gelbukh, A., Kuri Morales, Á.F. (eds.) MICAI 2007. LNCS, vol. 4827, pp. 19–29. Springer, Heidelberg (2007)
3. Bernardino, H.S., Barbosa, H.J., Lemonge, A.C.C.: Constraint Handling in Genetic Algorithms via Artificial Immune Systems. In: Grahl, J. (ed.) Late-breaking paper at Genetic and Evolutionary Computation Conference (GECCO 2006) (2006)
4. Bernardino, H.S., Barbosa, H.J., Lemonge, A.C.C.: A Hybrid Genetic Algorithm for Constrained Optimization Problems in Mechanical Engineering. In: IEEE Congress on Evolutionary Computation (CEC 2007), pp. 646–653. IEEE Press, Los Alamitos (2007)
5. Burnet, F.M.: The Clonal Selection Theory of Acquiered Immunity. Cambridge University Press, Cambridge (1959)
6. Cantú-Paz, E.: Efficient and Accurate Parallel Genetic Algorithms. Kluwer Academic Publishers, Dordrecht (2000)
7. Coelho, G.P., Zuben, F.J.V.: omni-aiNet: An Immune-Inspired Approach for Omni Optimization. In: Bersini, H., Carneiro, J. (eds.) ICARIS 2006. LNCS, vol. 4163, pp. 294–308. Springer, Heidelberg (2006)
8. Coello, C.A.C., Cruz-Cortés, N.: A Parallel Implementation of an Artificial Immune System to Handle Constraints in Genetic Algorithms: Preliminary Results. In: Congress on Evolutionary Computation (CEC 2002), vol. 1, pp. 819–824. IEEE Service Center (May 2002)
9. Coello, C.A.C., Cruz-Cortés, N.: Hybridizing a Genetic Algorithm with an Artificial Immune System for Global Optimization. Engineering Optimization 36(5), 607–634 (2004)
10. Coello, C.A.C., Cruz-Cortés, N.: Solving Multiobjective Optimization Problems Using an Artificial Immune System. Genetic Programming and Evolvable Machines 6(2), 163–190 (2005)

11. Cruz-Cortés, N., Trejo-Pérez, D., Coello, C.A.C.: Handling Constraints in Global Optimization Using an Artificial Immune System. In: Jacob, C., Pilat, M.L., Bentley, P.J., Timmis, J.I. (eds.) ICARIS 2005. LNCS, vol. 3627, pp. 234–247. Springer, Heidelberg (2005)
12. Dasgupta, D. (ed.): Artificial Immune Systems and Their Applications. Springer, Heidelberg (1998)
13. de Castro, L.N., Timmis, J.: Artificial Immune Systems: A New Computational Intelligence Approach. Springer, Heidelberg (2002)
14. de Castro, L.N., Zuben, F.J.V.: Learning and Optimization Using the Clonal Selection Principle. IEEE Transactions on Evolutionary Computation 6(3), 239–251 (2002)
15. Freschi, F., Repetto, M.: Multiobjective Optimization by a Modified Artificial Immune System. In: Jacob, C., Pilat, M.L., Bentley, P.J., Timmis, J.I. (eds.) ICARIS 2005. LNCS, vol. 3627, pp. 248–261. Springer, Heidelberg (2005)
16. Hajela, P., Lee, J.: Constrained Genetic Search via Schema Adaptation: An Immune Network Solution. Structural Optimization 12(1), 11–15 (1996)
17. Hajela, P., Yoo, J.S.: Immune Network Modeling in Design Optimization. In: Corne, D., Dorigo, M., Glover, F. (eds.) New Ideas in Optimization, pp. 203–215. McGraw-Hill, New York (1999)
18. Jerne, N.K.: Towards a Network Theory of the Immune System. Ann. Immunol. Inst. Luis Pasteur. 125C, 373–389 (1974)
19. Jiao, L., Gong, M., Shang, R., Du, H., Lu, B.: Clonal Selection with Immune Dominance and Anergy Based Multiobjective Optimization. In: Coello Coello, C.A., Hernández Aguirre, A., Zitzler, E. (eds.) EMO 2005. LNCS, vol. 3410, pp. 474–489. Springer, Heidelberg (2005)
20. Kelsey, J., Timmis, J.: Immune Inspired Somatic Contiguous Hypermutation for Function Optimisation. In: Cantú-Paz, E., Foster, J.A., Deb, K., Davis, L., Roy, R., O'Reilly, U.-M., Beyer, H.-G., Kendall, G., Wilson, S.W., Harman, M., Wegener, J., Dasgupta, D., Potter, M.A., Schultz, A., Dowsland, K.A., Jonoska, N., Miller, J., Standish, R.K. (eds.) GECCO 2003. LNCS, vol. 2724, pp. 207–218. Springer, Heidelberg (2003)
21. Luh, G.C., Chueh, C.H., Liu, W.: MOIA: Multi-Objective Immune Algorithm. Engineering Optimization 35(2), 143–164 (2003)
22. Mezura-Montes, E., Coello, C.A., Landa, R.: Engineering Optimization Using a Simple Evolutionary Algorithm. In: Proceedings of the Fiftheenth International Conference on Tools with Artificial Intelligence (ICTAI 2003), pp. 149–156. IEEE Computer Society, Los Alamitos (2003)
23. Michalewicz, Z., Schoenauer, M.: Evolutionary Algorithms for Constrained Parameter Optimization problems. Evolutionary Computation 4(1), 1–32 (1996)
24. Olivetti, F., Zuben, F.J.V., de Castro, L.N.: An Artificial Immune Network for Multimodal Function Optimization on Dynamic Environments. In: GECCO 2005: Proceedings of the 2005 conference on Genetic and evolutionary computation, pp. 289–296. ACM, New York (2005)
25. Petrowski, A.: A Clearing Procedure as a Niching Method for Genetic Algorithms. In: Third IEEE International Conference on Evolutionary Optimization, Proceedings, pp. 798–803. IEEE Press, Los Alamitos (1996)
26. Rajasekaran, S., Lavanya, S.: Hybridization of Genetic Algorithms with Immune System for Optimization Problems in Structural Engineering. Structural and Multidisciplinary Optimization 34(5), 415–429 (2007)

27. Runarsson, T.P., Yao, X.: Stochastic Ranking for Constrained Evolutionary Optimization. IEEE Transactions on Evolutionary Computation 4(3), 284–294 (2000)
28. Tan, K., Goh, C., Mamun, A., Ei, E.: An Evolutionary Artificial Immune System for Multi-objective Optimization. European Journal of Operational Research 127(2), 371–392 (2008)
29. Wu, J.Y.: Artificial Immune System for Solving Constrained Global Optimization Problems. In: IEEE Symposium on Artificial Life (Ci-ALife 2007), pp. 92–99 (2007)

Index

The manufacturer's authorised representative in the EU is Springer Nature Customer Service Centre GmbH, Europaplatz 3, 69115 Heidelberg, Germany. If you have any concerns regarding our products, please contact ProductSafety@springernature.com

Printed and bound by CPI Group (UK) Ltd, Croydon, CR0 4YY

29/07/2026

02180609-0001